SUSTAINABILITY IN THE DESIGN, SYNTHESIS AND ANALYSIS OF CHEMICAL ENGINEERING PROCESSES

SUSTAINABILITY IN THE DESIGN, SYNTHESIS AND ANALYSIS OF CHEMICAL ENGINEERING PROCESSES

Edited by

GERARDO RUIZ-MERCADO
HERIBERTO CABEZAS

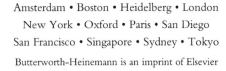

Amsterdam • Boston • Heidelberg • London
New York • Oxford • Paris • San Diego
San Francisco • Singapore • Sydney • Tokyo
Butterworth-Heinemann is an imprint of Elsevier

British Library Cataloguing-in-Publication Data
A catalogue record for this book is available from the British Library

Library of Congress Cataloging-in-Publication Data
A catalog record for this book is available from the Library of Congress

ISBN: 978-0-12-802032-6

For information on all Butterworth-Heinemann publications
visit our website at https://www.elsevier.com/

Working together
to grow libraries in
developing countries

www.elsevier.com • www.bookaid.org

Publisher: Joe Hayton
Acquisition Editor: Fiona Geraghty
Editorial Project Manager: Maria Convey
Production Project Manager: Nicky Carter
Designer: Maria Inês Cruz

Typeset by TNQ Books and Journals

DEDICATION

To my mother and father, Ana and Gerardo, to my wonderful wife, Carola, to my daughters, Itzel and Eluney, and to my sisters, Ana M. and Yenis.

Gerardo Ruiz-Mercado

To my parents, Heriberto Cabezas y Clávelo and Ana Rosa B. Fernández y Sanchez de Cabezas, and to my sister, Ana C. Cabezas y Fernández de Gonzalez, because without their love and support so many years ago, nothing would have been possible.

Heriberto Cabezas

CONTENTS

About the Authors   *xiii*

Acknowledgment   *xxiv*

Foreword   *xxv*

Preface   *xxvii*

1. Towards More Sustainable Chemical Engineering Processes: Integrating Sustainable and Green Chemistry Into the Engineering Design Process   **1**

D.J.C. Constable, M. Gonzalez, S.A. Morton

Underpinnings of Green Chemistry	1
The Principles and Implications	5
Problems With Chemicals and Reaction Spaces	12
Thinking About What More Sustainable Chemistry and Chemical Manufacturing Might Look Like	17
Tying It All Together	28
Disclaimer	31
References	31

2. Separations Versus Sustainability: There Is No Such Thing As a Free Lunch   **35**

L.M. Vane

The Separations Dilemma and Imperative	35
Methods of Analysis	38
Separation Alternatives	41
Examples	53
Concluding Thoughts	61
Disclaimer	62
References	62

3. Conceptual Chemical Process Design for Sustainability   **67**

R.L. Smith

Conceptual Chemical Process Design	67
Sustainability Approach for Chemical Processes	70
Example: Chlor-Alkali Production with Human Toxicity Potential Analysis	72
Discussion	81
Conclusions	82
Disclaimer	82
References	82

4. Process Integration for Sustainable Design   **87**
    M.M. El-Halwagi

    Introduction   87
    Mass Integration   88
    Property Integration   92
    Energy Integration   96
    Multiscale Approaches   99
    Conclusions   110
    References   111

5. Modeling and Advanced Control for Sustainable Process Systems   **115**
    F.V. Lima, S. Li, G.V. Mirlekar, L.N. Sridhar, G. Ruiz-Mercado

    Introduction to Sustainable Process Systems   115
    Proposed Approach: Modeling, Advanced Control, and Sustainability Assessment   118
    Case Study: Fermentation for Bioethanol Production System   127
    Sustainability Assessment and Process Control   135
    Conclusions and Future Directions   136
    Nomenclature   136
    Acknowledgments   138
    Disclaimer   138
    References   138

6. Sustainable Engineering Economic and Profitability Analysis   **141**
    Y. Jiang, D. Bhattacharyya

    Introduction   141
    Economic Sustainability Analysis   142
    Environmental Sustainability Analysis   145
    Social Sustainability Analysis   148
    Evaluation of Design Alternatives by Considering Various Sustainability Measures   151
    Example: Bioethanol Process   151
    Concluding Remarks   164
    Nomenclature   164
    References   165

**7. Managing Conflicts Among Decision Makers in Multiobjective
    Design and Operations**   **169**
    V.M. Zavala

    Introduction   169
    Approach   170

Illustrative Examples   173
Conclusions   179
Acknowledgments   179
References   180

8. Sustainable System Dynamics: A Complex Network Analysis   181
U. Diwekar

Introduction   181
Sustainable System Dynamic Models   182
Controllability Analysis   189
Optimal Control for Deriving Techno-Socio-Economic Policies   195
Summary   201
References   201

9. Process Synthesis by the P-Graph Framework Involving Sustainability   203
B. Bertok, I. Heckl

Introduction   203
Illustrative Example   206
Basics of the P-Graph Framework   208
Software: PNS Draw and PNS Studio   211
Summary   224
References   224

**10. Sustainability Assessment and Performance Improvement of Electroplating
Process Systems   227**
H. Song, N. Bhadbhade, Y. Huang

Introduction   227
Fundamentals for Process Sustainability   229
Sustainability Metrics System   230
Sustainability Assessment Framework   233
Case Study   237
Concluding Remarks   246
Acknowledgment   247
References   247

**11. Strategic Sustainable Assessment of Retrofit Design for Process
Performance Evaluation   249**
A. Carvalho

Introduction   249
State of the Art   250

Framework for Assessment of Retrofit Design Alternatives   259
Case Study: β-Galactosidase Production   265
Conclusions   270
References   271

12. Chemical Engineering and Biogeochemical Cycles: A Techno-Ecological Approach to Industry Sustainability   **275**

S. Singh, B.R. Bakshi

Motivation   275
Life Cycle Analysis for Chemical Industry Interaction with Carbon and Nitrogen Cycles   279
Chemical Industry Profile for Carbon   283
Chemical Industry Profile for Nitrogen   287
Techno-Ecological Approach and Chemical Industry Sustainability   290
References   293

13. Challenges for Model-Based Life Cycle Inventories and Impact Assessment in Early to Basic Process Design Stages   **295**

S. Papadokonstantakis, P. Karka, Y. Kikuchi, A. Kokossis

Introduction   295
LCI Aspects in Early to Basic Process Design Stages   300
Case Studies   305
Case Study 1: LCA Aspects of Solvent Selection Postcombustion CO_2 Capture   305
Case Study 2: LCA Aspects in the Design of Lignocellulosic Biorefineries   308
Case Study 3: Poly(methyl methacrylate) Recycling Process   317
Conclusions and Outlook   321
References   323

14. Life Cycle Sustainability Assessment: A Holistic Evaluation of Social, Economic, and Environmental Impacts   **327**

L.Q. Luu, A. Halog

Introduction   327
Methodologies for Assessing Life Cycle Sustainability   327
Three Pillars of Sustainability and the Need for Life Cycle Sustainability Assessment   333
Case Study of Rice Husk-Based Electricity and Coal-Fired Electricity   335
Goal and Scope Definition   337
Inventory Analysis   340
Results   342
Integrated Results of Sustainability Assessment   346

Some Remarks on the Methodology   348
Conclusion   349
References   350

**15. Embedding Sustainability in Product and Process Development—The Role
        of Process Systems Engineers   353**

C. Jiménez-González

Introduction   353
Material Selection   354
Process Design, Synthesis, and Integration   360
Process Intensification   366
Hazard Assessments and Inherent Safety   367
Impacts Throughout the Supply Chain—Life Cycle Thinking   370
Modeling and Computer-Aided Tools   371
Future Outlook   373
Acknowledgments   375
References   375

Index   379

ABOUT THE AUTHORS

BHAVIK R. BAKSHI

Bhavik R. Bakshi is a Professor of Chemical and Biomolecular Engineering at Ohio State University. He also holds appointments in civil, environmental, and geodetic engineering at Ohio State University and as a Visiting Professor at the Indian Institute of Technology in Mumbai, India. His research is developing methods and applications for assessing and designing sustainable systems while taking into account interactions between technological and ecological systems. He received his chemical engineering degrees from the University of Bombay and the Massachusetts Institute of Technology (MIT), with a minor in technology and environmental policy from MIT and Harvard.

BOTOND BERTOK

Botond Bertok graduated as Master of Engineering in Information Technology in 1999 at the University of Pannonia (UP), Veszprem, Hungary, and got his PhD in IT in 2004. Currently he is associate professor at the Faculty of Information Technology at the UP. His research area is development of methods and software for computer-aided process design and optimization based on mathematical modeling. In the last decade he has developed courses and educational programs on process modeling and optimization including a graduate specialization program on supply chain optimization in the petroleum industry in cooperation with MOL, the Hungarian oil and gas company. He has led numerous R&D projects resulting in software for improving efficacy, safety, and sustainability of complex production systems, logistic networks, and supply chains. He has more than 120 international publications and more than 380 citations according to Google Scholar.

NAVDEEP BHADBHADE

Navdeep Bhadbhade has a master's degree from Wayne State University. Before joining the master's degree program he earned his Bachelor's degree in Chemical Engineering from the University of Pune (India). His research interests are mathematical modeling, simulation, optimization, and process control for dynamic processes including electroplating processes.

DEBANGSU BHATTACHARYYA

Professor Bhattacharyya is currently an Associate Professor in the Department of Chemical Engineering at West Virginia University. Prior to this appointment he spent 3 years as a Research Associate Professor at West Virginia University and more than 10 years in the Refineries Division of Indian Oil Corporation Limited. His broad research interests are in the area of process systems engineering.

Professor Bhattacharyya received his PhD in Chemical Engineering from the Clarkson University, New York, in 2008. He has published more than 50 research papers and authored/coauthored more than 130 international presentations and 50 posters. He is a coauthor of the popular chemical engineering design book *Analysis, Synthesis, and Design of Chemical Processes.* He is a member of the prestigious 2015 R&D 100 (widely recognized as the "Oscars of Invention") winning team in the Software/Services category.

ANA CARVALHO

Ana Isabel Carvalho obtained her degree in Chemical Engineering in 2005. In 2009 she finished her PhD in Chemical Engineering. She is currently an Assistant Professor at Instituto Superior Técnico in Portugal. She is a researcher at the Centre for Management Studies of Instituto Superior Técnico (CEG-IST). Her research interests are related to sustainability, more specifically applied to process design and retrofit. She has also been working in sustainable supply chains, definition assessment, and optimization. Life cycle assessment for environmental and social evaluation are also on her research work. She has published several papers in peer-reviewed international journals and has extensively presented her work at national and international conferences. She has been awarded several national and international awards in pedagogical and scientific activities (e.g., Sustainable Engineering Forum Student Paper Award—AIChE).

DAVID J.C. CONSTABLE

David J.C. Constable is the Director of the American Chemical Society's Green Chemistry Institute®. In this role he works to catalyze and enable the implementation of green chemistry and engineering throughout the global chemistry enterprise. David has worked in a variety of industrial positions in aerospace and defense, pharmaceuticals, and the chemical industry. In those roles he has largely been involved in assisting companies with the development and implementation of programs, systems, tools, and methodologies that integrated sustainability, life cycle inventory assessment, green chemistry, green technology, energy, and environmental, health and safety activities into existing business processes.

URMILA DIWEKAR

Dr. Urmila Diwekar is currently the President of the Vishwamitra Research Institute (VRI, www.vri-custom.org), a nonprofit research organization that she founded in 2004 to pursue multidisciplinary research in the areas of optimization under uncertainty and computer-aided design applied to energy, environment, and sustainability. From 2002 to 2004 she was a Professor in the Departments of Chemical Engineering, Bio Engineering, and Industrial Engineering, and in the Institute for Environmental Science and Policy, at the University of Illinois at Chicago. From 1991 to 2002 she was on the faculty of the Carnegie Mellon University. She is the author of more than 155 peer-reviewed research papers, has given over 325 presentations and seminars, and has chaired numerous sessions in national and international meetings. She has been the principal adviser to 40 PhD and MS students, and has advised 13 postdoctoral fellows and researchers.

MAHMOUD M. EL-HALWAGI

Dr. Mahmoud M. El-Halwagi is the McFerrin Professor of Chemical Engineering at Texas A&M University. He received his PhD in Chemical Engineering from the University of California, Los Angeles, and his MS and BS from Cairo University. Dr. El-Halwagi has about 30 years of experience in the areas of process integration, sustainable design, and optimization. He has served as a consultant to a wide variety of gas, fuel, chemical, petrochemical, and pharmaceutical industries. He is the coauthor of about 200 papers and 60 book chapters. He is also the author/coauthor/coeditor of nine books including three textbooks on sustainable process design and integration. He is the recipient of several awards including the American Institute of Chemical Engineers Sustainable Engineering Forum (AIChE SEF) Research Excellence Award, the DuPont Excellence Award in Safety, Health and the Environment, and the National Science Foundation's National Young Investigator Award.

MICHAEL GONZALEZ

Michael is currently Chief of the Systems Analysis Branch in the US Environmental Protection Agency's (US EPA's) Office of Research and Development. His branch conducts life cycle assessment, impact assessment, and sustainable chemistry research. Michael has also served as the Senior Advisor of Green Chemistry to the Office of Research and Development (ORD) and was responsible for integrating green chemistry and engineering into ORD's research portfolio.

Michael's research interests focus on the design of green chemical synthesis routes, with an emphasis on the intersection of chemistry and chemical engineering. Michael is also a coinventor and developer of the Environmental Protection Agency GREEN-SCOPE tool, which evaluates a chemical process for its sustainability value in the areas of environment, efficiency, energy, and economics.

Michael earned his BS in Chemistry from the University of Texas—El Paso in 1992 and his PhD in Inorganic Chemistry from the University of Florida in 1998. Michael has published several articles in the areas of sustainability, green chemistry and engineering, and sustainability indicators for chemical processes.

ANTHONY HALOG

Dr. Anthony Halog is a Lecturer in Industrial Ecology at the University of Queensland, Australia. His research focuses on the sustainability of the human—nature complexity through understanding the nexus of material and energy systems. Dr. Halog is interested in the life cycle of manufactured goods and, ultimately, wastes—and in the environmental benefits and economic potential of circular and green economy. His Research Group in Industrial Ecology and Circular Economy at the University of Queensland endeavors to provide service/expertise for industry clients in transforming existing linear system-based value chains and operations toward circularity, which enhances their resource efficiency and productivity, creates added value products from wastes, and reduces their emissions into the environment for more sustainable consumption and production in the future.

ISTVÁN HECKL

István Heckl studied Information Technologies at the University of Pannonia in Veszprém, Hungary. He got his PhD with the supervision of Professor Ferenc Friedler in 2007. He works as an associate professor at the University of Pannonia. He is a lecturer in the subjects of C programming, C++ programming, and elements of the theory of computation, among others. He was the supervisor of 11 BSc and MSc theses and is the leader of several research and development projects. For example, the ProdSim project developed a product pipeline simulator to MOL (the major Hungarian oil company). ProdSim is capable of validating the short-term scheduling of the pipelines. As a result, planning periods could be increased from 5 to 6 days to up to a month. His research area includes process synthesis, optimization, separation network synthesis, and sustainability.

YINLUN HUANG

Dr. Yinlun Huang is a Professor of Chemical Engineering and Materials Science at Wayne State University, where he directs the Laboratory for Multiscale Complex

Systems Science and Engineering. His research has been mainly focused on the fundamental study of multiscale complex systems science and sustainability science, with applied study on engineering sustainability, including sustainable nanomaterial development, integrated design of sustainable product and process systems, and manufacturing sustainability. Dr. Huang is currently directing the National Science Foundation–funded Sustainable Manufacturing Advances in Research and Technology Coordination Network (SMART CN), which involves many domestic and foreign universities and national organizations and university centers. Among many honors, Dr. Huang was a recipient of the Michigan Green Chemistry Governor's Award in 2009, the AIChE Research Excellence in Sustainable Engineering Award in 2010, and the NASF Scientific Achievement Award in 2013. He is an elected AIChE Fellow.

YUAN JIANG

Ms Jiang is currently a graduate research assistant and PhD candidate of the Department of Chemical Engineering at West Virginia University, working on techno-economic analysis of coal biomass to liquids plants with carbon capture and storage and sustainable process design. Before that she received her BS degree in Chemical Engineering from the East China University of Science and Technology. Jiang's research focuses on energy conversion and process systems engineering, including unit and system-level simulation, design, analysis, and optimization of processes utilizing coal biomass and shale gas/natural gas.

CONCEPCIÓN "CONCHITA" JIMÉNEZ-GONZÁLEZ

Dr. Concepción Jiménez-González is a Program Lead for Global Manufacturing and Supply at GlaxoSmithKline (GSK). Prior to this role, she had a series of roles of increasing responsibility, including Director of New Product Development and Director of Operational Sustainability, among others. She has lived and worked in Mexico, the United Kingdom, and the United States. Before GSK she worked for GeoEnvironmental Consultants, Instituto Tecnológico y de Estudios Superiores de Monterrey (ITESM) (Mexico), and as a visiting researcher at Pfizer.

In addition to her job at GSK she is an adjunct professor in Chemical and Biomolecular Engineering at North Carolina State University (NCSU), and serves on the Governing Board of the American Chemistry Society's Green Chemistry Institute.

She holds a PhD in Chemical Engineering from NCSU, an MS in Environmental Engineering from ITESM, Mexico, and a BS in Chemical and Industrial Engineering at the Chihuahua Institute of Technology, Mexico.

PARASKEVI KARKA

Paraskevi Karka graduated from the National Technical University of Athens (NTUA), School of Chemical Engineering, in 2006. Currently, she is a PhD candidate at the NTUA, School of Chemical Engineering in the Department of Process Analysis and Plant Design. She holds two MSc degrees in the fields of Energy Production and Management (School of Electrical and Computer Engineering, NTUA, 2010) and in Systems of Energy and Environmental Management (School of Chemical Engineering, NTUA, and Department of Industrial Management of the University of Piraeus, 2010). Her research interests include life cycle analysis and sustainability assessment of industrial processes. She has participated in various national and European research projects related to biomass utilization processes and the design of biorefineries.

YASUNORI KIKUCHI

Yasunori Kikuchi is a project associate professor, Presidential Endowed Chair for the "Platinum Society" and Department of Chemical System Engineering at the University of Tokyo (UTokyo), Tokyo, Japan. He is concurrently a visiting associate professor at Chiba University in Kashiwa, Japan, a visiting scholar, International Institute for Carbon-Neutral Energy Research at Kyushu University in Fukuoka, Japan, and a project fellow at the Center for Research and Development Strategy in the Japan Science and Technology Agency in Tokyo. He received a PhD degree from the Department of Chemical System Engineering at UTokyo. He has been honored by the Society of Chemical Engineers, Japan Award for Outstanding Young Researcher (2016) and the Institute of Life Cycle Assessment, Japan Award for Outstanding Paper (2011).

ANTONIS KOKOSSIS

Dr. Kokossis, FIChemE, FIEE, FRSA, and FIET, is a Professor of Process Systems Engineering at the National Technical University of Athens (NTUA). He holds a Diploma in Chemical Engineering from NTUA and a PhD from Princeton University. He has expertise in process systems design and integration, renewable energy systems, polygeneration, biorefineries, and industrial symbiosis. His research has addressed the design of multiphase reactors, complex separation and reactive-separation systems, energy and power networks, and environmental problems across a wide spectrum of applications. He holds 142 communications in international conferences, 129 publications in peer-reviewed journals, and 41 invited lectures in conferences at universities and multinational companies. He is a National Representative of the International Energy

Agency, the Greek Secretary for Research and Technology in Climate Change, and the Computer Aided Process Engineering Group at the European Federation of Chemical Engineering.

SHUYUN LI

Shuyun received her BS degree from Northeast Petroleum University in 2007 and MS degree from the China University of Petroleum in 2010, both in Chemical Engineering. Upon her master's degree graduation, she worked as a process engineer at Beijing Petrochemical Engineering Co., Ltd. on simulation and design of refinery processes from 2010 to 2013. She is currently pursuing a PhD degree in Chemical Engineering at West Virginia University. Her research mainly focuses on integrating advanced process control with sustainability assessment tools for optimization and evaluation of energy systems.

FERNANDO V. LIMA

Dr. Lima joined the faculty as an Assistant Professor of Chemical Engineering at West Virginia University in 2013. His research interests are in the areas of process design and optimization, advanced control and state estimation, emerging energy systems, and sustainability. His current research focuses on modeling, optimization, and control of advanced energy plants for maximized efficiency and sustainability.

He received his BS degree from the University of São Paulo in 2003 and his PhD from Tufts University in 2007, both in Chemical Engineering. Upon completion of his PhD he was a Research Associate at the University of Wisconsin—Madison and a Postdoctoral Associate at the University of Minnesota. Dr. Lima has authored and coauthored numerous publications and presentations in process systems engineering, energy, and sustainability.

LE QUYEN LUU

Ms Quyen Luu is a 5-year experienced researcher of sustainable energy development from the Institute of Energy Science, Vietnam Academy of Science and Technology. Quyen did her master's on Environment Management at the University of Queensland, Australia, with the support from an Australia Awards Scholarship. She got the Dean's Commendation for her achievements during her study, and received the Young Southeast Asian Leaders Initiative scholarship to work with the US local government on their environmental sustainability programs. Quyen participated in several projects of bringing equity in energy access to disadvantaged people, and developing a renewable energy mix for isolated areas. She is now working on the upscaling of renewable energy consumption and promoting the closed loop of energy/resource production and consumption chain and zero waste.

GAURAV V. MIRLEKAR

Gaurav received the B.Tech. degree in Chemical Technology (Oils) from the Institute of Chemical Technology, Mumbai, India, in 2012. He is currently pursuing the PhD degree in Chemical Engineering at West Virginia University, Morgantown, USA. Before joining West Virginia University he held a position of manufacturing engineer with Gulf Extrusion Co. LLC, Dubai, UAE, in an aluminum anodizing facility from 2012 to 2013. His current research interests include model-based control and optimization, data-based process modeling, bio-inspired optimal control design, and sustainability.

SAMUEL A. MORTON III

Samuel A. Morton III is an Associate Professor in the Department of Engineering at James Madison University in Harrisonburg, Virginia. Dr. Morton has more than 15 years of engineering research and development experience. He has expertise in the areas of conceptual process development and life cycle assessment with much of his recent activities in the area of bioenergy development. His research activities focused on the development of sustainable bioenergy resources through improved use of lignocellulosic biomass, the development of industrial hemp as a biorefinery feedstock, and the use of microalgae as phytoremediation agents. His other interests include a broad range of green separation techniques utilizing ionic liquids as alternatives to energy-intensive separation processes.

STAVROS PAPADOKONSTANTAKIS

Stavros Papadokonstantakis studied Chemical Engineering (National Technical University of Athens) and holds a PhD in the field of "Modeling Chemical Engineering Processes with Neural Networks." He has worked as an engineering consultant at American Process Inc. (Atlanta, USA—Athens, Greece) in the field of "Energy Monitoring in Pulp and Paper Industries." Since 2007 he has been conducting research in sustainability aspects of process design (Swiss Federal Institute of Technology, Zurich). Since 2014, as an Associate Professor in the Department of Energy and Environment at Chalmers University of Technology (CUT), he focuses on energy efficiency of chemical production, hazard and life cycle assessment, biorefineries, and industrial symbiosis. He has held a professional chemical engineering license since 1999 and has been a senior member of the American Institute of Chemical Engineers since 2014. He is the author of 40 papers in books and peer-reviewed journals.

GERARDO RUIZ-MERCADO

Dr. Ruiz-Mercado holds a PhD from the University of Puerto Rico—Mayaguez (2009) and a BS (magna cum laude) from the University of Atlántico—Colombia (2002) all in Chemical Engineering. Before joining the US Environmental Protection Agency he continued his research studies as a research associate in the Department of Chemical Engineering and Bioengineering at the University of Illinois at Chicago. He is an active member of the AIChE at the Environmental Division and the Sustainable Engineering Forum.

Dr. Ruiz-Mercado is currently leading and developing research projects in areas of sustainable development. He is a coinventor and developer of the GREENSCOPE process sustainability methodology and tool. In addition he has a published record of contributions in research areas such as sustainable product and process design, energy efficient multicomponent distillation, reactive separation processes, sustainability evaluation, and life cycle approaches.

SHWETA SINGH

Shweta Singh is an Assistant Professor in the Department of Agricultural and Biological Engineering and Environmental and Ecological Engineering at Purdue University, USA. Before joining Purdue she held postdoctoral fellowships at the University of Toronto, Canada (2013—14), and a National Research Council Postdoctoral Fellowship by the National Academy of Sciences and the US Environmental Protection Agency (2012—13). She received a B.Tech. in Chemical Engineering from the Indian Institute of Technology (IIT)—Banaras Hindu University, India (2006), a Master's in Applied Statistics, and a PhD in Chemical Engineering from Ohio State University, USA (2012).

Her research is focused on sustainable engineering and systems modeling for the interactions of engineered systems with ecological systems. Particular focus is on studying the interaction of the nitrogen biogeochemical cycle with input—output models, biodiversity loss drivers such as nutrient pollution, and urban metabolism. She also specializes in life cycle assessment, industrial ecology, and applied statistics.

RAYMOND L. SMITH

Ray Smith has been a chemical engineer in the US Environmental Protection Agency's (US EPA's) Office of Research and Development for over 15 years, after earning his PhD in Chemical Engineering from the University of Massachusetts, Amherst. During his tenure with the EPA, Ray has established an expertise in sustainability by performing research and publishing in the areas of life cycle assessment, biofuels, industrial ecology, process design, sustainability indicators, optimization, and decision making. Over the course of his career he has also been employed in industry and collaborated with domestic and international academic, industrial, and consultant groups. He is a coinventor and developer of the GREENSCOPE process sustainability methodology and tool. Ray has volunteered his time to the American Institute of Chemical Engineers, serving as the first Chair of both the Environmental Division and the Sustainable Engineering Forum.

HAO SONG

Hao Song is a PhD student working at Wayne State University. Prior to joining the PhD program he received his master's degree in Chemistry from Auburn University. During his PhD study at Wayne State University he also earned a master's degree in Materials Science. His main research interests are industrial process design, modeling, and optimization, systematic sustainability assessment, and decision making for industrial systems including automotive coating systems and electroplating systems.

LAKSHMI N. SRIDHAR

Dr. Lakshmi N. Sridhar has been a faculty member at the University of Puerto Rico at Mayaguez since 1993. He completed his B.Tech at the Indian Institute of Technology—Madras and his PhD at Clarkson University in Chemical Engineering. He works in analysis and optimization of chemical engineering process problems.

LELAND M. VANE

Dr. Vane leads a separations research team at the US Environmental Protection Agency's (US EPA's) National Risk Management Research Laboratory. His team has been active in developing green separations technologies, particularly membrane-based pervaporation and vapor permeation, for volatile organic contaminant removal from aqueous wastes, alcohol recovery from water, solvent recovery/reuse, and brine management. Dr. Vane received his Bachelor of Chemical Engineering degree with distinction and summa cum laude from the University of Delaware in 1987 and his PhD, also in chemical engineering, from Cornell University in 1992. He has been involved in separations research since joining the US EPA in 1992.

VICTOR M. ZAVALA

Dr. Zavala is the Richard H. Soit Assistant Professor in the Department of Chemical and Biological Engineering at the University of Wisconsin–Madison. Before joining University of Wisconsin–Madison in 2015 he was a computational mathematician in the Mathematics and Computer Science Division at Argonne National Laboratory. He received his BSc degree from Universidad Iberoamericana (2003) and his PhD degree from Carnegie Mellon University (2008), both in Chemical Engineering. He is currently a recipient of a Department of Energy Early Career Award under which he develops scalable optimization algorithms. He is also a technical editor of the *Mathematical Programming Computation* journal. His research interests are in the areas of mathematical modeling of energy systems, high-performance computing, stochastic optimization, and predictive control.

ACKNOWLEDGMENT

The quality of a book is only as good as that of the authors and the referees who so kindly conducted thorough reviews of the component chapters. Hence the editors would like to gratefully acknowledge the contribution of the referees, who are listed in alphabetical order along with their institutional affiliations: Alireza Banimostafa (Schneider Electric, USA), Donald J. Chmielewski (Illinois Institute of Technology, USA), Mario Richard Eden (Auburn University, USA), David Ferguson (US Environmental Protection Agency, USA), Raymond Girard Tan (De La Salle University, Philippines), Gonzalo Guillen-Gosalbez (Imperial College of Science, Technology and Medicine, UK), David Habeych (Avebe, the Netherlands), Wesley Ingwersen (US Environmental Protection Agency, USA), Viatcheslav Kafarov (Universidad Industrial de Santander, Colombia), Zdravko Kravanja (University of Maribor, Slovenia), Rafael Luque (Universidad de Cordoba, Spain), Dharik Mallapragada (ExxonMobil, USA), Michael Narodoslawsky (Technical University of Graz, Austria), Weiyi Ouyang (Universidad de Cordoba, Spain), Jeffrey Seay (University of Kentucky, USA), Yogendra Shastri (Indian Institute of Technology Bombay, India), and Petar Varbanov (Peter Pazmany Catholic University, Hungary).

FOREWORD

The next generation of chemical processes requires a transition to new methods for analysis, synthesis, and design. This body of work is a compilation of research by professionals in the field and addresses approaches to make this transition focused on sustainable chemical processes. Editors Dr. Gerardo Ruiz-Mercado and Dr. Heriberto Cabezas have assembled international contributions that inform the path forward to achieve sustainable design of chemical processes. Gerardo's and Heriberto's experience and oversight of research in developing methods, metrics, and tools related to sustainable chemical product and process design provide a sound foundation for producing this body of work.

Thus sustainability requires a holistic approach, which treats the challenges as inherently connected. Sustainable chemical processes require being sustainable within the context of the larger societal integrated system. A central theme for the book is the need for a holistic approach to chemical process design and operation that expands the traditional system boundaries and explores the transition from today's chemical process "building blocks" to a future set of sustainable "building blocks." This holistic approach includes methodologies for process integration and incorporates the need to integrate science, technology, economic, environmental, and societal needs and constraints into the design process. The recognition that there are diverse stakeholders, with different objectives and priorities, in the decision-making process for designing processes is an important illustration of including expanded system boundaries. The inclusion of a chapter on an alternative approach to modeling conflict recognizes the importance of incorporating stakeholder perspectives and the need to think beyond traditional boundaries when developing sustainable products and processes. Another example of incorporating important decision drivers is the work reported on ecosystem models.

To emphasize practical application, the work includes specific case studies that provide examples for addressing sustainable design and process performance. For example, a methodology is presented to assess the sustainability of an electroplating process system. The challenge of reducing the environmental impact from separations technologies is addressed. A chapter is included that presents a framework for the sustainable assessment of retrofit designs. How to effectively utilize the existing asset investments has sustainable economic and resource use implications.

Chemical process design choices are coupled with changing resource constraints—for example, feed materials and water resource. The need to incorporate sustainable resources as part of a methodology for sustainable chemical process design and operation is also included in this work. While this work is focused on sustainable chemical process design, it is important to see it as part of the larger sustainability challenge to provide quality access to energy, water, and food for an increasing global population with lifestyles based on increasing material consumption. Sustainable chemical product and process designs that are consistent with a sustainable integrated system are essential to achieving this goal.

Chemical engineering has a key role to play in sustainability. An integrated systems approach, as reflected in the contributions, is central to achieving the goal. The analyses, methodologies, and case studies presented inform future research and opportunities for advances in education.

Dale L. Keairns, PhD
Past President American Institute of Chemical Engineers (2008)
Executive Advisor
Booz | Allen | Hamilton

PREFACE

WHY THIS BOOK?

The genesis of this volume was a desire by the editors to assemble as much as possible the state of the art of technical and scientific knowledge relevant to chemical engineering and sustainability, and to do so in an accessible and coherent manner. This is important because sustainability is a profoundly complex problem, and addressing it effectively will require knowledge from many different disciplines and subdisciplines. The practitioners of many of these disciplines and even subdisciplines often travel in separate circles, and there is therefore a need for a volume such as this one to tie together the different important threads of knowledge in a compact form in one place for practical application.

HISTORICAL DEVELOPMENT OF SUSTAINABILITY

In the United States the concept of sustainable development implicitly originated with the National Environmental Policy Act of 1969 (NEPA), one of the first major federal laws establishing a broad framework for the protection of the environment. It declared as a tenet of national policy the need "to foster and promote the general welfare, to create and maintain conditions under which man and nature can exist in productive harmony and fulfill the social, economic and other requirements of present and future generations." However, the embryonic sustainable development approach described by NEPA in 1969 did not contain clear details that could be applied in practice.

In 1970, as a result of the need and the public demand for cleaner water, air, and land, the US Environmental Protection Agency (US EPA) was established as an independent federal agency. This was accomplished by bringing together 15 components (e.g., the Clean Water Act, the Clean Air Act, the Safe Drinking Water Act) from five different executive departments and independent agencies (e.g., US Department of Interior, US Department of Agriculture). The US EPA has responsibility for administering federal laws that protect human health and the environment.

A second relevant event occurred during the 1970 Stockholm conference. This event was the recognition that the detriment of the environment from human activities, widespread poverty, economic growth, etc. was interrelated to the worldwide efforts for achieving economic (quality of life) and social (freedom and opportunity) development. However, this concept of development was missing the important aspect of environment conservation, and it therefore required further expansion since all human existence is dependent on the environment.

A decade after the US EPA was established, a National Research Council report entitled "Risk Assessment in the Federal Government: Managing the Process" was released in response to a directive from the US Congress to perform a study of the practices of risk assessment (RA) and risk management (RM). It mainly dealt with controversies related to RA techniques (scientific basis) and their influence on regulatory actions and particularly policy decisions (RM). At the time, there were many public concerns regarding RA/RM related to the risk of cancer from exposure to chemicals present in the environment. There were also issues with the RA procedures supporting the control of asbestos, formaldehyde, and saccharin.

The concept of sustainable development was finally placed on the global stage by the 1987 report "Our Common Future" from the World Commission on Environment and Development (WCED). The WCED defined sustainable development as "development that meets the needs of the present without compromising the ability of future generations to meet their own needs." This is the most widely accepted definition of sustainable development across the world. In addition, a general framework of environmental strategies for achieving sustainable development by the year 2000 and beyond was established. Nevertheless, both the WCED and the resulting framework are still general concepts leaving much detail for implementation to be developed.

The next major event in sustainable development was the 1992 UN Conference on Environment and Development (UNCED) that took place in Rio de Janeiro, Brazil, also known as the 1992 Rio Earth Summit. This was the second worldwide meeting for discussing the nexus of environment and development. During this important event, the triple bottom line of Environment—Society—Economy (the three pillars of sustainable development) was adopted into international agreements related to climate change, biodiversity, and a unified declaration of environment and development. This was summarized in a global sustainable development action plan called "Agenda 21" and in the Rio Declaration (27 principles for sustainable development). Some of the content of US conservation and environmental law coincided with those described by many of the principles of the Rio Declaration.

Several conferences and meetings on sustainable development hosted by government, nongovernmental, and private organizations have followed the 1992 Rio Earth Summit. These include two main UNCED conferences: the 2002 World Summit on Sustainable Development in Johannesburg (South Africa) and the Rio+20 or Earth Summit 2012 hosted in Rio de Janeiro, Brazil. The latter was intended to reconcile the sustainable goals of our planet and to develop a political framework outlining global sustainability policies. In addition, at the US level, there have been some initiatives, reports, and studies for addressing and achieving sustainable development (PCSD 1996; NRC; Environmental Law Institute), but these have not transcended into mainstream law.

However, all of these sustainable development and sustainability approaches are good starting points for creating frameworks within some government sectors and for implementation as part of environmental, social, and economic laws and regulations. The main goal is achieving sustainability or sustainable development to meet the needs of a much larger but stabilizing human population, to sustain the support systems of the planet, and to reduce and minimize hunger and poverty (Our Common Journey, 1999 NRC report). Sustainable development should be the framework used for solving challenging problems and needs by any business sector including the chemical industry. But bringing it to practical fruition will require a major and sustained effort by many scientists and most importantly engineers. This again brings us back to the purpose of this book, which is the practical application of sustainability in engineering.

SUSTAINABILITY AND SUSTAINABLE DEVELOPMENT

One important issue that merits careful discussion is the commonality and the contrast between sustainability and sustainable development. Conceptually, both sustainability and sustainable development aim to preserve human existence on planet Earth over the long term. However, the focus on sustainability is roughly more toward the preservation of Earth systems that make human life possible without directly addressing any particular development trajectory. Sustainable development, on the other hand, adds an element of social development, which favors some level of equity in terms of social and economic opportunity. For instance, a working definition of sustainability that we have often used is "Sustainability, at its core, is an effort to create and maintain a dynamic regime of the Earth under which the human population and its necessary material and energy consumption can be supported indefinitely by the biological system of the Earth." One can see that here the focus is on maintaining or promoting the conditions that make "civilized" human existence possible on Earth over the long term. Sustainability under this particular definition can possibly be viewed as a precondition to sustainable development, i.e., sustainable development is unlikely to be feasible if some form of "civilized" human existence is not possible. Sustainability is a natural extension of the evolution of ideas from environmental protection to pollution prevention to sustainability. The normative content here is relatively minor, other than the expressed desirability of preserving human existence.

The definition of sustainable development is contained in the WCED report, and it basically states that sustainable development is "development that meets the needs of the present without compromising the ability of future generations to meet their own needs." This may seem superficially simpler, but it is implicitly a far more complex endeavor. Here the aim is not only to make "civilized" human existence possible, but to create conditions favorable to the social and economic development of people across the globe. This implies elements of social equity and justice, which while desirable in general do contain a significant normative component. One can certainly argue that these normative components are necessary for sustainability or sustainable development to succeed because they provide a social and, perhaps, political foundation. However, care must be taken to ensure that the development concept is not too deeply embedded in a culturally biased context, which may or may not be applicable across the many human cultures of Earth.

Consistent with the core mission of the US EPA, the focus of the present work is technological developments that may aid progress toward sustainability with the social, economic, and developmental aspects as secondary considerations. The present work is therefore more about sustainability than sustainable development.

BOOK CONTENT AND PURPOSE

A perusal of the book's table of contents will reveal a wide coverage of important and practical topics. These address a wide range of themes from green chemistry and engineering design to life cycle assessment, network design and control, conflict management, social and economic impacts, and many other central topics. Each chapter is richly referenced with the most important contributions from the literature, and each chapter has been rigorously reviewed by both the editors and independent reviewers. The editors would respectfully contend that this volume represents a comprehensive primer suitable for anyone interested sustainable process engineering. The volume is suitable for use as a reference by advanced undergraduate students, graduate students, and practicing professionals.

A first set of chapters describes the concepts of sustainability based on chemical engineering and process systems engineering foundations. Chapter "Toward More Sustainable Chemical Engineering Processes: Integrating Sustainable and Green Chemistry Into the Engineering Design Process" describes the integration of sustainability and green chemistry into the engineering design procedure. Sustainable processes are best designed, implemented, and realized by taking green chemistry and engineering design approaches from project inception. The foundations of green chemistry principles and their implications in chemical reaction development are described.

After a chemical reaction occurs, separation is the next unit operation that needs to be addressed to obtain a usable product. Therefore, the chapter "Separations Versus Sustainability: There Is No Such Thing As a Free Lunch" deals with the needs of integrating sustainability with process separations. Since separation operations involve a large environmental footprint, progress in reducing this footprint is needed using sustainable approaches. However, the high energy demand of separation operations remains a challenge for sustainable solutions.

Next, chapter "Conceptual Chemical Process Design for Sustainability" explores and proposes a methodology for incorporating sustainability analysis throughout all conceptual process design stages. This is done by using a hierarchical framework, short-cut solutions, and sustainability evaluation of different design alternatives for decision making. In addition, the importance of incorporating sustainability at early process design stages is discussed.

After discussing sustainability in conceptual chemical process design, chapter "Process Integration for Sustainable Design" brings key concepts, tools, and applications in the area of sustainable design through process integration. Mass, energy, property, and functionality integration approaches are discussed. These have the benefit of increasing not only sustainability but also efficiency. In addition, the application of sustainable process integration at multiple scales (molecular and life cycle levels) is offered.

Chapter "Modeling and Advanced Control for Sustainable Process Systems" starts with the inclusion of sustainability in process systems engineering by integrating process control with sustainability assessment tools for the simultaneous evaluation and optimization of process operations. The sustainability evaluation results provide information on whether the implementation of control systems is moving process performance toward a more sustainable operation.

Since a sustainable economic outcome must be achieved for any new process technology or proposed modification for commercial-scale use, chapter "Sustainable Engineering Economic and Profitability Analysis" addresses how sustainability analysis strongly impacts process economics and vice versa. In addition, some methodologies and approaches for capturing these impacts and interactions during process design are proposed.

Practically, any decision-making activity in the design of a more sustainable process system involves multiple decision makers. Therefore stakeholders must trade off many economic, environmental, and societal objectives. Chapter "Managing Conflicts Among Decision Makers in Multiobjective Design and Operations" proposes a novel multicriteria decision-making framework by computing a more sustainable compromise solution that minimizes disagreement among the stakeholders without the need of a Pareto front.

Because systems and processes are not always operating at steady state, dynamic approaches should be also accounted for when discussing sustainable systems. Chapter "Sustainable System Dynamics: A Complex Network Analysis" describes the implementation of complex network analysis to examine some integrated dynamic models of sustainability. In addition, it shows by using optimal techno-socio-economic policies how the system can be brought toward stable sustainability.

When there are many technology options for process synthesis, there is a need to determine the most sustainable flowsheet structure among the available and feasible alternatives. Chapter "Process Synthesis by the P-Graph Framework Involving Sustainability" describes process synthesis by the P-graph framework employing sustainability as an alternative objective for process synthesis running structural, economical, and ecological analysis simultaneously.

After combining sustainability concepts with process design and systems engineering foundations, it is necessary to implement process performance evaluation assessments in order to determine areas of improvement toward sustainability. Therefore practical sustainability assessment and performance improvement for chemical processes is described in chapter "Sustainability Assessment and Performance Improvement of Electroplating Process Systems." An electroplating process case study is employed to outline the proposed evaluation method, which prioritizes improvement measures to guide advances toward sustainability. In addition, chapter "Strategic Sustainable Assessment of Retrofit Design for Process Performance Evaluation" shows a new retrofit design framework for developing sustainability assessment at operational and supply chain levels in which economic, environmental, and social aspects are identified and classified according to retrofit actions.

The implementation of sustainability requires a holistic approach, one encompassing an entire system beyond the manufacturing facility. In addition, chemical processes have a huge reliance on ecological systems. Therefore life cycle considerations beyond the process to decide which design alternative is more sustainable are needed. For example, chapter "Chemical Engineering and Biogeochemical Cycles: A Techno-Ecological Approach to Industry Sustainability" describes a model for capturing the connection of natural systems with industrial systems for providing resources and dissipation of emissions, and the needs of including these links in decision making for sustainability. However, interconnecting chemical processes and their corresponding life cycle stages in early to basic process design phases is not an easy task. Therefore, chapter "Challenges for Model-Based Life Cycle Inventories and Impact Assessment in Early to Basic Process Design Stages" shows some challenges and solutions on how to account for life cycle inventory (LCI) and impact assessment in conceptual chemical process design. Some of these challenges are related to an early estimation of process LCIs, LCI data gaps, design decisions, and process scaling-up.

Since life cycle assessment (LCA) can be more than an environmental impact assessment approach, chapter "Life Cycle Sustainability Assessment: A Holistic Evaluation of Social, Economic, and Environmental Impacts" discusses life cycle sustainability assessment. This extends the holistic environmental LCA to account for the economic and social pillars of sustainability. Lastly, chapter "Embedding Sustainability in Product and Process Development—The Role of Process Systems Engineers" describes the practical role of process systems engineers in the implementation of sustainability in product and process development. It shows some key aspects and tools that practitioners should take into account to design and develop more sustainable products and processes during material selection, process design, process and product modeling, and supply chain implications.

▷ IN SUMMARY

The contribution of this book aims to demonstrate the development of new process design, process systems engineering, and life cycle perspective that have high influence, are easier to adopt, and are less complex, making it easier to create and design for sustainability. Invited well-recognized contributors from academia, industry, and government submitted chapters related to many of the concepts, tools, and applications in the areas of sustainable material management, social responsibility, product development, LCA/LCA tools, material reuse, material and energy efficiency, performance evaluation, and decision making. This book serves as an invaluable source of information and advances as society continues to implement the concepts associated with sustainable development in the analysis, synthesis, and design of chemical processes.

Gerardo Ruiz–Mercado
Heriberto Cabezas
Cincinnati, Ohio
June, 2016

Towards More Sustainable Chemical Engineering Processes: Integrating Sustainable and Green Chemistry Into the Engineering Design Process

D.J.C. Constable
American Chemical Society, Green Chemistry Institute, Washington, DC, United States

M. Gonzalez
US Environmental Protection Agency, Cincinnati, OH, United States

S.A. Morton
James Madison University, Harrisonburg, VA, United States

UNDERPINNINGS OF GREEN CHEMISTRY

With the benefit of hindsight, it may be easy to say that green chemistry and engineering began with the simple idea that chemistry, as it is generally practiced, needs to change. Green chemistry and engineering evolved from the ideas and practices of pollution prevention and waste minimization that were established in the 1980s. It also had additional help because industry had to respond to an array of legislation that governments promulgated to reduce and eliminate the generation of toxics and their emissions to the air, water, and on to the land [1]. Green chemistry and engineering, if it is practiced correctly, is the ultimate form of resource consumption minimization and pollution source reduction and control. It should not be a surprise that if one is reducing resource consumption and minimizing pollution, there will be great economic benefits that result.

From a chemist's perspective, one generally writes a simple equation (Eq. [1.1]) to depict a chemical reaction:

$$A + B \rightarrow C \qquad [1.1]$$

The problem in writing a reaction this way is that only a small part of the overall chemical story is represented. Very rarely can a chemist take two chemicals, mix them in their pure states, and expect them to react quantitatively to form a single product. Chemistry, like most of life, is messy and rarely this simple. To start with, chemicals

1

have to be contained in some way; they need some kind and type of reaction space that keeps the reactants close to one another so they can react and at the desired reaction temperature and pressure. Chemists do not often think about the types and places where these reactions will proceed and how to optimize thermal conditions (heating and cooling) and mass transfer phenomena (e.g., mixing, flow, etc.); that is the province of the chemical engineer. But chemists do know that for most reactions to proceed, reactants need to be dissolved in a solvent with unique properties that are beneficial to the reaction and along with other reagents (e.g., acids, bases, salts, phase transfer agents, etc.), catalysts, or the need for other chemicals to be added to get A and B to react in a timely fashion. Then there is a need to either isolate the product C or contain it as an intermediate that is to be used in the next step of a chemical synthesis or manufacturing process. Isolation of the desired product C is not a trivial step and many times will require the use of a filter aid, temperature swing, or an antisolvent to shift the chemical equilibrium and force the formation of a product-containing precipitate. Sometimes, as a result of these actions, an emulsion is formed that requires the chemist to add another chemical, like a surfactant, or another solvent or salt to break the emulsion. This results in another step or series of steps to the already growing process sequence.

While the resulting end product is obtained, all these additional chemicals, catalysts, solvents, regents, etc. generally end up as waste, with each needing additional process step(s) to separate, purify, recycle, remediate, or dispose of in some appropriate fashion. These additional actions require and add to the overall energy use and environmental footprint of that once simple chemical reaction, as identified in Eq. [1.1]. It is for these reasons that people started thinking more about how pollution could be minimized and prevented through changes in chemistry and chemical processes. What this typically meant prior to the 1990s was that in practice no one changed the chemistry occurring within the chemical processes; they just attempted to treat the waste to make it less impactful before discharging into the environment as a gas, vapor, liquid, or solid. As a strategy, it is certainly possible to continue managing and treating wastes of all kinds, but this approach bears significant costs, and even if something is less harmful, it may still have a variety of potential impacts to humans and the environment. It is also true that there are valuable resources within these waste streams that are being irreversibly lost to use by these management and treatment methods.

How Did We Get Here?

If one thinks about the varieties of chemicals used in many chemical processes, it is likely that a majority of them have the propensity to cause some type of toxic effect to humans [2] and to a variety of organisms if they are released into the surrounding environment. The inherent toxicity associated with chemicals has the potential to result in considerable economic costs, as well as environmental and human health impacts as chemicals are manufactured, transported, and stored prior to their use in a chemical process [3]. Mixing these process chemicals together, i.e., chemicals that are not reactants, in a manufacturing process usually does not reduce their toxicity. In many cases these process chemicals retain their toxic effects and this may lead to potentially multiple toxic effects to be managed within a waste stream.

It could be argued that most chemists generally take it for granted that the chemicals and materials they work with are hazardous and very reactive. They learn to control this reactivity in a variety of ways, usually through controlling conditions in a laboratory and by using small quantities of each chemical, activities that are feasible at small scale for minimal economic cost. It is that tendency for chemicals to react, combined with other specific and unique properties, that can lead to many potentially harmful effects or impacts that create problems for human and environmental organisms. Chemists rely on the fact that chemicals react with each other, and they purposefully choose chemicals that react quickly, robustly, predictably, and quantitatively [4–6]. But, for the nonchemists in the world, the idea of working with materials that can cause explosions, fires, or which in most cases have a variety of toxicity concerns is not something most people aspire to do. Nor are many consumers very comfortable with the idea of these chemicals ending up in a commercial product they use in their home or wear. Chemists may be amused by the naivety most of society has about chemicals, but the fact is that many consumers have an unhealthy fear of chemicals [7], despite the fact that they are heavily dependent on chemicals and the products made from chemicals.

This fear of chemicals in recent years has spurred considerable talk among government regulators and industry about the design and use of safer chemical alternatives [8]. On the surface of it, this sounds like something everyone can get behind. In practice it is not so easy to implement, and basing chemical use on the inherent hazard properties of a chemical compound or a mixture of chemicals creates a major problem for chemistry, as it is practiced today. Following this line of thinking to its logical extreme, a company could not make or sell a great number of commercial products that are critical to maintaining the lifestyles of those people living in the developed and developing world, and critical to the continuance of modern society. Nor could this approach allow a company to maintain their market position within their sector and be economically sustainable. Go one or two steps upstream in the value chain of any product and you will encounter a reasonable number of hazardous chemicals or by-products that are present as a consequence of the chemical reactions used in making the product.

Hazard and Risk

There are at least two strategies to help chemists address this particular problem. The first and traditional way industry approaches this problem is to merely say that industry needs to do a better job of communicating about the benefits of chemicals or products made from chemicals. A portion of communicating about these benefits is for industry to better help people understand the concept of chemical risk assessment [9–12]. This will aid in understanding that there is not a strict reliance upon basing acceptable chemical use and selection solely upon considering the inherent hazards of any given chemical. Risk is generally understood as follows:

$$\text{Risk} = f(\text{hazard, exposure}) \qquad [1.2]$$

where exposure is determined by the physical properties of the compound and the frequency and duration of the exposure to that compound. In the case of physical hazards like explosivity, exposure is further characterized by severity and probability. Developing a better understanding of risk allows one to accept that a person can work with hazardous chemicals, and exposures can be controlled to reduce or eliminate the impacts of hazardous chemicals in products.

The second strategy is to look at hazard as an opportunity. This tactic may sound a bit strange at first, but all that manufacturers want to do is produce a product or a chemical that performs its desired function and make a profit doing just that. For example, consumers want a material or product to have a particular color, or perhaps it is a protective coating, or a car that has good gas mileage, so a new structural composite that reduces the vehicle weight while providing greater crash protection would be desirable. If one thinks about color for a moment, color is imparted to different products sometimes as a dye (e.g., textiles), in other applications as a pigment (e.g., printing or coatings), or in other applications as a physical structure (e.g., how a butterfly imparts color to its wings). Each method a product developer might use to impart a color can be done in a way that either uses hazardous or nonhazardous chemicals. By avoiding the use of hazardous chemicals many potential problems may be avoided. For example, there are costs associated with purchasing the chemical, managing the chemical during a manufacturing process (i.e., engineering controls, personal protective equipment, emission and effluent permits, etc.), treatment of the process waste, and finally disposal of the chemical. And these are just the primary considerations associated with the use of a hazardous chemical. Also to be considered with housing and using chemicals, especially hazardous chemicals, are the insurance premiums, environmental permits and inspection along with their associated fees, security needs, zoning considerations, and public perception, to name but a few. Perhaps, it might be argued, that the greatest risk (economically speaking) is to those who are still using hazardous chemicals to impart color to a product, in this example.

Waste and Hazard Are Insufficient

Getting back to the multitude of chemicals used in chemical processing and general manufacturing, if those chemicals are not part of the final product, they in most cases become waste. Historically, say before the 1970s in the United States and the European

Union, waste was not very heavily scrutinized. It was certainly not uncommon for waste streams to be minimally treated, if treated at all, before being discharged into the environment as a gas or vapor, a liquid, or as solid waste. In 1969, after a river caught on fire [13] and a plethora of environmental and human health impacts were encountered with increasing regularity, industry was forced to pay greater attention to waste generation and treatment. Its first approach was to treat waste at the end of the pipe and make it less harmful prior to discharging it to the environment. Then, as mentioned previously, industry began to think about how it might change its manufacturing processes to produce less waste because they discovered that treatment and disposing of wastes was becoming increasingly expensive. Additionally, increased cost came about as more regulations were created that specified what could and could not be discharged, and increasingly complex treatment technologies were implemented to eliminate and/or reduce the overall risk associated with waste treatment, storage, and disposal.

THE PRINCIPLES AND IMPLICATIONS

In the early 1990s it struck many within the chemistry enterprise that not producing waste, especially waste containing toxic chemicals, would be best accomplished by doing more than just changing chemical process operations and/or treating wastes with end-of-the-pipe technologies. Additionally, the steady accumulation of environmentally related legislation was bringing greater attention to the fact that there were a considerable number of toxic chemicals commonly used in high volumes. Not surprisingly there was a desire, especially on the part of the US Environmental Protection Agency's (US EPA) Office of Pollution Prevention and Toxics, to reduce or eliminate the use and generation of hazardous chemicals. Two individuals, Paul Anastas and John Warner, were among the first to coalesce a number of ideas about how this might be done into a set of principles [1]. The 12 principles of green chemistry were published in 1998, and these were followed by the generation of 12 more green chemistry principles [14]. Paul Anastas and Julie Zimmerman [15] then extended these concepts and were the first to publish a set of green engineering principles followed by another set known as the San Destin Principles of Green Engineering [16].

In practice the Anastas and Warner 12 principles of green chemistry receive all of the attention, and this is most unfortunate since all the other principles expand the scope of green chemistry and engineering. Among the Anastas and Warner principles, most chemists promote principles 3 and 4, which have to do with molecular design and designing out toxicity in new molecular entities developed for commerce. Because catalysis is used in such a vast number of chemical processes, principle 9, a principle that promotes catalysis, has also received considerable research and development attention in fields of green chemistry and engineering. The remaining nine Anastas and Warner principles, along with the 12 by Winterton, and all the green engineering principles rarely are popularized. This is an important point because it is these "forgotten" principles that help one to move the practice of chemistry and engineering to be more sustainable.

Knowing that most people, including most business executives, like no more than three or four ideas in their heads at any given moment, all of the principles from the people mentioned previously were grouped into three general areas [17]:

- Maximize resource efficiency
- Eliminate and minimize hazards and pollution
- Design systems holistically and using life cycle thinking

While no one set of principles is sufficient to encompass the breadth of sustainability, these three categories represent a reasonable jumping-off point. Another general point to make about principles is that a principle is not a metric. For many in the public, and certainly among scientists and engineers considering the green chemistry and engineering principles, there is a tendency to become permanently mired in the details of questions like: "What does maximize resource efficiency mean in practice?" Then if one makes an interpretation of what they think that might mean, they then ask: "How do I know when I have the 'right' metric or set of metrics?" and "How do I measure progress against it/them?" This is where many allow the pursuit of perfection to become the enemy of the good. One can make considerable progress in green chemistry and engineering by simply considering the many ways in which these principles may be applied. It is better to make progress than to maintain the status quo or get caught up in extensive discussions or investigations seeking the "best" metric before acting.

A final point to make regarding the principles is that in order to truly make progress and understand how to make chemistry, chemicals, and manufacturing processes greener and more sustainable, one must bring a range of scientific and engineering disciplines to bear. It is simply not sufficient, for example, to know only synthetic organic chemistry, or thermodynamics and kinetics, or some other subdiscipline of chemistry or engineering. It could be argued that one's inability to embrace the inherent complexity in sustainability is the biggest reason for resistance among scientists and engineers in moving sustainable chemistry and engineering practices forward.

Maximizing Resource Efficiency

Thinking on a global scale, something that most people rarely do, one might ask where things come from, how the vast array of chemicals are produced, or ask how these chemicals are used to produce the products and services in commerce. Making materials and products in recent years has driven industry to adopt increasingly complex global value chains in their relentless drive to remain competitive and profitable. Progressively over the next 10–25 years, access to key raw materials for chemical production, alternative energy, automotive applications, and electronics, to name just a few sectors, will come at significantly increased cost and environmental impact. There are a variety of reasons for this trend, but basically, known reserves of high grade ores from which we obtain key elements are diminishing or access to these markets is becoming more difficult [18]. These constraints facilitate the shift to the exploration for new sources of key elements from lower-grade ores and ores that are invariably less accessible. The more difficult it is to access the ore or the source of the element, the more energy, materials, and wastes that will be associated with the desired refined metal or element.

As society comes to terms with increasing costs and demands, it is likely to be driven toward increasing material and energy efficiencies, i.e., decreasing the quantity of material and energy utilized for every product produced. The consequence of high material and energy intensity (low efficiency) is, of course, the production of large amounts of waste. In 1992 Roger Sheldon published a paper on the relative amount of waste in different industrial sectors and coined the phrase known as E-factor [19]. E-factor is related to mass intensity (MI) as follows:

$$E_{factor} = MI - 1$$

$$\text{Or in expanded form: } \frac{kg_{waste}}{kg_{product}} = \frac{kg_{input}}{kg_{product}} - \frac{kg_{product}}{kg_{product}} \qquad [1.3]$$

As can be seen from Table 1.1 the further one is removed from petroleum extraction the greater the quantity of waste produced. The challenge for making chemical processes more sustainable and greener, in the light of resource efficiency, is threefold. First, we must significantly decrease the material and energy intensities associated with chemical manufacturing observed in all industrial sectors; for example, decrease the mass intensity by at least an order of magnitude, if not more. The second issue, which is slightly more intractable at the moment, is transitioning from fossil sources for our carbon frameworks, i.e., petroleum and natural gas, to renewable sources. Third, the transition to alternative frameworks must be done in a way that reduces toxic chemical usage, not merely obscures their use in early stages of the system.

The magnitude of this challenge should not be underestimated. To make a switch to renewably derived feedstocks, our economy would need to change a deeply entrenched and interdependent system for making chemicals that took more than 100 years (plus or minus a few 100 million years to form fossilized carbon such as that in petroleum) to

Table 1.1 Mass Intensity of Different Sectors of the Chemical Industry [19]

Industry	Mass Intensity (MI) $kg_{total}/kg_{product}$ Excluding H_2O
Oil refining	c. 0.1
Bulk chemicals	1.1–5
Fine chemicals	5–50
Pharmaceuticals	25–100

develop and continually optimize. This switch must also be done in a shorter timeframe without major social or economic disruptions and with decreased environmental impacts, if society is to preserve the earth's biodiversity and maintain our current large human and nonhuman populations. Using renewable sources to develop new chemicals and the products that we make from those chemicals, while not causing environmental degradation, will also require a considerable amount of research and work.

The major obstacles to increased waste reduction are institutional and behavioural rather than technical.

Serious Reduction of Hazardous Waste. US Congress; 1986.

Historically, governments have developed policies and created extensive regulatory frameworks or voluntary initiatives to force or induce industry to focus on waste emissions to air, land, and water. They have also focused on the reduction and elimination of toxics as opposed to preventing waste and toxics generation through chemistry and chemical technology innovations. As a result, and somewhat predictably, there is now an extensive industry to help companies identify hazardous properties of the chemicals routinely found in commercial products. Sadly there continues to be far less focus on, and certainly far less is understood, about how to transition from current unsustainable practices of making chemicals and the myriad of products made from chemicals.

Eliminating and Minimizing Hazards and Pollution

Thanks to the US EPA, and most everyone else who is involved in green chemistry and engineering, eliminating or reducing the generation of hazardous substances and waste is typically what most people associate with green chemistry [20]. There is considerable debate about exactly how many chemicals are in common use, and numbers range from 80,000 to two or three times that. However, if you look at chemicals currently used in high volume, there are about 4000 routinely handled. Many of these compounds have one or more hazardous properties, and a smaller number are extremely toxic and therefore present challenges in their use and handling. But they are used because they are extremely expedient for effecting key chemical transformations. Without them, many of the products people use routinely throughout the world and have come to rely on simply would not exist.

All substances are poisons; there is none which is not a poison. The right dose differentiates a poison from a remedy.

Paracelsus (1493–1541)

If one thinks about the remaining 76,000 or more chemicals that are not part of routine high-volume chemical production and where these are routinely used, one need only to look to the world of research and development and to a lesser extent into small specialty chemical or niche chemical applications. There are now many people who passionately believe that it is critically important to understand all of the potentially hazardous properties of all chemicals that are made and used. Unfortunately that scale of understanding comes at a significant cost. The cost to perform hazardous property testing on the large number of chemicals the global chemistry enterprise uses on a daily basis is probably for all practical purposes well beyond society's ability to pay.

While there is an increasing body of chemicals legislation being promulgated throughout the world [21], it will be some time before a significant number of chemicals are evaluated. In the meanwhile, at least in Europe, and in some states within the United States [22], as well as a few other countries, the use of selected chemicals has been banned. Interestingly, chemicals legislation may be superseded by customer demands in the not-too-distant future. A great example of this has been the reduction in the use of bisphenol A, a chemical used as a plasticizer in plastics like polycarbonate, adhesives, can linings, and in other applications. Concerns regarding bisphenol A and its potential for endocrine disruption has moved manufacturers to remove it from many products, despite rulings by the US Food and Drug Administration that its intended use in a range of common products is considered to be safe for humans [23]. In the future this tendency toward corporate risk aversion is more likely to be a driving force in reducing or eliminating chemicals in some products than through the means of targeted legislation.

While there are green chemistry and engineering principles focused on reducing or eliminating hazardous chemicals, and these principles may challenge some chemists to design new chemical entities and basic building blocks or framework molecules, this collection of principles is arguably among the hardest for chemists to make progress on. This is not surprising given the fact that for a majority of chemists there is usually no formal training in traditional chemistry degree programs to help them understand the relationship between chemical properties or molecular structures, and how chemical properties and molecular structures affect the toxicity or hazard potential of any given chemical. Moreover, a particular molecular structure is synthesized because it is known to react in a certain way, or to exert a desired effect during a reaction, or to affect a desired property in a particular end use application, as in the case of pesticides, herbicides, and drug substances. Even in instances where chemists are trained or develop experience over many years in structure—activity relationships, it remains a considerable challenge to predict how a particular chemical may affect one organism or another.

To address this there are a variety of researchers working to better understand chemical structure—activity relationships in the hopes of developing practical molecular design guidance [24]. Given that the pharmaceutical and agrichemical industries have been working on developing a better understanding of structure—activity for over 60 years, and it is something that remains a challenge today, it is likely to be some time before this tactic yields a general set of molecular design principles to guide chemists as they make new chemicals.

It is well known that within the current US economy the majority of the commodities purchased, on the basis of mass, only remain in use for approximately 90 days before they are disposed of as waste. This is certainly a sobering fact and one should not therefore forget that there are hazards and associated risks with a chemical's or product's end of life. A share of pollution prevention and elimination must therefore include a consideration of the chemical or product design for recycle, reuse, or biodegradation. Chemicals produced from waste are something that historically has received little or no attention; society simply wants to make things, use them, and dispose of them when no longer needed without much thought about what happens to them once they leave the consumers' hands. This concern is an area that is now receiving greater attention, especially in Europe, and there is the promise of great progress in converting waste into chemicals in the not-too-distant future [25,26].

For materials that cannot be recycled or reused, one should think about designing them for biodegradation. Implicit in any consideration of biodegradation or any chemical form of degradation is chemical fate, i.e., where does a molecule, material, or product and in which form does it end up once it is considered to be at the end of its useful life and/or released to the environment? Where something ends up, whether it makes its way into the air, or water, or on land, there will be different degradation pathways and rates associated with that chemical, material, or product. And, as a chemical, material, or product degrades, the many degradation products that are formed need to be nontoxic. While this is considered for products like active pharmaceuticals and agricultural chemicals, it is generally not considered for any of the many consumer compounds that are made every year. The downside of chemical innovation is the potential creation of a plethora of unintended consequences.

Design Systems Holistically and Using Life Cycle Thinking

Perhaps it is a consequence of how chemists are trained, but they tend to mainly think about creating a new molecule or running an experiment that proves a certain chemical phenomenon or mechanism in a particular reaction. This falls in line with the understanding that humans, in general, are reductionist thinkers; they break things down into smaller pieces so they can better understand a complex problem. This is a concern when expanding this rationale into the concepts of sustainability. Sustainability is inherently complex with a large number of interacting variables and it is difficult, to say the least, to control many of the variables. However, if one is to succeed at sustainable design, one must learn to embrace complexity, employ systems thinking, and learn how to use tools like life cycle assessment and identify how their research contributes to this system perspective.

There are three main aspects to this set of principles. The first is that one must employ the best designs; that is to say, a person needs to think about design, but not just design of a chemical or reaction in a beaker or a mini-test tube or flow cell. This is becoming known as "Thinking beyond the bench." The second aspect we need to consider is design at a systems level; it is no longer just sufficient to create a new chemical entity or use a particular reaction sequence that has the desired effect or makes a specific chemical as if that effect or that chemical will only ever be observed in the confines of a laboratory experiment. Third, chemical design needs to be informed by knowledge of where things come from, how the desired chemical will be used, and where it will end up after it becomes part of a product. This is what it means to think holistically or to employ life cycle thinking.

It could be argued that some researchers in certain scientific disciplines or areas of study are more amenable to systems thinking than most researchers in the chemical sciences given their work to understand entire ecosystems; the interplay of, for example, microorganisms, plants, invertebrates, vertebrates (animals), and, of course, humans across space and time. Or perhaps it is a climate scientist who attempts to model the complex interactions of atmospheric chemicals at differing concentrations, aerosols, cloud coverage, albedo, wind, temperature, humidity, etc. It is this ability to look at the big picture and discern key interactions, responses, and impacts that is critically important to furthering the implementation of green chemistry and engineering in the global chemistry enterprise.

Among the disciplines that should be taught to chemists and engineers is life cycle inventory (LCI) and assessment (LCI/A). This has been explained elsewhere and the reader is encouraged to develop a good understanding of LCI/A [27,28]. Very simply, LCI/A is a rigorous analytical methodology that systematically inventories and evaluates the environmental impacts of a product or activity. Starting with the function or functional unit of interest, for example, a computer, a single application of a pesticide, the coating on a car, a service, etc., the analyst works their way through the unit operations, chemicals, and services all the way back to raw material extraction. For each unit operation there is a detailed input/output inventory or an accounting of all the mass and energy used, the by-products formed and sold, emissions, and what is disposed of as waste and in which form (i.e., gas, liquid, solid). LCI/A also looks at the product as it is used, and what happens to it when it no longer performs the service or function for which it was originally intended. LCI/A forces one to look across entire systems because most products are not merely raw material extractions followed by immediate use with no attendant emissions. LCI/A broadens one's perspective as one seeks to understand material and energy flows through and across systems and accounts for the associated impacts.

Upon obtaining the LCI representing the system under evaluation, the next phase is to assess these inputs and outputs for their potential impact. The impact assessment (IA) phase of an LCI/A is the evaluation of potential human health and environmental impacts of the environmental resources and releases identified during the LCI. IA addresses ecological and human health effects as well as resource depletion (e.g., water, land, minerals). A life cycle IA attempts to establish a linkage between the product or process and its potential environmental impacts.

The tie-in those LCI/A principles seek is to have scientists and engineers look across systems holistically and consider life cycle impacts with green chemistry and green engineering approaches, which should be readily apparent. A chemist only considering or optimizing a single reaction or a reaction sequence that is part of a process will easily miss the use of a very nasty chemical in another part of the value chain. By looking across the entire value chain, it is possible to optimize the use of chemicals and energy so that only those chemicals having the fewest impacts, together with the most appropriate chemistries, synthetic strategies, and the best possible processes, are selected.

PROBLEMS WITH CHEMICALS AND REACTION SPACES

In general, chemists do not often think about the actual environment where a reaction takes place in a chemical process. In a laboratory setting they focus on the type of chemistry they are doing, for example, traditional analytical wet chemistry (e.g., titrations or C, H, N analysis), materials chemistry (composites, ceramics, nano, etc.), natural products synthesis, perhaps traditional synthetic organic chemistry, etc., and chemists make use of the glassware, reaction vessels (e.g., round bottom flask, high pressure cell, a micro-test tube, etc.), or a lab apparatus that is readily available to them. The chemical reaction occurs in some kind of matrix (liquid, gas, solid, and variations on these physical states) and that matrix is contained in something that is readily available, which generally bears remarkable similarity to vessels and lab apparatus that have been in use for hundreds of years. It is now time for chemists to expand their reaction space to represent that which occurs at the chemical process level.

Chemical Reactivity

As was pointed out previously, chemists also select chemicals they are familiar with, are confident will react and react predictably (e.g., no large exotherms, endotherms, rapid gas evolution, etc.), are robust, and, ideally, will react within a rapid timeframe. Another way to say this is that chemists like well-behaved and robust reactions they can set up, walk away from, and come back to at a later time to see if the desired reaction has taken place. The chemist's focus is solely on the reactants and the desired product. And, most of the time, their focus is only on the desired product. Very rarely is it on the reagents, solvents, number of steps, reaction conditions, purification needs, or the vessels or apparatus that contain the reaction. All this other "stuff" is more typically seen either as agents to promote the desired reaction or as being a part of the chemical engineer's domain, and it is the chemical engineer that is tasked with turning the reaction or a series of reactions into a chemical process.

From a sustainable processing standpoint, this state of affairs has many implications. Because discovery chemists do not generally worry about making large amounts of any given molecule or product of interest, there is a tendency to pay the greatest attention to ensuring that chemical reactions proceed rapidly to completion. In this context, energy demand is ignored, so heating, cooling, pressures, or mixing for long periods are common, as are complex isolation protocols, multiple salt washes, etc. These practices are seen as a natural part of attaining the highest possible yields of the desired product. It is usually very late in compound discovery or in the initial multikilo runs where chemists begin to think about all the other components that may be part of a chemical process. And if this is the case, it really is too late to do too much about the particular chemistry route that is chosen. It is at this point that a process chemist receiving a route from a discovery chemist will very quickly change it (i.e., proceed to the final product through a different set of intermediates) to avoid these common excesses.

However, commercial development of new chemical entities typically requires that the time to identify and demonstrate commercial viability is minimized and products are pushed to market rapidly and as soon as an economically viable route and manufacturing process can be established or retrofitted into an existing process. This tends to lock in a less optimized process containing excesses in chemical, solvent, and energy, etc. because the chemist is merely optimizing what is now seen to be the commercial process and is focusing on important matters like throughput, quality, and yield. It is beyond the scope of this chapter to go into great depth on all the potential areas of latent excess associated with reagents, solvents, salts, etc., so solvents will be chosen to illustrate a few of the key issues to be resolved more generally in moving toward more sustainable processing. In the next few paragraphs where you see solvents you would also find the same to be true for many reagents, reactants, catalysts, salts, etc.

Solvents—Why We Use Them and Can We Eliminate Them?

Solvents are extremely important to the chemical processing industries both as an enabler and as a major contributor to large waste volumes. Solvents are used in many different industries to process, manufacture, and formulate products. They are also used in various unit operations like separations (gas/liquid, liquid/liquid, solid/liquid), in situations where the solvent facilitates the reaction medium, or is only there to dissolve the reactants so each may come into intimate contact with the other and the reaction may proceed more rapidly to completion. Solvents are used to control reactivity through moderating heat transfer, to clean equipment like reactors and overheads, and as part of product formulations such as in paints, textiles, rubber, adhesives, and many other products. In fact, it is quite hard to imagine chemistry and chemical processing without solvents in one part of a process or another.

While solvent use is ubiquitous, most solvents have a collection of environmental, health, and safety (EHS) hazards associated with them. These include human and eco-toxicity hazards, process safety hazards, and waste management issues. Many solvents are also facing increasing and continuous regulatory scrutiny, especially those like chlorinated solvents (e.g., CCl_4) or the dipolar aprotics (e.g., CH_3CN). These issues and others underline the importance of eliminating, minimizing, or optimizing the use of solvents in chemical processing and products. Chemists and engineers generally find it challenging to avoid using large volumes of solvent. When it is not possible to dramatically reduce or eliminate solvent use, they should pay close attention to solvent selection to ensure that EHS and sustainability impacts across the value chain are minimized. This approach is an excellent example of applying life cycle thinking to quantify and visualize how a solvent change or reduction can impact the entire value chain. Once a solvent is selected, green chemistry and engineering principles should help shape optimization and minimization decisions in solvent use. Finally, recycle and reuse of solvents is critical and there are many methodologies provided in this book to enable this.

While it has been well established and documented that regulation, legislation, and technology development have played an important role in the advancement of policies and methods for pollution prevention, it can be viewed by some that the resulting impact of new regulations or legislation leads to the advent of new technologies to support, enable, or defend these new policies. Whereas it can also be viewed by others that the development of new innovative technologies spurs the need for and enacting of new regulations and legislation to allow these technologies to be utilized in the commercial sector with the end goal of protecting human health and the environment. Regardless of which path is taken, it is a fact that technology development and regulation/legislation are working together to ensure that more sustainable pathways are taken. In each path the intent is to ensure resource longevity and that the products and services we produce have as minimal an adverse impact as possible to human health and the environment. One such example is the recent passage in the United States of the 2015 definition of solid waste (DSW) Final Rule [29]. This new redefinition of the existing 2008 DSW rule promotes responsible recycling of hazardous secondary materials. The modifications to the 2008 rule prevent the mismanagement of hazardous secondary materials intended for recycling, while promoting and encouraging safe and environmentally responsible recycling of such materials, including solvents (industrial solvents in four industrial sectors). This new ruling will facilitate solvent reuse within the industrial sector and encourage the use of material substitution and/or reclamation in the chemical industry. As more and more companies begin to utilize this ruling, the results, of this reuse option, will lead to the quantification of the environmental, economic, and material benefits gained with the implementation of new rules and regulations that lead to alternative material life cycles.

To further elaborate on the potential of a new policy to promote sustainability, let us take a more in-depth look into the benefits gained from solvent reuse. By reusing a solvent, previously destined for incineration, we are extending the useful life of the solvent, offsetting the need for virgin solvent, while generating significant financial and life cycle environmental benefits and lowering overall risk. Another benefit is the reduction of short-lived climate pollutants (SLCPs), known as super pollutants, such as ozone precursors, which are the result of a solvent's incineration. Because of their short lifetimes (when compared to CO_2, which remains in the atmosphere for approximately a century), actions to reduce emissions of SLCPs will quickly lower their atmospheric concentrations, yielding a relatively rapid climate response. This approach will provide a better understanding of avoiding the significant life cycle impacts associated with creating solvents, using them only once, and then destroying them.

Reaction Spaces

This section began by discussing chemical reactions and where they take place. Following a brief discussion of the reaction environment from a chemical perspective, it is time to revisit this topic. From a sustainable processing perspective, where reactions take place can dramatically influence the reaction chemistry and the use of the other reagents, solvents, catalysts, etc., or "stuff." In the global chemistry enterprise there are basically two types and scales, or variations of these scales, that are routinely used for chemical manufacture. The first is large petrochemical processing characterized by high volume continuous flow regimes. The second is batch chemical processing characterized by much smaller volumes and discrete batches. Petrochemical manufacturing processes have been extensively modified over the past 100 years and many of them are catalytic, so they are among the most mass and energy efficient chemical processes available. Batch chemical processing tends to be considerably less efficient, and much has been published about just how inefficient these processes are.

Recently there has been a considerable amount of research and development into alternative technologies to traditional batch chemical processing [30—34]. It would be useful to talk about these innovations in light of their potential to impact the type of chemistries and chemistry strategies a chemist might employ to make batch chemical processing more mass and energy efficient. It would also be useful to think about using different chemistry strategies that influence how reactions can be carried out in nontraditional media as, for example, in the case of using micelles as minireactors. Perhaps the best way to illustrate these approaches is with a few examples.

Imagine for a moment that a chemist wants to use a reaction that is highly exothermic, or is energetic in some fashion (say rapid CO_2 evolution), or maybe it is exothermic with a rapid evolution of hydrogen (explosion hazard). In the laboratory, as most chemists are trained, this can be handled in one of several ways. The chemist might precisely control the order and/or rate of addition of a particular reagent or

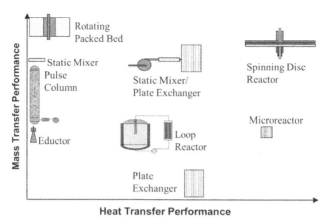

Figure 1.1 Alternative reactors not generally used by chemists in the laboratory.

reactant. Alternatively they may use a dip tube or some other mechanism to deliver the reagent or reactant into the reaction vessel under conditions of high mixing. Another option might be to run the reaction in a large excess of solvent that is cooled to a low temperature. One or more of these strategies could be used to safely and productively run the reaction.

In the laboratory any of these approaches may be used, but there are alternatives that most chemists would not consider because of their lack of training and experience to use alternate technologies to run a reaction. Fig. 1.1 provides an example of different types of reactors that are generally not used by chemists but which would provide alternatives to conventional reaction spaces.

In fact the best way to proceed with this reaction is maybe to run it in a microreactor. The microreactor could be constructed to have a large mass relative to the introduced mass of reactant that could absorb a rapid temperature rise. The reactor size and thermal mass could be combined with a clever configuration of different channels and pumping rates to achieve cooling through heat transfer, or one could precisely control the order and rate of reagent addition in one part of the reactor to ensure that there were no hot spots and good heat and mass transfer. It would also be possible to achieve rapid and intense mixing in a tiny volumetric space, and in the case of rapid gas evolution one could include a vent for gas emission, or a back-pressure regulator to ensure that the gas remains in solution. The engineer would then only have to number up a series of reactors as opposed to scale up to a larger reactor volume where safety and operating concerns caused by the exothermic and gas evolving nature of the reaction would present significant operational and engineering challenges. A reaction does not have to be problematic as in the case of a gas-evolving and exothermic reaction to take advantage of the many features a mini- or microreactor affords. For example, reactions requiring very precise temperature control to avoid by-product formation, or high-pressure reactions, or multicomponent reactions with short-lived intermediates, etc. are reasonably and productively accommodated. Such environments open up the chemist's toolkit to allow a high degree of thermodynamic and kinetic control over reactions that are not available in a large batch environment. One can also think of expanding the process opportunities to permit a series of reactions without the need to isolate intermediate or continuous separation operations employing in-process recycle of reagents and solvents. Accessibility to these operation and control opportunities are commonly known as "process windows" [35].

A second area of recent interest has been in creating reaction environments using nontraditional media, such as water [36–38]. In some cases, for example, surfactants are used to create micelles, which provide a unique opportunity to promote reactions that were not believed possible in water [39]. Such reaction environments also challenge long-held beliefs by chemists that reactants need to be completely dissolved or could only be dissolved in organic solvents in order for reactions to proceed. The point here is that unchallenged beliefs by chemists about the conditions required to run reactions should be challenged not only for their potential sustainable and green chemistry benefit, but because a significant number of important chemical reactions can have improved chemical reactivity (i.e., lower mass and energy intensity).

THINKING ABOUT WHAT MORE SUSTAINABLE CHEMISTRY AND CHEMICAL MANUFACTURING MIGHT LOOK LIKE

Earlier in this chapter there was a brief discussion regarding the green chemistry and engineering principles and the idea of those principles associated with maximizing resource efficiency. Further consideration of the concept of maximizing resource efficiency is worthy of additional attention here, as one considers the implications of moving society closer toward living in a style that ensures future generations can live as well as or better than the current generation. When you look at the global chemistry enterprise it is overwhelmingly dependent on fossil carbon sources for its organic building blocks, and the rise of the chemical industry closely tracks the rise of petroleum as a source of transportation fuels. It is interesting to note that in 2014 only 1.3% of petroleum used in the United States was naphtha for petrochemical feedstock with an additional 0.7% as other oils for petrochemical feedstock use. This compares to 45% as finished motor gasoline, 29.9% as distillate fuel oil, and 9.6% as jet fuel [40]. This means that needing to make more sustainable changes in the chemical value chain is not likely to be a major driver for change in the petroleum industry, and changes to the petroleum industry are only likely to occur as the world considers heating and transportation, or, more generally, our desire for mobility.

Implications of Different Chemical Feedstocks

The chemicals industry is also heavily dependent on a variety of metals and other elements mined or extracted from the earth. As the human population has increased, the rate at which key mineral resources used in the chemistry enterprise are extracted from the earth continues to escalate [41,42]. A largely unanswered question in this scenario is: "Can the rate at which resources are being extracted continue to rise as the human population also continues to grow, and can an increasing number of existing and new products be supplied to a population desirous of obtaining greater access to those products that are associated with modern Western society and a higher standard of living?"

Currently the petrochemical industry, as noted, produces key building blocks that are used in high volumes, for example, alkanes, alkenes, olefins, aromatics like benzene, aniline, terephthalic acid, etc. Fig. 1.2 shows an illustration of the supply chain for some of these compounds. These molecules are straight chains or aromatic compounds of varying saturation where the carbon atoms are in a highly reduced state and they have very little functionalization, i.e., they do not have varying heteroatoms like oxygen, nitrogen, sulfur, etc., which are sites for further reaction. The global chemistry enterprise has been built over the past 150 years to deliver these chemicals in high volume and it does so with considerable mass and energy efficiency, which has been, and continues to be, under constant improvement. In fact, it has been reported that there are about 120 molecules that account for a majority of the chemical frameworks or building blocks used by chemists to build new molecules [43,44]. Given the large number of chemicals

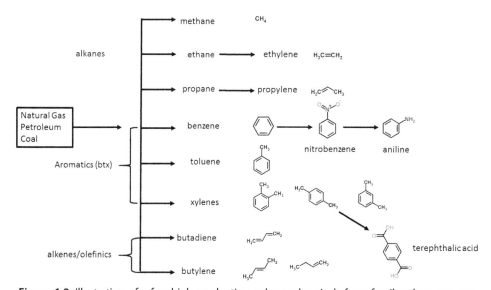

Figure 1.2 Illustration of a few high production volume chemicals from fossil carbon sources.

that have been discovered and the potential chemical diversity available in the world, this is a surprisingly small number of compounds that chemists work with.

In addition, chemists have developed a variety of chemistries, i.e., methodologies or protocols for activating, functionalizing, and otherwise chemically transforming these framework molecules or basic organic chemical building blocks. These framework molecules need to be functionalized (adding heteroatoms like N, S, O, etc.) or coupled with other building blocks in specific configurations to synthesize the many different types of molecules used in almost everything that humans use for food, shelter, and clothing. These framework molecules derived from petroleum generally look very dissimilar to what are made by plants and other living organisms. Therefore if the global chemistry enterprise is to transition from the petrochemical building blocks used today to a different set of building blocks, the ways that most chemists used to chemically transform building blocks will necessarily have to transition to a different set of chemistries or methods to chemically transform molecules. In general, chemists are not currently trained to as great a degree to effect chemical transformations on chemicals derived from biological processes nor do they have the experience with reactions and synthetic strategies required to construct new molecules based on these types of building blocks.

Framework Molecules—Moving From Petroleum to Sugars, Lignocellulosics, and Proteins

Transitioning to biologically based and renewable carbon framework molecules is not going to be an easy transition to achieve. In the first instance, sustainably or renewably sourced framework molecules will have to be produced as cost competitively as those produced via the petrochemical complex if they are going to displace petroleum-derived molecules. At this point one might reasonably ask where else might the chemical manufacturing enterprise find or develop the basic building blocks for chemicals. There are a variety of strategies that have been proposed over the past 10—15 years and many of these are under advanced development.

For example, there has been a tremendous amount of research and development effort directed toward using many different types of biomass and an even greater number of chemical, thermal, and biological processes to obtain desirable chemical building blocks or drop-in replacements. Fig. 1.3 contains the top value-added chemicals derived from biomass as envisioned by the US Department of Energy National Renewable Laboratory in 2014. As is the case with the modern petrochemical complex, the vision is for a highly mass and energy efficient integrated biorefinery [46—49]. As was mentioned in the previous section, these molecules are very different than those directly obtained from petroleum sources as they contain significant amounts of oxygen and, in a few cases, nitrogen atoms, but none of them have any aromaticity.

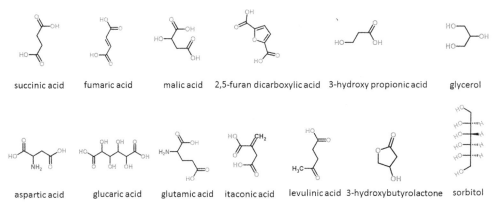

Figure 1.3 US Department of Energy National Renewable Laboratory top value-added chemicals from biomass [45].

It is beyond the scope of this chapter to cover all of the potential biomass and processing strategies, but a few general strategies are worth mentioning here. The first is represented by feedstocks from the forest products industry like cellulose, hemicellulose, lignin, sugar, pine chemicals, etc. The second is the development of feedstocks from a substance chemically similar to lignin, and that is chitin. Chitin is the material found in many marine organisms like shrimp, lobster, etc. The third is the development of feedstocks from waste biomass like food wastes, for example, citrus peels and pulp, potato skins, etc.

Considering the forest products industry, there are two general strategies that account for a majority of the activity to develop useful chemical products. The first is looking for opportunities in lignocellulosics and the second is opportunities for high value-added chemicals derived from the leaves, roots, seeds, etc. of trees, bushes, etc. Within the lignocellulosic arena there have been many advances in separating cellulose, hemicellulose, and lignin, with multiple products made from cellulose and hemicellulose, derivatives of cellulose, and sugars derived from cellulose. The work to successfully break down lignin into various component parts continues, but to date there is no strong viable commercial process for turning lignin into useful chemical products. Finding a use for lignin is a similar dilemma to that for biodiesel production where a considerable amount of biodiesel is being produced, but the by-product glycerol remains a problem when it comes to finding a viable high value-added commercial use. The chemical composition of lignin is quite interesting given its high aromaticity and the potential for using the component parts as chemical building blocks. The problem, as with most things in chemistry, is how to break and make new bonds precisely where it makes the most sense while doing it in a mass and energy efficient manner.

To illustrate another component of the forest products industry it may be instructive to take a brief look at the pine chemicals industry. The pine chemicals industry is not a new industry, but it is a very small portion of what is now known as the specialty chemicals industry, despite the fact that pine chemicals have been in active use for longer than the modern chemical industry era that arose in the early part of the 20th century. The pine chemicals industry has been extracting useful products such as turpentine and other simple materials for literally hundreds of years. With the rise of the pulp and paper industry, chemicals have in the majority been extracted from two waste streams: crude tall oil and crude sulfate turpentine. Crude tall oil can be further separated into a fatty acid fraction, a tall oil fraction, a tall oil rosin fraction, and a tall oil pitch fraction. The crude sulfate fraction is separated into a variety of terpene monomers that can be further transformed into a variety of terpene resins. All of these streams can be used as raw materials for coatings, various oil applications, surfactants, adhesives, inks, etc. [50].

While the pine chemicals industry is not currently as technically sophisticated an industry as the petrochemical industry from a chemical engineering perspective, it does have many similar unit operations like distillation as the primary means of separating complex mixtures. It does, however, represent a model for the integrated biorefinery of the future where the various distillate fractions are further separated into additional chemicals that may be transformed into useful framework molecules.

One company that has achieved success in recent years in separating the various components of woody biomass is Renmatix [51]. Renmatix has succeeded in developing and licensing a near-critical water process that precisely delivers streams of cellulose, hemicellulose, and lignin. It can further separate the cellulose and hemicellulose fractions into their component sugars. Once separated, it is then possible to use these sugars for a variety of fermentation processes to produce fuels (ethanol or biodiesel) or chemicals, depending on the microorganism that is used. Sugar is seen by some as the next oil, i.e., a common substance that can be used in a variety of industrial fermentation processes to deliver chemicals of interest. Lignin at this point is usually burned for its energy value, but this is arguably a less than optimal use of biomass.

A great example of companies making the transition to sugar as a source for chemicals may be found in companies like BioAmber [52] and Myriant [53]. These companies have identified succinic acid, one of the compounds identified in Fig. 1.3 as a market opportunity. Succinic acid is a 4-carbon chain molecule that is more cost effectively obtained via a sugar fermentation process than it is obtained from petrochemical sources. Additionally, succinic acid is a useful building block for a variety of commodities (e.g., 1,4-butanediol) and specialty chemical end use applications. So market demand for it is assured. Sadly there are currently not many other success stories like these.

An example of the difficulty in making a transition to a large volume chemical or chemicals from a sugar source might be the example of polylactic acid (PLA), pictured in the following page. DuPont initially discovered this molecule in the 1950s. But it was not commercialized to a great extent because it was difficult to produce from the chemicals readily available from petrochemical companies. In the early 2000s a joint venture between Dow and Cargill known as NatureWorks [54] was formed. As a large diversified agricultural company looking for additional sources of income from its corn milling operations, Cargill continues to be an enthusiastic backer of chemicals from agriculture. Dow contributed a significant expertise in making polymers to the joint venture, but eventually withdrew because of the slow progress to realize commercial success. The venture was focused on obtaining lactic acid from the fermentation of sugar (obtained through enzymatic conversion of corn starch), followed by conversion to the polymer PLA. PLA was seen as a desirable polymer since it could be consistently obtained from agricultural sources and is biodegradable. The caveat to PLA biodegradability, however, is that it must be degraded in industrial composting facilities, i.e., where the temperature of composting can be maintained above $50°C$ for an extended time. If one does not compost at elevated temperatures, PLA has pretty much the same biodegradability as a variety of other plastics.

$$\left(O - \underset{\underset{CH_3}{|}}{\overset{\overset{H}{|}}{C}} - \overset{\overset{O}{\|}}{C} \right)$$

The fact that lactic acid could be reliably obtained via fermentation from sugar and its product is biodegradable was seen by many as an unbeatable combination. However, PLA is not without its own set of problems. From a performance perspective and customer acceptability standpoint, it is not as good as many other already well-established and entrenched polymers like polyvinyl chloride, polyethylene, polycarbonate, polystyrene, or other polyesters. It has taken over 10 years and the consistent financial backing of Cargill for NatureWorks to make its production profitable and to develop a strong market for PLA. A key to commercial success has been in NatureWorks' developing end uses where waste streams can be tightly managed and diverted to industrial composting facilities, as is the case with professional sports arenas where team owners like to be seen as "going green" and the waste streams are largely food product wastes.

Another good example of what the future may hold in transitioning to more sustainably derived chemical building blocks from sugar may be found in companies like Solazyme and Amyris, two of the winners of the 2014 US Presidential Green Chemistry Challenge Awards [55,56]. In the case of Solazyme the company is using an algae platform to make tailored oils for a variety of end use applications. Because the basic biological systems for making oils in higher plants are present in algae, it was possible for Solazyme to genetically modify the organisms to produce a range of oils containing specific fatty acid ratios. So it is possible to exactly replicate oils like olive oil, sunflower oil, soybean oil, or palm oil or to create or "tailor" oils with the exact ratio of fatty acids and triglycerides a chemist may want. This offers a unique opportunity to create framework molecules that are not currently available, or to create them without geographic restrictions. In other words, making these oils does not require conditions that are suitable for growing a palm or olive tree, but conditions that look more like a biobased chemical plant.

In the case of Amyris a different biological platform, in this case yeast, was modified to produce a completely different molecule as a biodiesel drop-in replacement when compared to the usual type of biodiesel molecule. The advantages for a molecule like farnesane are that it is a true drop-in replacement; upon combustion it does not have emissions like SO_x or excessive NO_x, and burns more cleanly than traditional biodiesel. Interestingly, at the moment, Amyris finds greater demand and profitability for molecules like squalene, a molecule used in cosmetics and personal care products. For the next few years at least, it is likely that Amyris will not be able to compete with petroleum-derived diesel given the recent precipitous drop in oil prices and this will push them into the higher value-added compounds like squalene. Both Solazyme and Amyris, like companies making biosuccinic acid (Bioamber, Myriant, etc.), are competing on cost and performance in addition to having a sustainability benefit. This is an important point since the hurdle for sustainability, from a commercial perspective, is higher than the hurdle for a company whose feedstocks come from petrochemicals.

As mentioned previously, chitin (pictured below) is a material that has captured significant academic interest in recent years as a potential source for polymers and other applications in medicine [57], coatings, membranes, agriculture, etc. [58,59]. Chitin is a polysaccharide composed of β-(1,4)-linked units of the amino sugar N-acetyl-glucosamine. Interest in chitin is fueled by its ready availability from large amounts of natural and farmed shrimp, lobsters, crabs, many insects, its presence in the internal support structures of a variety of invertebrates, and in the cell walls of many fungi. A majority of chitin is currently disposed of as waste. As is the case with lignin, there are considerable difficulties in depolymerizing chitin, or in further functionalizing the polymer to obtain a variety of different products like fibers, membranes, etc. The most common treatment of chitin is to treat it with sodium hydroxide to form chitosan, a chain of randomly distributed β-(1,4)-linked units of D-glucosamine (deacetylated) and N-acetyl-glucosamine. Because chitosan has been deacetylated the amine group is readily available for further functionalization.

A very interesting twist on using naturally formed polymers like chitin and chitosan is worth mentioning here to illustrate a different way of thinking about the delivery of a high-performance product through routine biological systems. A relatively new company, Ecovative [60], is using agricultural waste materials like corn stover or other similar cellulosic materials, inoculating them with fungus, and promoting mycelial (fungal) growth to form a tightly linked polysaccharide composite that is readily biodegradable. By growing the composite in defined shapes, it is possible to create packaging materials, insulation, building composites like fiber board or oriented strand board, etc., which can be used in many different industries. This is an excellent example of sustainable processing of a different nature, where biological processes are harnessed to deliver critical performance characteristics of great benefit in the absence of significant environmental impact across the life cycle of the product. The chemical plants are those that exist on a molecular level, and the manufacturing plant looks decidedly different than a standard chemical plant.

The final example of chemicals from nonpetrochemical sources may be found in another waste stream from the food processing industry, as exemplified by a company like Florida Chemical [61]. Apart from what might be expected as potential products extracted from citrus peels, like flavors and fragrances, the citrus industry has been seeking to develop a greater number of end uses for products like extractable solvents, oils, and other chemicals. There are large volumes of waste citrus peels available from juicing operations (over 8 million tons per year in Brazil alone [62]), which represents a ready supply of useful chemicals given that about 14% of the peel is a mixture of oils, terpenes, etc. While there are a variety of oils, waxes, and other alkanes that can be isolated from citrus peels, the largest marketed chemical is D-limonene. Limonene can be a useful building block for a variety of products like polyethylene terephthalate, or it may be converted into chemicals like cresol or p-α-dimethylstyrene [63]. While it is unlikely to displace current supplies of these chemicals there are those companies that are interested in obtaining chemicals from nonpetroleum, renewable sources.

Catalysis

Catalysis is arguably one of the most important processes in the chemical industry [64] with over 50% of commercial chemical processes incorporating at least one catalytic step. While the use of catalysis and catalytic reagents is included in the principles of green chemistry and engineering, catalysis is not without significant challenges from a sustainability perspective. When synthetic organic chemists think about catalysis, they usually think about using organometallic catalysts, i.e., metals that are coordinated with a variety of generally very complex organic ligands. The overwhelming majority of these catalysts are based on the use of platinum group metals like Pt, Pd, Ru, Au, etc., and the resulting synthetic schemes are typically complex, oftentimes employing very hazardous reagents. From the academic research chemist's perspective the more difficult the synthesis, the more complex the ligand, and the more rare the metal, for example, iridium, the better the catalyst. The drivers here are for scientific novelty so that publication is ensured. As for efficacy of the designed catalyst, that is another story.

The difficulty from a practical perspective is that most of these catalysts will never perform sufficiently well at a commercial scale to be useful. Many catalysts developed in academia are also homogeneous catalysts and recovery is often not possible, but when it is possible, the recovery process can be problematic. From a sustainability perspective the use of the platinum group metals creates a number of challenges [65,66]. Most of these metals come from only two regions of the world: Russia and South Africa. Russia has recently had political and economic conflicts with Europe and the United States, while South Africa has a continued history of labor unrest related to the conditions under which miners are working. Obtaining pure metals requires that tons of ore be mined, crushed, and extracted either as a mercury amalgam (less common) or extracted by acids and a cyanide complex, with further downstream processing to isolate the smaller quantities of rarer metals (Ru, Rh, Re, Ir, etc.) from the Pt, Au, and Ag. All methods of extraction and purification have significant environmental impacts [67–69].

While there are other issues with the use of Pt group metals, the ones cited previously should suggest that their extraction and use is not currently sustainable. As these metals are used ubiquitously for many end use applications in electronics, jewelry, and chemical manufacturing, the cost of these metals will continue to rise over time as they become increasingly harder to find, mine, and process. It would be reasonable to think that increasing costs will over time drive chemists toward the development of organometallic catalysts from comparatively more earth abundant metals like Fe, Ni, or perhaps Co [70,71]. Another strategy is for chemists to make greater use of biocatalysis in their synthesis strategies [72]. The use of biocatalysis is growing given the emergence of robust tools and methods for directed evolution and genetically modifying microorganisms [73].

Implications of Biocatalysis

Biocatalysis, and, more broadly, the use of what some call synthetic biology, has enabled the routine implementation of genetically modified organisms at scale, and this has become widespread in industry for both single transformations and for the production of chemicals [74]. Whole cell fermentations and isolated enzymes are more routinely being investigated and incorporated as part of synthetic strategies, especially in the pharma-ceutical industry [75]. Harnessing large libraries of organisms capable of producing a variety of enzymes has enabled scientists to use or evolve the very specific and chemically selective properties of enzymes to perform a variety of asymmetric syntheses, coupling reactions, functionalizations, and other commonly desired chemical synthesis endpoints. Biocatalysis and related technologies have the added advantage of not requiring highly toxic reagents and solvents, and reactions can be performed at room temperature and neutral pH.

This is not to say all that attends the implementation of synthetic biology is without issues. Fermentations are generally batch operations, they require a carbon source (e.g., sugar), phosphorus (e.g., phosphate buffers), nitrogen (e.g., NH_3 for both pH control and N for protein), and a variety of trace minerals (e.g., K, Mg, Ca, etc.) and significant amounts of time to produce the desired enzyme or framework molecule. Keeping a large number of microorganisms in an exponential growth phase means that there is considerable respiration (e.g., sugar converted to CO_2 or cell substructures) and even-tually the desired compound needs to be separated from the fermentation broth. This is typically done with modified resins, solvents, or chromatographically, and whatever technology is used for separation, there are mass and energy consequences. With continued application of biocatalysis and maturation of the field current, impediments of lengthy reaction times, dilute reaction concentrations, enzyme preparation, and use in alternative reactor configurations (e.g., continuous flow) will be overcome. As is true of any technology or chemistry, there are always benefits and impacts and the question to be answered is: "Which impacts are less bad across the system?"

Reducing the Number of Steps

The concept of using mini- or microreactors to address chemical synthesis routes with a large number of steps, many of which are superfluous, was discussed previously in this chapter. To further expand on them as well as demonstrate how their contributions influence the integration of chemistry and engineering, the authors would like to include the additional concepts of process intensification and multicomponent reactions. Each of these concepts has the end goal of reducing the number of steps in a chemical synthesis sequence along with the mass, energy, and number of inputs and outputs to the synthesis or process.

In the late 1970s the term process intensification (PI) was coined [76]. At the time this phrase was used to describe a novel rotating packed bed reactor whose added benefit was to promote an increase in the mass transfer occurring between the gas—liquid phases within the reactor. The phrase was also loosely used to express that while there was a miniaturization in the physical size of the equipment, the smaller reactor could retain an identical, if not greater, throughput and performance when compared to the larger reactor. The use of PI approaches is a means to implement process substitution. PI minimizes the environmental footprint (resource and energy consumption) of processes and greatly reduces the time, costs, and operational needs when compared with standard chemical production processes. This is accomplished by minimizing the processing steps required for material synthesis through the use of next-generation equipment configurations that support simultaneous material synthesis and separation. In addition to mitigating environmental impacts, these process changes can reduce safety risks to workers and have the potential to impact insurance rates, zoning requirements, and transportation needs. All these benefits have the potential to immediately and drastically affect the presence of a chemical facility in a community.

Mini- or microreactors are examples of technology that can be used in PI. These continuous-flow reactors are designed to decouple the mass and energy transport dependencies from the reactor's size. This is achieved in a mini- or microreactor because they possess a high volume-to-surface ratio and a well-defined and controlled flow regime, which increases mass and energy transfer within the reactor. It is possible to create steady-state conditions in these reactors through excellent reactant stoichiometric and reaction condition control. This ability to create very well-controlled and defined reaction conditions enables one to "tune" reaction conditions for a specific product, thereby resulting in optimal process conditions. These reactors also offer an excellent example of application to many of the green chemistry principles including: atom economy, reaction mass efficiency, energy efficiency, less hazardous conditions, catalysis, and real-time analytics, all of which prevent waste.

An additional opportunity that arises from the unique design and properties of mini- and microreactors is the ability to perform a series of sequential reactions in a single continuous-flow pass as the reactant proceeds across the reactor bed. This is known as a multicomponent reaction. Unlike the batch reactor, the flow reactor regime, combined with some chemical synthesis route design foresight, allows one to introduce the desired reagents that allow subsequent reactions to proceed to completion. In a batch reactor environment, only a single step of a multistep reaction is usually performed. This single step includes charging the reactor with the reactants, establishing and maintaining the desired reaction conditions, waiting for the reaction to reach completion, followed by separation of the products from any unreacted reactants. Many times the resulting product is purified to prepare it for the next step of the sequence, only to go through this entire process again. In a continuous-flow reactor all the solvents, reagents, and reactants are rapidly and completely mixed together and the desired product is obtained in a single pass.

The synthesis of 2-aminochromes by Vaddula and Gonzalez [77] is an example of a multicomponent reaction performed within a single pass of a continuous-flow ThalesNano H-Cube Pro reactor. Using this reactor the authors demonstrated that various chromene derivatives could be obtained with a simple and rapid one-step continuous-flow synthesis route from the reaction of aromatic aldehydes, α-cyano-methylene compounds, and naphthols. The efficient, safer, faster, and modular reaction proceeded to completion with very high yields and residence times of less than 2 min at a slightly elevated pressure of approximately 25 bar.

This merging of disciplines, expertise, and concepts provides just a few examples of the novel and innovative research being conducted in the name of green chemistry and engineering. As the reader proceeds through this book, he or she will be introduced to a significant number of topics, concepts, and examples that either directly contribute to the design of a chemical or are supportive of using process design to influence the underlying chemistry of the process. But the overall goal is to utilize the combined experience and knowledge to advance the sustainability of chemicals and chemical processes.

TYING IT ALL TOGETHER

The introduction, development, integration, and practice of sustainability concepts in many sectors of society has been steadily increasing over the past few decades. Within each sector the definition and practice of sustainability is honed for a specific path, with each having different goals and constraints. While the path to sustainability is usually sector specific, the common intention of each sector is to provide goods and services in a way that advances society, economics, and the environment, not only for our current generation, but also for future generations. In particular, advancing environmental sustainability includes reducing harmful pollution and maintaining adequate resources while satisfying other needs.

As industries move toward sustainability, they must consider how to identify and reduce potential sources of environmental impact within their realm of influence, such as products designed, chemicals used, chemistries performed, process technologies used, waste and water infrastructure, transportation systems, and industrial commerce. These decisions are made with the understanding that effective and sustainable environmental protection is linked to human health and quality of life, economic opportunity, and community vitality. This approach is meeting the needs of society and prospering while using less material, reducing toxics, and recovering more of the material used.

Many examples of environmental sustainability are found in the chemical and manufacturing sectors. Individual companies are developing and implementing methods, tools, and practices that provide comprehensive solutions to address and increase their sustainability performance. With the concept of sustainability providing the foundation for these methods and tools, there is a strong desire that their application and the information (inputs/outputs) from them contribute to the advancement and use of quantitative assessments that inform the future practice of sustainability. See Table 1.2 for a summary of the current and future state of chemical manufacturing. The methods and tools developed in this sector include, but are not limited to, sustainable molecular design [78], green chemistry [79], green engineering [80], process design and operation [81,82], industrial ecology [78], sustainable supply chain design and management [83], and life cycle assessment [84,85].

Throughout this chapter many concepts have been introduced, detailed, and demonstrated for their ability to contribute to increasing the sustainability of the chemical enterprise. Also expressed through these concepts are the overarching goals of "dematerialization," "detoxification," and "design for value recovery," leading to reduced system-wide environmental impacts while preserving natural capital. These strategies include the application of sustainable and green chemistry and engineering to promote material substitution, process substitution, cleaner technology options, product or process redesign, and material reclamation from process or waste streams for beneficial reuse. The key constraint when designing and applying these strategies is in preserving a material's function and the good or service the resulting product provides to society. Once potential alternative strategies have been created, new product or service life cycles based on these alternatives are developed and compared with the original product life cycle to evaluate whether or not the proposed alternative improves the sustainability of the incorporated/replacement/alternative materials. Also, as previously discussed, there will be the critical need for material and energy flow data (i.e., inventories of resources used and pollutants emitted) to assess the product or service and their alternatives. This

Table 1.2 Current and Future State of Chemical Manufacturing

	Current State	Future State
Feedstocks	Fossil carbon sources (e.g., natural gas, natural gas liquids, oil, coal)	• Systems thinking promotes innovations that bridge societal/market needs and health, safety, and environmental impacts • Feedstocks are diversified and renewable to avoid resource depletion • Renewable carbon sources (e.g., biomass (e.g., plants, trees, etc.) converted to sugars or other simpler molecules, • Framework molecules derived from genetically modified single cell or multicellular organisms
Chemicals	• Integrated petrochemical refinery • 120 framework molecules • Rapid depletion of readily available elements critical to modern society	• Integrated biorefinery • Specific chemicals from industrial-scale fermentation processes • Use of greater molecular diversity • Diverse and renewable chemical building blocks beyond 120 framework molecules • Less inherently hazardous chemicals • Design rules, guidance, and tools available to predict and optimize new chemical entities for integrated performance, cost, and life cycle safety, health, and environmental impacts
Chemistries	• Transformation of chemicals in highly reduced state, functionalization (introduction of heteroatoms like N, S, O, halogens, etc. into aromatic compounds), organometallic catalysts using platinum group metals (Pt, Ru, Rh, Pd, etc.), multistep synthetic routes • Use of mass and energy inefficient chemistries developed over the past 150 years	• Removal of functionality • Multicomponent, single pot cascade reactions, coupling and cyclization reactions, new reaction pathways • Enzymes (biocatalysis) • Highly efficient chemistries that unite biological and chemical transformations as part of synthetic route design • Artificial intelligence-enabled synthetic route design • Development and implementation of mass and energy efficient reactions
Processes	• Petrochemical: large scale, continuous flow • Specialties: multipurpose chemical plants, batch reactors, etc. • Software exists for scaling from 1 L to multithousands of liters per hour	• Integrated biorefinery: large scale, continuous flow • Specialties: continuous flow, micro- and minireactors, heat exchangers, mixers, etc. numbered up to meet volume requirements • On-demand synthesis using skid-mounted equipment • Process design software for micro- and minireactors • State-of-the-art metrics and methodologies for assessing process efficiency, hazard, risk, and sustainability impacts

information can be obtained through the use of process engineering and operations data. To support a more sustainable outcome, this information can be extended to tools for the evaluation and incorporation of socioeconomic impact data during this alternative assessment step.

The application of accepted sustainable and green chemistry and engineering design principles helps the scientist or engineer think about or reframe any given problem, process, or service they are planning to develop and implement. The intent of this chapter was to demonstrate that sustainable chemical engineering processes are best designed, implemented, and realized by taking a holistic and comprehensive sustainable and green chemistry and engineering design approach from project inception. It is during the design phase that the chemist and chemical engineer have the greatest ability to positively influence and/or prevent environmental impacts that are ultimately associated with a product or service from cradle to grave or cradle to cradle. It is hoped that the argument for design was clearly made and that the chemist and chemical engineer are inspired to spend sufficient time to make their chemical manufacturing processes more sustainable.

DISCLAIMER

The views expressed in this contribution are those of the authors solely and do not necessarily reflect the views or policies of the US EPA.

REFERENCES

[1] Anastas PT, Warner JC. Green chemistry: theory and practice. Oxford Univ. Press; 1998.
[2] US Dept of Health and Human Services, CDC, NIOSH. NIOSH pocket guide to chemical hazards. 1997.
[3] Chemical Management Services (CMS). http://www.epa.gov/osw/hazard/wastemin/minimize/cms.htm [accessed 31.07.15].
[4] Corey EJ, Cheng XM. The logic of chemical synthesis. John Wiley; 1989.
[5] Pross A. Theoretical and physical principles of organic reactivity. John Wiley; 1995.
[6] Smith MB, March J. March's advanced organic chemistry. John Wiley; 2001.
[7] National Conversation on Public Health and Chemical Exposures. http://www.atsdr.cdc.gov/nationalconversation/ [accessed 31.07.15].
[8] Selecting safer chemical alternatives, http://www.ecy.wa.gov/programs/hwtr/ChemAlternatives/ [accessed 31.07.15].
[9] Chemical Alternatives Assessment. http://www.epa.gov/sustainability/analytics/chem-alt.htm [accessed 31.07.15].
[10] Transitioning to safer chemicals: a toolkit for employers and workers, https://www.osha.gov/dsg/safer_chemicals/ [accessed 31.07.15].
[11] Becker M. The right chemistry − how collaboration can lead to better decisions on safer chemical alternatives, http://www.greenbiz.com/blog/2012/10/26/how-collaboration-can-lead-better-decisions-safer-chemical-alternatives [accessed 31.07.15].
[12] IPCS harmonization project document; no.8 WHO human health risk assessment toolkit: chemical hazards. 2010. IPCS project on the Harmonization of Approaches to the Assessment of Risk from Exposure to Chemicals.
[13] Adler JH. Fables of the Cuyahoga: reconstructing a history of environmental protection. Fordham Environ Law J 2002;XIV. http://law.case.edu/faculty/adler_jonathan/publications/fables_of_the_cuyahoga.pdf [accessed 31.07.15].

[14] Winterton N. Twelve more green chemistry principles? Green Chem 2001;3:G72−5.

[15] Anastas PT, Zimmerman JB. Design through the twelve principles of green engineering. Environ Sci Technol 2003;37(5):94A−101A.

[16] Abraham MA, Nguyen N. Green engineering: defining the principles − results from the San Destin conference. Environmental Progress 2003;22(4):223−36.

[17] Jimenez-Gonzalez C, Constable DJC. Green chemistry and engineering, a practical design approach. John Wiley; 2011.

[18] Hanson DJ. C&E news. Crit Mater Probl Contin 2011;89(43):28−31.

[19] Sheldon RA. Consider the environmental quotient. Chemtech 1994;24(3):38−46.

[20] http://www.epa.gov/greenchemistry/index.html [accessed 31.07.15].

[21] http://echa.europa.eu/web/guest/regulations/reach [accessed 31.07.15].

[22] http://www.dtsc.ca.gov/PollutionPrevention/GreenChemistryInitiative/safer_products_regs_outline.cfm [accessed 31.07.15].

[23] http://www.fda.gov/Food/IngredientsPackagingLabeling/ucm166145.htm [accessed 31.07.15].

[24] Molecular Design Research Network (MoDRN). http://modrn.yale.edu/ [accessed 31.07.15].

[25] Luque R, Clark JH. Sustain Chem Process 2013;1(10). http://www.sustainablechemicalprocesses. com/content/1/1/10 [accessed 31.07.15].

[26] Arancon RAD, Lin CSK, Chan KM, Kwan TH, Luque R. Energy Sci Eng 2013;1(2):53−71.

[27] US Environmental Protection Agency. Life cycle assessment: principles and practice. EPA 600-R-06−060. Cincinnati, OH: National Risk Management Research Laboratory; 2006. Available from: http://www.epa.gov/nrmrl/std/lca/lca.html [accessed 31.07.15].

[28] ISO 14000 − Environmental Management. Available from: http://www.iso.org/iso/home/standards/ management-standards/iso14000.htm [accessed 31.07.15].

[29] US EPA. Definition of solid waste ruling. 2015. http://www.gpo.gov/fdsys/pkg/FR-2015-01-13/ pdf/2014-30382.pdf [accessed 31.07.15].

[30] Haswell SJ, Watts P. Green chemistry: synthesis in micro reactor. Green Chem 2003;5:240−9.

[31] Razzaq T, Kappe CO. Continuous flow organic synthesis under high-temperature/pressure conditions. Chem Asian J 2010;5:1274−89.

[32] Kockmann N, Roberge DM. Scale-up concept for modular microstructured reactors based on mixing, heat transfer, and reactor safety. Chem Eng Process 2011;50:1017−26.

[33] Pennemann H, Watts P, Haswell SJ, Hessel V, Löwe H. Benchmarking of microreactor applications. Org Process Res Dev 2004;8:422−39.

[34] Elvira KS, Solvas XC, Wootton RCR, deMello AJ. The past, present and potential for microfluidic reactor technology in chemical synthesis. Nat Chem 2013;5:905−15.

[35] Hessel, Kralisch D, Kockmann N, Noel T, Wang Q. ChemSusChem 2013;6:746−89.

[36] Lipshutz BH, Ghorai S. Transitioning organic synthesis from organic solvents to water. What's your e factor? Green Chem 2014;16:3660.

[37] Li C-J, Chen L. Organic reactions in water. Chem Soc Rev 2006;5:68.

[38] Uhlig N, Li C-J. Alkynes as an eco-compatible "on-call" functionality orthogonal to biological conditions in water. Chem Sci 2011;2:1241−9.

[39] Minkler SRK, Isley NA, Lippincott DJ, Krause N, Lipshutz BH. Leveraging the micellar effect: gold-catalyzed dehydrative cyclizations in water at room temperature. Org Lett 2014;16(3):724.

[40] Petroleum and Other Liquids, U.S. Energy Information Administration, http://www.eia.gov/dnav/ pet/PET_PNP_PCT_DC_NUS_PCT_A.htm [accessed 26.09.15].

[41] Wagner LA. Materials in the economy—material flows, scarcity, and the environment. US Geol Surv Circ 2002;1221. http://pubs.usgs.gov/circ/2002/c1221/ [accessed 31.07.15].

[42] Rogich MD, Cassara A, Wernick I, Miranda M. Material flows in the United States: a physical accounting of the U.S. industrial economy. World Resources Institute; 2008.

[43] Bishop KJM, Klajn R, Grzybowski BA. Angew Chem Int Ed 2006;45:5348−54.

[44] Grzybowski BA, Bishop KJM, Kowalczykm B, Wilmer CE. The 'wired' universe of organic chemistry. Nat Chem 2009;1:31−6.

[45] Werpy T, Petersen G, editors. Top value added chemicals from biomass. Results of screening for potential candidates from sugars and synthesis gas, vol. I. NREL; 2004. Report No. NREL/ TP-510−35523.

[46] Integrated Biorefinery Research Facility. Partnering with industry to advance biofuels and bio-products. NREL; 2011. Report No. FS-5100-53429, http://www.nrel.gov/biomass/publications. html [accessed 31.07.15].

[47] Davis R, Tao L, Tan ECD, Biddy MJ, Beckham GT, Scarlata C, et al. Schoen, process design and economics for the conversion of lignocellulosic biomass to hydrocarbons: dilute-acid and enzymatic deconstruction of biomass to sugars and biological conversion of sugars to hydrocarbons. NREL; 2013. Report No. NREL-TP-5100—60223.

[48] Biorefinery Developments for Europe. http://www.biosynergy.eu/fileadmin/biosynergy/user/docs/ Results_of_the_Integrated_Project_BIOSYNERGY_2007-2010.pdf [accessed 31.07.15].

[49] Stuart PR, El-Halwagi MM. Integrated biorefineries: design, analysis, and optimization. CRC Press; 2012.

[50] http://www.arizonachemical.com/Global/Graphs/Company_Materials_Tree_of_Life.pdf [accessed 31.07.15].

[51] http://renmatix.com/ [accessed 31.07.15].

[52] http://www.bio-amber.com/ [accessed 31.07.15].

[53] http://www.myriant.com/ [accessed 31.07.15].

[54] http://www.natureworksllc.com/ [accessed 31.07.15].

[55] http://www2.epa.gov/green-chemistry/2014-greener-synthetic-pathways-award [accessed 31.07.15].

[56] http://www2.epa.gov/green-chemistry/2014-small-business-award [accessed 31.07.15].

[57] Pillai CKS, Paul W, Sharma CP. Chitin and chitosan polymers: chemistry, solubility and fiber formation. Prog Polym Sci 2009;34(7):641—78.

[58] Yeul VS, Rayalu SS. Unprecedented chitin and chitosan: a chemical overview. J Polym Environ 2013;21(2):606—14.

[59] Khor E, Wan ACA. Chitin: fulfilling a biomaterials promise. Waltham, MA: Elsevier; 2014.

[60] http://www.ecovativedesign.com/ [accessed 31.07.15].

[61] http://www.floridachemical.com/ [accessed 31.07.15].

[62] Ferreira-Leitão V, Gottschalk LMF, Ferrara MA, Nepomuceno AL, Molinari HBC, Bon EPS. Biomass residues in Brazil: availability and potential uses. Waste Biomass Valorization 2010;1(1):65—76.

[63] Clark JH, Pfaltzgraff LA, Budarin VL, Hunt AJ, Gronnow M, Matharu AS, et al. From waste to wealth using green chemistry. Pure Appl Chem 2013;85(8):1625—31.

[64] Bartholomew CH, Farrauto RJ. Fundamentals of industrial catalytic processes. 2nd ed. John Wiley; 2006.

[65] Scott A. Chem Eng News 2014;92(35):30—3.

[66] Ritter SK. Chem Eng News 2013;91(13):41—3.

[67] Izatt RM, Izatt SR, Bruening RL, Izatt NE, Moyer BA. Chem Soc Rev 2014;43:2451—75.

[68] Mudd GM. Plat Met Rev 2012;56:2—19.

[69] Norgate TE. In: Graedel TE, van der Voet E, editors. Linkages of sustainability. Cambridge, MA: The MIT Press; 2010. p. 131—48.

[70] van der Vlugt JI. Cooperative catalysis with first-row late transition metals. Eur J Inorg Chem 2012;3:363—75.

[71] Bullock RM, editor. Catalysis without precious metals. Weinheim, Germany: Wiley-VCH; 2010.

[72] Clouthierab CM, Pelletie JN. Expanding the organic toolbox: a guide to integrating biocatalysis in synthesis. Chem Soc Rev 2012;41:1585—605.

[73] Davids T, Schmidt M, Böttcher D, Bornscheuer UT. Strategies for the discovery and engineering of enzymes for biocatalysis. Curr Opin Chem Biol 2013;17(2):215—20.

[74] Buchholz K, Kasche V, Bornscheuer UT. Biocatalysts and enzyme technology. Wiley; 2012. 626 pp.

[75] Tao J, Xu JH. Biocatalysis in development of green pharmaceutical processes. Curr Opin Chem Biol 2009;13(1):43—50.

[76] Cross WT, Ramshaw C. Chem Eng Res Des 1986;64:293—301.

[77] Vaddula BR, Yalla S, Gonzalez MA. J Flow Chem July 11, 2015. http://dx.doi.org/10.1556/ 1846.2015.00015.

[78] Voutchkova-Kostal AM, Kostal J, Connors KA, Brooks BW, Anastas PT, Zimmerman JB. Green Chem 2012;14:1001—8.

[79] Ruiz-Mercado GJ, Gonzalez MA, Smith RL. Clean Tech Environ Policy 2014;16:703—17.
[80] McDonough W, Braungart M, Anastas PT, Zimmerman JB. Environ Sci Technol 2003;37(23): 434A—41A.
[81] Gonzalez MA, Smith RL. A methodology to evaluate process sustainability. Environ Prog 2003;22(4):269—76.
[82] Smith RL. Hierarchical design and evaluation of processes to generate waste-recycled feeds. Ind Eng Chem Res 2004;43:2508—15.
[83] Wesley I, Cabezas H, Weisbrod A, Eason T, Demeke B, Ma X, et al. Sustainability 2014;6(3): 1386—413.
[84] Meyer DE, Curran MA, Gonzalez MA. Environ Sci Technol 2009;43(5):1256—63.
[85] Meyer DE, Curran MA, Gonzalez MA. J Nanopart Res 2011;13(1):147—56.

CHAPTER TWO

Separations Versus Sustainability: There Is No Such Thing As a Free Lunch

L.M. Vane
US Environmental Protection Agency, Cincinnati, OH, United States

THE SEPARATIONS DILEMMA AND IMPERATIVE

Industrial separation processes have a significant energy and environmental footprint. In the 2005 report "Materials for Separation Technologies: Energy and Emission Reduction Opportunities," the US Department of Energy's Oak Ridge National Laboratory (ORNL) estimated that separation processes account for 22% of all in-plant energy usage in the United States [1]. The thermally driven processes of distillation, evaporation, and drying accounted for 49%, 20%, and 11%, respectively, of the separation energy usage. In the chemicals, petroleum refining, forest products, and mining industries, separation processes account for an even higher fraction of in-plant energy usage, about 47%, and consume over 98% of the separation energy used by US industries. The ORNL report concluded the most significant reductions in energy usage could be achieved by replacing energy-intensive processes like distillation with low-energy separation systems such as membranes, extraction, sorption, or synergistic hybrid systems of low- and high-energy systems.

The ORNL report highlighted the following separation problems as being "High-Energy Distillation Processes with Potential for Replacement with Lower-Energy Alternatives":

1. Olefin/paraffin separations: ethylene/ethane, propylene/propane, etc.
2. Removal of organic compounds from water where azeotropes are formed: ethanol, isopropanol, sec-butanol, etc.
3. Recovery of dilute organics from water: acetic acid, ethylene glycol, methanol, many high-boiling polar organic compounds
4. Cryogenic air separation
5. Polyol separations: ethylene glycol/diethylene glycol, ethylene glycol/propylene glycol
6. Isomer separations

Also in 2005 the National Research Council (NRC) of the National Academies issued the report "Sustainability in the Chemical Industry: Grand Challenges and Research

Sustainability in the Design, Synthesis and Analysis of Chemical Engineering Processes   2016 Published by Elsevier Inc.

35

Needs," identifying eight challenge areas for chemical industry sustainability [2]. The report highlighted the need for advances in efficient chemical separations, especially the development of effective alternatives to distillation. The recovery of chemicals from relatively dilute aqueous solutions, such as fermentation broths or wastewaters, and the separation of carbon dioxide were specifically highlighted. As with the ORNL report, the need for alternatives to distillation, the workhorse separation technology of the chemical and biofuels industries, was emphasized in the NRC report. In addition to the energy-related environmental impact of separation processes, the ability to achieve environmental objectives is hindered by the inherent energy demand of conventional separation technologies. For example, the US Environmental Protection Agency finalized a new definition of solid waste intended to promote the reuse/reprocessing of 18 solvents to reduce the use of virgin solvents, cutting energy and waste, particularly in the pharmaceutical, paint and coating, plastic and resin, and basic organic chemicals sectors [3]. However, replacement of virgin solvents requires the application of separation technologies to recover those solvents from their mixtures with other solvents and/ or water used in the industrial process and to purify the solvents to meet process specifications.

The greatest potential advancement in separation processes would be to minimize their usage. That is to say, begin with a raw material or alter the production system preceding the separation step such that the separation requirement is eliminated or reduced. These concepts are embodied in several of Anastas and Zimmerman's 12 Principles of Green Engineering, particularly Principle #3: "Design for Separation" [4]. Intuitively, we understand that a higher product concentration will make recovery of a unit of that product less demanding. Unfortunately, the separation step is often considered later in the product/process development timeline. The task of just designing a workable chemical or biological reaction scheme is often daunting enough. However, ignoring or delaying design of the separation step imperils both the economic and environmental viability of the enterprise.

The economic impact of product concentration is clearly demonstrated in what has been termed a "Sherwood plot," relating the selling price of a variety of purified materials to the concentration of the materials in the initial matrices from which they are being separated [5,6−10]. The most widely referenced Sherwood plot is from Grübler's book *Technology and Global Change* [5], shown in Fig. 2.1. Grübler's Sherwood plot illuminates the dramatic effect initial concentration has on the selling price of a commodity or the cost of contaminant removal across a range of material types. The prices in Fig. 2.1 embody more than just the cost of separation steps, but, because of the significance of separation processes in total cost, it is easy to imagine that separation costs are similarly correlated with initial concentration. For example, the capital cost of separations and supporting facilities as a fraction of total capital costs has been estimated to range from 33% for a typical chemical plant to 70% for a refinery or bioprocess facility [9]. As a

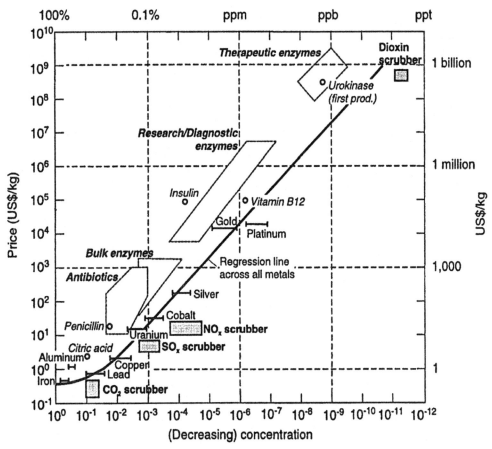

Figure 2.1 Sherwood plot of material selling price as a function of the concentration of the material in the initial matrix. *Reproduced from Grübler A. Technology and global change. Cambridge (UK): Cambridge University Press; 1998.*

result, improvements in the energy usage, material usage, and capital cost of separation processes would go a long way to improving the financial and environmental bottom lines for chemicals and materials.

Unfortunately, a chemical separation requires energy to accomplish because, in general, Mother Nature wants to mix things up. In thermodynamics, this means the Gibbs free energy of mixing is usually negative. In fact the minimum work (w_{min}) required to accomplish a separation is equal to the negative of the Gibbs free energy of mixing (Δg^*_{mix}):

$$w_{min} = -\Delta g^*_{mix} = \Delta h_{mix} - T_0 \Delta s_{mix} \qquad [2.1]$$

where Δh_{mix} is the heat of mixing, Δs_{mix} is the entropy of mixing, and T_0 is the temperature. The minimum work to separate a mixture of n components into pure streams of each is calculated as:

$$w_{min} = -T_0 R \sum_{i=1}^{n} x_i \ln(\gamma_i x_i) \qquad [2.2]$$

where γ_i and x_i are the activity coefficient and mole fraction, respectively, of component i in the feed to the separation unit and R is the gas constant ($8.314\,J/mol \cdot K$). The product ($\gamma_i x_i$) is the activity (a_i) of the component. For ideal binary mixtures of compounds A and B (i.e., $\gamma_i = 1$), Eq. [2.2] simplifies to:

$$w_{min} = -T_0 R [x_A \ln x_A + (1 - x_A) \ln(1 - x_A)] \qquad [2.3]$$

In cases where pure component A is recovered with a specific degree of recovery (Y_A) from an ideal binary mixture, the minimum work required per mole of pure component A recovered is calculated as [11]:

$$w_{min} = \left[\frac{1}{Y_A} \ln(1 - Y_A) \right] R T_0 \ln(x_A) \qquad [2.4]$$

This relationship, plotted in Fig. 2.2, demonstrates the rapid increase in minimum work as the concentration decreases or as the fraction of recovery increases. Later in this chapter the actual work for several separations will be discussed. The energy required for some separations is near the thermodynamic minimum, while most are far, far higher than the minimum. However, for all, the general behavior illustrated in Fig. 2.2 is retained.

METHODS OF ANALYSIS

The separation processes within a chemical process are selected and designed using a variety of criteria, with the criterion used in a specific situation established from the information set available and the priorities of the organization. In most cases the almighty bottom line, cost, drives process decisions, including separation decisions. Ideally, sufficient information and organizational motivation would exist to determine the life cycle impact of separation process options. Less refined analyses rely on comparisons of energy intensity, material intensity, or waste intensity—the amount of each category per unit of product recovered or purified. This may be appropriate when evaluating the effect of design variables within a given separation technology rather than between technology platforms. Because each form of energy, material, and waste has a different value and impact, other refinements are included. For example, comparing a process that relies on thermal energy to one driven by mechanical work, like a natural

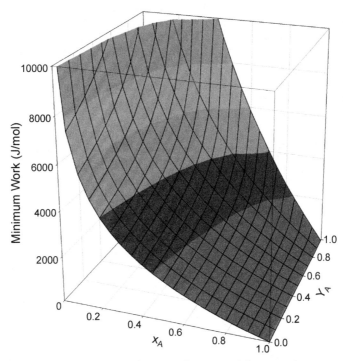

Figure 2.2 Effect of concentration (mole fraction of A, X_A) and fraction of A recovered (Y_A) on the minimum work required per unit of pure component of A produced from an ideal binary mixture.

gas-fired boiler versus an electrically powered heat pump. This requires more than a simple energy input comparison. To this end, energy intensity may be further scrutinized by converting all energy inputs into a common unit, such as the primary energy or fuel equivalents required (i.e., MJ-primary/kg-product or MJ-fuel/kg-product) or greenhouse gas equivalents released (kg-CO_2-equivalents/kg-product) [12,13].

Exergy destruction has been proposed as a metric of process sustainability. Exergy is the maximum useful shaft work that could be obtained from a system at a given state in a specified environment without violating any laws of thermodynamics [14]. Irreversibilities in a process result in a loss in work potential, referred to as exergy destruction. As noted by Çengel, the difference between the exergy and the actual work performed "represents the room engineers have for improvement" [14]. Although valuable, exergy analyses have not been commonly used to evaluate separation schemes [15,16]. In a 2013 mini-review, Luis reported on the evolution of exergy analyses in chemical engineering literature and highlighted the past and future potential for using exergy analyses in separation process design [15]. She concluded that 10—20% of exergy losses arise in the separation stages of chemical manufacturing while 65—90% were because of the irreversibilities in chemical reactions [15]. Nevertheless, there are opportunities for reducing

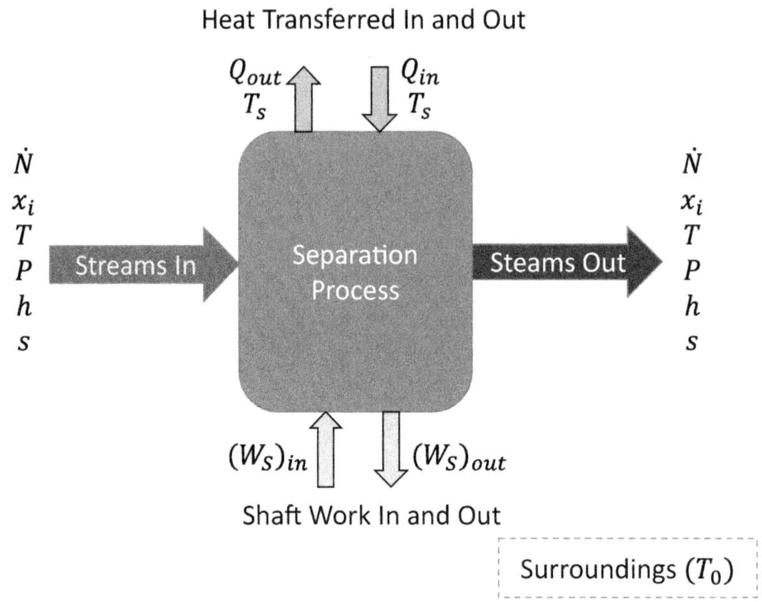

Figure 2.3 Schematic of a generic separation process.

the exergy losses in the separation stages caused by the high thermodynamic inefficiency of many distillation systems, particularly those with close-boiling mixtures [17].

This leads to the definition of a "Second Law efficiency" (η) for the generic steady state separation process depicted in Fig. 2.3 as the ratio of the minimum work to the actual work:

$$\eta = \frac{w_{\min}}{LW + w_{\min}} \tag{2.5}$$

where LW is the lost work, or destroyed exergy, defined as [18]:

$$LW = \sum_{\substack{\text{streams,} \\ \text{heat,} \\ \text{work} \\ \text{in}}} \left[\dot{N}(h - T_0 s) + Q\left(1 - \frac{T_0}{T_s}\right) + W_S \right] - \sum_{\substack{\text{streams,} \\ \text{heat,} \\ \text{work} \\ \text{out}}} \left[\dot{N}(h - T_0 s) + Q\left(1 - \frac{T_0}{T_s}\right) + W_S \right]$$

$$\tag{2.6}$$

where $\dot{N}$ is the stream molar flow rate, h and s the enthalpy and entropy of the stream, respectively, T_0 and T_s the temperature of the reference surroundings and heat source (or sink), respectively, Q the heat flow, and W_S the shaft work. As noted by Demirel, only zero lost work has no impact on the environment: "lost work causes the inefficient use of energy (loss of exergy), and environmental cost due to (1) discharging lost exergy into the environment, and (2) the depletion of natural resources because of inefficient use of fossil fuels" [17]. Thus, unless all of the environmental, risk, and societal costs can be accounted for, simple financial cost calculations cannot capture the true "cost" of a separation system.

SEPARATION ALTERNATIVES
Distillation

As noted earlier, distillation is the dominant separation process in the chemical industry. This is because of the "simplicity" of design and operation as well as the economies of scale offered by distillation relative to many of the alternative separation processes. Two books by Henry Kister, *Distillation Design* and *Distillation Operation*, are recommended reading for those seeking to understand both the fundamental and practical aspects of distillation systems [19,20]. The leading drawback to distillation is the reliance on high-quality energy to provide heat to the reboiler and the removal of heat in the condenser at a lower temperature (schematic shown in Fig. 2.4). This combination results in poor energy efficiencies and dreadful Second Law efficiencies, particularly for close-boiling mixtures. For example, η is only 5% for olefin/paraffin separations (e.g., ethylene and propylene production), 12% in crude oil units, and 18% in cryogenic air distillation [21]. As the cost of fossil fuels has increased, so too has the operating cost of distillation, creating opportunities for alternative separation technologies. Those alternatives include: liquid—liquid extraction, gas stripping, adsorption, absorption, membrane separation, crystallization, and combinations thereof.

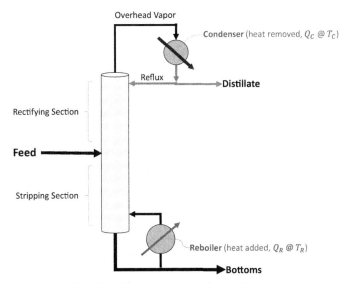

Figure 2.4 Simplified schematic of a distillation column.

Although distillation is an established technology, there are still opportunities for improvement. A review by Kiss provides an overview of many of these opportunities [21]. Thermally linking the reboiler and condenser is a popular concept for improving distillation efficiency. This requires the use of a heat pump to upgrade the heat released in the condenser to be useful in the reboiler. All of the heat pump concepts add complexity and capital cost with the goal of reducing energy usage and operating cost. The most straightforward heat pump concept for distillation, vapor compression, mimics the classic design commonly used in household heat pumps/air conditioners: a working fluid/ refrigerant is circulated in a closed loop using a compressor and expansion valve, where the evaporator of the heat pump is the distillation condenser and the heat pump condenser is the distillation reboiler. With this closed loop type of heat pump, the distillation process streams and the heat pump streams are physically separated, with only heat exchange.

Another popular heat pump concept, if not widely adopted, is the direct use of the overhead vapor as the heat pump working fluid, an open cycle heat pump termed mechanical vapor recompression (MVR). In MVR the compressor raises the pressure of the overhead vapor to the point where it will condense at a temperature sufficient to drive the reboiler. This has the advantage that the reboiler heat exchanger serves as both the reboiler and the condenser. One disadvantage is the process vapor may not be the most efficient working fluid for the temperatures involved. Nevertheless, primary energy savings over 50% are anticipated [21].

The coefficient of performance (COP) for an ideal heat pump is defined as the amount of heat moved divided by the shaft work required:

$$COP_{ideal} = \frac{Q_C}{W_S} = \left[\frac{T_{hp,C}}{T_{hp,C} - T_{hp,E}} \right] \qquad [2.7]$$

While the actual COP observed is:

$$COP_{actual} = \frac{Q_C}{E_{hp}} = \eta_{hp} \frac{Q_C}{W_S} = \eta_{hp} \left[\frac{T_{hp,C}}{T_{hp,C} - T_{hp,E}} \right] \qquad [2.8]$$

where Q_C is the heat released in the condenser of the distillation column (J/s), W_S is the work performed by an ideal heat pump (J/s), E_{hp} is the electricity consumed by an actual heat pump (J/s), $T_{hp,C}$ and $T_{hp,E}$ are the temperatures (in kelvin) of the heat pump condenser and evaporator, respectively, and η_{hp} is the heat pump efficiency factor (η_{hp} ranges from 0.5 to 0.75) [22–24]. The quantity in the brackets in Eq. [2.7] is the Carnot efficiency. The difference between $T_{hp,C}$ and $T_{hp,E}$ is the temperature lift. Heat transfer resistances result in a heat pump temperature lift higher than simply the difference between the bottoms and overhead temperatures in the distillation column, thereby reducing the COP.

So, just how much energy can a heat pump save? Take the distillation separation of a mixture of methanol and water at atmospheric pressure. Assuming almost pure methanol in the overhead and almost pure water in the bottoms, the overhead temperature is about 65°C (338K) and the bottoms temperature is 100°C (373K) for a temperature lift of 35K. According to Eq. [2.7], COP_{ideal} is 10.7. Assuming $\eta_{hp} = 0.7$ and that there is a 5°C minimum approach in each heat exchanger (i.e., $T_{hp,C} = 378K$ and lift $= 35 + 2 \times 5 = 45K$), COP_{actual} is 5.9, 45% lower than the ideal. Thus, 1 unit of electrical energy results in 5.9 units of thermal heat being upgraded. The average efficiency of the fossil-fuel powered electrical power grid results in 0.37 units of electrical energy delivered per unit of higher heating value of primary energy source consumed (average for US fossil fuel-powered grid in 2011 [25,26]). This means the heat pump moves 2.18 units of heat to the reboiler per unit of primary energy consumed. For a normal steam generator/reboiler system, 0.85 units of heat are delivered to the reboiler per unit of primary energy converted. As a result, the heat pump primary energy efficiency is 2.6 times that of a steam-heated reboiler. The bottom line, then, for this example of methanol/water separation is that primary energy usage could be reduced 61% by using a closed loop heat pump.

A number of additional factors must be considered to determine if the energy savings would result in financial savings [27,28]. With such high energy savings, it would seem that heat pumps would be more widely utilized in the chemical industry. One reason why they are not is that most heat pump concepts add a piece of rotating equipment (the compressor), with all of the added operational costs and downtime risks, to a unit operation that did not already have that type of equipment. On top of that, this added equipment is relatively pricey, meaning the capital cost versus operation cost payback period may be longer than with other improvements. Humans, particular corporate boards and investors, are not so keen on delayed gratification.

Other distillation process options for improved energy efficiency range from simple to advanced concepts. On the simpler end of the spectrum: adding a feed/bottoms heat exchanger if the feed is at a much lower temperature than the bottoms liquid. One step up is to split the feed into two streams, heating only one using the available bottoms heat and feeding the two feed streams to optimized points on the column [29]. The next leap in complexity is similar in concept to a multieffect evaporator, the feed stream would be split into "n" streams and each would be fed to a separate distillation column operated at a pressure so as to enable the exchange of heat between the reboiler and condenser of different columns in the cascade [28,30]. In another concept, termed "heat integrated distillation column," the stripping and rectifying sections of the distillation column are operated at different pressures by compressing the overhead vapor from the stripping column [31−33]. This is akin to an MVR heat pump, except that the rectification and stripping sections operate at different pressures and these sections are in direct heat exchange to most efficiently affect heat and mass transfer, as opposed to just the condenser and reboiler operating at different pressures. For multicomponent mixtures, an advancement receiving significant attention is combining the multiple distillation columns into a single, divided wall column allowing for efficient heat transfer [21,31,34,35]. Similarly, for processes involving a reaction step and a distillation step, combining the two into one reactive distillation unit may be more efficient if there is an intermediate bottleneck in one of the two steps. On the more exotic end of the spectrum is the replacement of the fixed vertical distillation column with a rotating bed/contactor to utilize centrifugal forces ("HiGee" technology) to enhance mass transfer [21,36].

Extraction: Liquid—Liquid Extraction, Gas Stripping, Adsorption, and Absorption

The separation processes of liquid—liquid extraction, gas stripping, adsorption, and absorption all have one thing in common: one or more components of the initial stream are transferred to a mass separating agent (MSA) and subsequently recovered from that MSA. The MSA may be regenerated in situ or in another device. The latter is depicted in a continuous mode of operation in Fig. 2.5. The form of the MSA—gas, liquid, or solid—determines how we refer to the process and how the separation scheme is designed. For example, a liquid or gas MSA is more likely to be operated continuously while a system based on a solid MSA will most likely involve an MSA-filled contactor undergoing sequential loading and regeneration cycles.

The following factors are considered important for selection of an appropriate MSA (adapted from factors for liquid—liquid extraction for ethanol/water separation [37—40]):

1. *Selectivity* for one component relative to another. Higher selectivity results in a more concentrated regeneration product stream. This is often presented as a separation

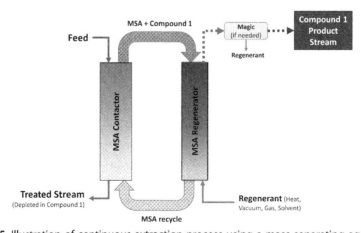

Figure 2.5 Illustration of continuous extraction process using a mass separating agent (MSA).

factor that is defined similarly to that of relative volatility, β, comparing the ratio of component concentrations in/on the MSA (y_i) to that in the feed material (x_i):

$$\beta = \frac{y_1/y_2}{x_1/x_2} \qquad [2.9]$$

2. *Equilibrium distribution coefficient*, K_D, is the ratio of the concentration in the MSA material to that in the feed phase (i.e., y_i/x_i). Higher values of K_D are desirable because it reduces the amount of MSA required to remove a given mass of product with concomitant reductions in the capital and operating costs for the separation system. The separation factor is the ratio of K_D values for two compounds in the feed.

3. *Mutual solubility*. Ideally the MSA and the bulk feed material would not be soluble in each other. Solubility of MSA in the bulk feed material results in loss of MSA from the system and may present downstream complications since the MSA could be transferred to other process steps, to the products, or to the waste treatment facility. Solubility of the bulk feed material into the MSA may alter the selectivity, capacity, and physical stability of the MSA.

4. *Ability to separate MSA and bulk stream*. Separation of the MSA and bulk feed phase is usually accomplished relying on density or other physical differences between the materials. In the cases of gas stripping and adsorption, the density differences are usually sufficient to enable easy separation of the MSA and the bulk stream. In liquid–liquid extraction with a limited density difference, gravity settling may be insufficient, requiring the use of a liquid–liquid centrifuge. In other cases, magnetic or electric forces may be used to enhance the separation.

5. *Interfacial tension*. When the two phases are in direct contact, intimate mixing is desirable, but the formation of stable or even metastable emulsions or mixed phases must be avoided. In liquid–liquid extraction where a porous membrane is used as an interfacial support, the wetting of the pores may be undesirable.

6. *MSA viscosity/flowability*. If the MSA is a liquid, a suspension of solids, or loose particles, then a low viscosity or high flowability is generally desirable since it improves mass transfer in the contactors and reduces energy required for moving and mixing the MSA.

7. *Volatility*. The desired vaporization characteristics of the MSA depend to a great extent on the regeneration scheme. If the component separated is to be recovered from the MSA by evaporation, either at ambient or at elevated temperatures, then a low-volatility MSA is desirable. However, if a distillation-based regeneration scheme is employed for a liquid MSA, then some degree of volatility and a low heat of vaporization may be advantageous.

8. *Cost.* Ideally, if the MSA was completely recovered and reused, only the initial charge of MSA would be needed. Because of normal losses or chemical reactions, some amount of fresh MSA must be added periodically. Thus the purchase of MSA represents both an upfront capital cost and an ongoing operating cost. The cost of end-of-life disposal or reprocessing of the MSA must also be considered as an operating expense.

9. *Stability/reactivity.* This factor links to several of the other factors in that degradation of the MSA necessitates replacement/reprocessing and may lead to leaching of degradation products into process streams.

10. *Life cycle health and safety.* Issues of health and safety during use of the MSA should be considered when selecting an MSA. So too should life cycle health and safety issues. For example, the MSA itself may be benign, but the method of producing the MSA or regenerating it may not. A full life cycle assessment would identify areas of concern related to both the use and manufacture of the MSA. In bioprocesses, the MSA may inhibit the microorganisms in the bioreactor or may negatively interact with other critical bioreactor components such as enzymes. In addition, the toxicity associated with the MSA in wastewater and air releases from the production facility must be considered. Further, the flammability, flash point, and reactivity with process chemicals of the MSA and of MSA—extractant mixtures present facility health and safety issues.

The concentration of a compound removed by the MSA in the regeneration product stream is directly dependent on the selectivity of the MSA for that compound. Although higher selectivities may be possible, the regeneration product stream may still require some level of purification, depending on product specifications and the concentration of the compound in the feed stream. Unfortunately, for a given class of MSA, K_D tends to decrease as β increases. As a result there is often a tradeoff between selectivity and amount of the MSA required to remove a given mass of a compound, particularly in the case of liquid—liquid extraction. Common methods of MSA regeneration include: vacuum flash vaporization, indirect heating, direct heating (i.e., steam, hot gas, or hot liquid), distillation, gas stripping, and membrane processing. Most of these methods simply transfer the recovered compound from the MSA to another phase or material, from which it must be separated, often by condensation, distillation, or another MSA-based method. In many scenarios the regeneration is performed at an elevated temperature in order to increase the tendency of the compound to leave the MSA. Ideally the distribution coefficient for compound between a gas phase (vacuum or gas) and the MSA increases significantly with increasing temperature in order to achieve as complete a regeneration as possible.

Neither the extraction nor the regeneration steps/cycles are run to completion. Thus when the regenerated MSA is returned to the extraction contactor, or the MSA column is cycled back to extraction mode, there will be residual extracted compound in/on the

MSA. This residual level must be low enough to ensure the desired level of the extracted compound in the treated stream can be reached. In this way, the extraction efficiency is linked to the regeneration efficiency, with energy demand and capital cost increasing as regeneration efficiency increases. Heat management may become a limiting design factor in contactors because of the heat of desorption or the need to heat or cool the material/ column at various stages of the process/cycle. Heat added in a thermal regeneration scheme may not be easily recoverable, especially when the heating source and the cooling sources are not naturally linked. Much of the energy and efficiency of an extraction process revolves around how efficiently the MSA can be regenerated, and this can be overlooked during the initial screening of MSA candidates.

Membrane-Based Separation Processes

Finally, the crème de la crème of separation alternatives—membrane processes. Sustainability achieved, Earth is saved! *Non?* Membranes are often viewed as the solution for many chemical process ills, and not just by those of us in the membrane community. The truth is that membranes may be a more efficient or even more sustainable alternative in many situations, but there are limitations to membrane processes, just as there are for traditional approaches.

While membrane processes have infiltrated many process schemes, probably the most extensive replacement has been in the area of water treatment. For example, reverse osmosis has largely replaced distillation/evaporation for the production of drinking water from seawater [41,42]. Clearly, with the right mix of material properties, efficient process designs, and economic drivers, energy-efficient membrane processes can replace conventional thermally driven separation processes. Later in this chapter, desalination will be a featured example.

The category of "membrane processes" covers a wide range of membrane-mediated processes with disparate governing principles, driving forces, and descriptive language. In general, membranes separate components in a process stream based on some combination of the size and chemical differences between those components. In the continuum of membrane processes, there are those that separate components based primarily on the size of the components and those that separate components based primarily on the molecular-scale chemical interactions between the components and the membrane material. For size-based separations the driving force for material transport through the membrane is an applied fluid pressure gradient. Particle filtration, microfiltration, and ultrafiltration are examples of size-based membrane separations. For chemically mediated membrane separations the driving force is the chemical potential difference between the upstream and downstream sides of the membrane. Gas separation, vapor permeation, pervaporation, reverse osmosis, and forward osmosis fall into this category. Nanofiltration spans the two. Electrodialysis adds electrical charge as a separating feature. For many of the chemically mediated membrane processes the selective layer is a dense material and the terms "solution—diffusion" or "sorption—diffusion" are commonly used to describe transmembrane transport. In these, feed components sorb into the upstream face of the dense membrane and then diffuse through the membrane to the downstream permeate side of the membrane. The term "dense" has been used to indicate the selective material only has molecular-scale pores or transport pathways with diffusion as the primary means of movement through the membrane. Sometimes, dense membranes are referred to as "nonporous", but this does not capture membranes with pores on the scale of small molecules, such as zeolite materials.

One of the leading voices for the opportunities and challenges for displacing thermally driven separation technologies with membrane-based processes for improved sustainability has been Professor William Koros of the Georgia Institute of Technology. His publications on the subject are recommended reading [41,43,44] as is a review on the subject of energy-efficient gas separation membranes authored by several leading membrane researchers [45]. As noted by Dr. Koros, chemically mediated membrane processes, and hybrid technologies thereof, have great potential to displace traditional separation processes. As a result, we will emphasize these here.

In chemically mediated membrane processes the throughput or productivity (the rate a compound passes through the membrane) is related to the chemical potential driving force as follows:

$$\text{Throughput of } i = \frac{\left(\mu_i^{\text{Feed}} - \mu_i^{\text{Perm}}\right)}{R_i} A = \frac{J_i A}{MW_i} \qquad [2.10]$$

where μ_i^{Feed} and μ_i^{Perm} are the chemical potentials of compound i on the feed and permeate sides of the membrane, respectively, R_i is the overall resistance to mass transfer for i, A is the membrane area, J_i is the flux of i through the membrane (e.g., $kg/m^2 \cdot s$), and MW_i is the molecular weight. The change in chemical potential and the mass transfer resistances present for a hypothetical membrane are depicted in Fig. 2.6. As indicated, the overall mass transfer resistance is composed of three main individual mass transfer resistances situated in series: a feed-side fluid boundary layer, the membrane, and a permeate-side fluid boundary layer. The relative importance of the individual mass transfer resistances is determined mainly by the permeability and thickness of the membrane, the design of the membrane module (turbulence promoters, feed and permeate spacer layers), the feed-side flow rate, and the permeate-side flow rate. The net result is that the permeate is enriched, relative to the feed, in species that are preferentially permeated through the membrane and the feed stream becomes depleted in those same compounds. Boundary layer resistances decrease both selectivity and throughput. As a result, membrane module designs and operations are chosen to minimize these. Ideally, the membrane represents the sole mass transfer resistance (i.e., $R_i \approx R_i^{\text{Mem}}$). For gas separation, vapor permeation, and pervaporation processes, the

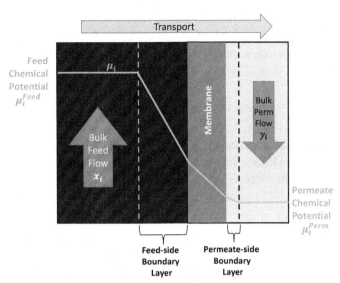

Figure 2.6 Mass transfer in a chemically mediated membrane process. The chemical potential gradient from the bulk feed to the bulk permeate streams is the driving force for mass transfer (shown as *yellow line*). Three main resistances to mass transfer are shown—fluid boundary layers on the feed and permeate sides of the membrane and diffusion through the membrane.

driving force is represented by the partial pressure gradient and the inverse of the resistance is represented by the permeance (Π_i). Flux is calculated as:

$$J_i = \Pi_i \left(p_i^{\text{Feed}} - p_i^{\text{Perm}} \right) MW_i \qquad [2.11]$$

where permeance is the membrane permeability divided by the thickness of the selective layer (i.e., $\Pi_i = P_i/\ell$) with units of kmol/m$^2\cdot$s$\cdot$kPa and p_i^{Feed} and p_i^{Perm} are the partial pressures of the compound on the feed and permeate sides of the membrane, respectively.

Membrane separation quality can be described in terms of a separation factor as defined in Eq. [2.9] with x_i and y_i representing the bulk feed-side and bulk permeate-side compositions, respectively, of two compounds being separated. However, this separation factor includes the driving forces for the two compounds and so is not an accurate reflection of the selectivity of the membrane alone. A better representation of membrane selectivity is to calculate the resistances for two compounds according to Eq. [2.10] and then calculate selectivity as the ratio of the inverse resistances. If the membrane resistance is dominant for both compounds, then the calculated selectivity would be the membrane selectivity, sometimes referred to as the permselectivity, α, defined as:

$$\alpha_{ij} = \frac{R_j^{\text{Mem}}}{R_i^{\text{Mem}}} = \frac{\Pi_i}{\Pi_j} = \frac{P_i}{P_j} \qquad [2.12]$$

Membrane permeability is defined as the product of the solubility and diffusivity of a compound in the selective membrane layer ($P_i = S_i D_i$). Note the selectivity offered by the membrane is independent of the phase equilibrium of the system, such as the vapor—liquid equilibrium (VLE) behavior.

Based on Eq. [2.10], the two ways to maximize flux are to maximize the chemical potential difference and to minimize the resistance. To minimize the membrane resistance, decrease the thickness of the selective layer and/or increase the inherent permeability of the selective material. At some point, however, improvements in membrane resistance will have a diminishing impact as the other mass transfer resistances become dominant. At the material level, attempts to control the selectivity or resistance of a polymer membrane commonly encounter a tradeoff between selectivity and permeability that has been often reported for permselective membranes. In other words, when researchers attempt to alter the formulation of a particular polymer to improve selectivity, it usually results in a decline in permeability, and vice versa. This is usually presented graphically as a log—log plot of selectivity versus permeability with the extreme data values shown as a line, termed the "upper bound." These are often referred to as "Robeson plots" and the "Robeson upper bound" in recognition of the trailblazing work of Professor Lloyd Robeson on this subject [46,47]. An example of a Robeson plot is given in Fig. 2.7 for the separation of carbon dioxide from nitrogen. This graph is a cautionary tale for anyone seeking the ideal membrane or even expecting to tweak a membrane to enhance both selectivity and permeability. The practical significance is that a high-flux membrane (aka low area/low capital cost) may produce a permeate stream requiring additional processing (aka higher costs) because of the low selectivity it achieves and a high-selectivity membrane producing high-purity streams may require an unaffordable amount of membrane area because of the low flux inherent in the selective material.

As illustrated in Fig. 2.7 the ideal membrane with infinite selectivity and no mass transfer resistance for the preferred species (the upper right corner of the figure) does not exist. In addition, the infinite membrane area required to completely remove a target species is unaffordable. As a result the separation performed by a real membrane system will be incomplete, yielding a permeate stream containing at least a portion of all species in the feed and a reject stream (aka "retentate") that has not been completely rid of the preferentially permeated species. The question then becomes: "Can imperfect membranes be utilized to improve upon traditional separation processes?" Sometimes the answer is that the membranes are too imperfect, at least at this time, to compete. However, in many situations standalone membrane units or combinations of membranes with the traditional technology can make a significant difference. The lowest hanging fruit for applying solution—diffusion membranes have been situations where the VLE behavior is not favorable for distillation-based schemes. Azeotropic mixtures and close-boiling components are examples. In these cases a membrane that can separate the

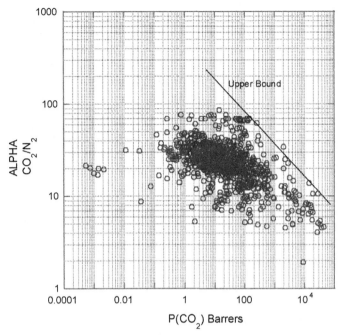

Figure 2.7 Robeson plot illustrating the tradeoff between selectivity (α, ALPHA) and permeability (P) for the separation of carbon dioxide from nitrogen with polymer membranes [47]. The *circles* indicate all literature data considered relevant. The *upper bound line* is an empirical judgment of the outermost range of reliable data. *Reprinted from Robeson LM. The upper bound revisited. J Membr Sci 2008;320(1—2):390—400. Copyright (2008), with permission from Elsevier.*

two species independent of any VLE thermodynamic limitation could be used to carry out the entire separation or to at least jump the azeotrope or limiting region. Examples include azeotropic solvent/water systems like the ethanol/water or isopropanol/water systems that have azeotropes at 5 and 15 wt% water, respectively, and the close-boiling olefin/paraffin systems, such as ethylene/ethane and propylene/propane.

Another tradeoff in membrane operations is the diminishing return of added membrane area. Each increment of membrane area added in series removes relatively less of the preferentially permeating species than the previous increment of membrane area and yields a reduction in the average permeate purity. Assume a gas separation membrane unit contains membrane area A_1 and removes 90% of compound 1 from the feed mixture. All other factors being equal, adding a second membrane unit with area A_1 will be able to remove 90% of compound 1, but that is 90% from the feed to that unit, or only an additional 9% based on the original concentration. Likewise, adding a third unit of area A_1 removes only 0.9% more of the original amount of compound 1. In this unconstrained scenario the natural log of the fraction of compound 1 remaining after treatment with a membrane system is proportional to the membrane area in the system.

The reason for this is the diminishing driving force as the feed-side concentration decreases through the membrane system. Unfortunately, at the same time, the concentration of the rejected compounds remains about the same or increases through the membrane system, meaning that the driving force for transport for the rejected compound increases, leading to a higher flux of the undesired permeating species. So, not only is added membrane area less efficient at removing the preferred permeating species (based on the initial amount present), it yields a permeate stream with relatively more of the rejected species than in the first parts of the membrane system. In other words, as membrane area is increased, the purity of the rejected species increases in the retentate but the purity of the preferentially permeating species in the permeate decreases. To compensate for this situation, multiple permeate streams may be withdrawn from a membrane system and processed according to their compositions to improve on overall efficiency. Such system-wide thinking is often necessary to make an alternative process, like a membrane process, functional. An example of this will be given later for the post-combustion capture of carbon dioxide.

Hybrids!

This brings us to hybrid processes. Hybrids are all the rage: be it cars, fruits, vegetables, pets, etc. But, as some curmudgeons might point out, a hybrid chemical process may just represent good process design. Rarely does a process consist of a single unit operation. So it should come as no surprise that the most efficient separation process just might be a combination of multiple technologies. The moniker "hybrid technology" may indicate that two disparate technologies have been united, but it may also indicate that an emerging technology has been combined with an established technology. In the latter case, it designates that an emerging technology has reached a point of maturity where process designers have enough performance information or models to intelligently link it to conventional technologies. It may also indicate a level of maturity on the part of both the champions of the new technology and the guardians of the established technology to work together. In some situations, hybrids might be a low risk way of introducing a new technology—by allowing the established technology to perform most of the separation, but more efficiently. As noted in the 2005 ORNL report: "Hybrids that can be retrofit to, and easily coupled and decoupled from existing production units would provide facilities with energy improvements (and debottlenecking opportunities) without risking the normal production" [1]. In other situations the hybrid approach may be necessary to enable the new technology to actually be used outside the idealized world of the laboratory.

Coupling two distinct technologies to make a more efficient integrated technology is the most common type of hybrid technology. Another type is to meld two technologies into one. For example, Professor Ed Cussler's group and his research collaborators have shown how hollow fibers can be used to significantly enhance mass transfer in a distillation column by using the hollow fibers as a nonselective structured packing [48–51]. This hybrid unites the most pertinent aspect of the hollow fibers, the high surface area per unit volume, with the VLE separation characteristics of a distillation column to significantly improve efficiency. Olefin/paraffin separation is the initial target of that work. Similarly, Dr. Koros' group has borrowed concepts from the field of hollow fiber membranes to develop high surface area adsorbents for a variety of separations including CO_2 removal from flue gas and desulfurization of natural gas [52,53]. In still another form of melded hybrid, Agrawal and Noble describe a "composite separation system" wherein two or more MSAs are combined in one process in a synergistic fashion [54]. The example given is of combining zeolite particles into a polymer membrane to yield a mixed matrix membrane having separation performance significantly greater than that of the polymer alone, but at a cost and ease of manufacturing close to that of a polymer membrane. Thus hybrids can take many forms, from a simple coupling of different technologies to the alloying of different materials/structures.

EXAMPLES

In the preceding sections a variety of separation processes and means of assessing energy usage or environmental impact have been presented. In this section, three specific separation challenges will be highlighted.

Example 1: Desalination

The first separation example is seawater desalination. Traditionally, desalination was done by distillation or simple evaporation/condensation [55]. Today, thermally driven desalination has been largely replaced by the membrane process reverse osmosis. In reverse osmosis an applied pressure exceeding the osmotic pressure of the salt solution causes water to permeate through a dense membrane. Hydrated salt ions are relatively large compared to water and have a lower permeability through the membrane resulting in relatively salt-free water being collected as the reverse osmosis permeate.

A quick analysis of energy usage will indicate why reverse osmosis has replaced thermal desalination. First, Eq. [2.2] can be used to calculate the minimum work required to produce the first drop of pure water from seawater as [11]:

$$w_{min} = -T_0 R\ln(\gamma_w x_w) = \Pi_s \overline{V}_w \qquad [2.13]$$

where Π_s is the osmotic pressure of the salt solution and $\overline{V}_w$ is the molar volume of water. For seawater with 3.5 wt% NaCl ($x_w = 0.98894$) in water at 15°C, water activity $(\gamma_w x_w)$ is 0.980 [11,56]. Thus, according to Eq. [2.13], the minimum work required to recover the first drop of water is 48.4 J/mol, more commonly reported as 2.74 kJ/kg or 0.76 kWh/m^3. (Note: There is some variability in this value of minimum desalination energy reported in the literature because of different assumptions regarding the water activity. For example, if an ideal solution is assumed [i.e., $\gamma_w = 1$], then the minimum work is 45% lower [26.6 J/mol] [57]). For higher degrees of water recovery the minimum work increases. For example, for 50% water recovery from the same seawater the theoretical minimum increases 39% to 1.06 kWh/m^3 [11]. Simple thermal desalination requires 40,700 J/mol of heat to evaporate the first drop of water. Even considering the conversion of heat to work, that is a huge inefficiency for evaporation—indicating why thermal desalination is done in a multistage or multieffect process, essentially a cascade of evaporation and condensation steps at successively lower pressures to reuse the thermal energy several times over. However, if the thermal energy could be reused five times, the amount of thermal energy is still on the order of 8000 J/mol. Conversely, the latest seawater reverse osmosis (SWRO) units require about 2 kWh/m^3 for 50% recovery or about 127 J/mol of electrical energy, about twice the theoretical minimum for 50% recovery [11,58]. Using the 37% fuel-to-electricity efficiency referenced earlier, this translates to a primary energy usage of 344 J/mol. Adding in another 1—2 kWh/m^3 for the intake, pretreatment, posttreatment, and brine discharge stages of the SWRO plant [58] still yields an SWRO primary energy requirement that is an order of magnitude lower than that of a thermally driven process.

This largely explains why SWRO has supplanted distillation/evaporation for seawater desalination. In situations where "waste" heat is available, thermal desalination may still be economically attractive. It should be noted that, despite an eightfold reduction in energy required for SWRO separation since the 1970s, additional SWRO energy reductions will be harder to achieve because of proximity of current SWRO systems to the theoretical minimum [11,58].

Example 2: CO$_2$ Capture

Earlier in this chapter, in the discussion of solution—diffusion-controlled membrane processes, it was noted that as membrane area is added to increase the product purity of the compound rejected by the membrane, the purity of the preferentially permeating compound in the permeate stream decreases. This results in the need for clever management of multiple permeate streams and system-wide thinking. An example of this is the post-combustion capture of CO$_2$. The conventional CO$_2$ capture process is amine scrubbing where the flue gas is contacted with an aqueous amine solution to sorb/complex the CO$_2$. The spent amine is then thermally regenerated in a stripping column, producing a concentrated CO$_2$ stream. The conventional amine process leads to a loss in power plant efficiency of between 15% and 29%, depending on the type of fuel and plant design [59]. The low end of the range is for natural gas combined cycle plants and the high end is for subcritical pulverized coal power plants. A concerted effort is under way to develop alternatives with a lower parasitic power loss and lower cost than the conventional scrubbing process [60]. In 2014 the US Department of Energy's National Energy Technology Laboratory estimated the power penalty, per unit of CO_2 captured, for the addition of several technologies to existing pulverized coal power plants [61]. At 0.249 kWh/kg CO_2 a membrane process was 38% lower than the 2005 benchmark amine process and 22% lower than both an updated (2012) amine process and a sorbent-based process.

The membrane process for post-combustion CO_2 capture uses CO_2-selective membranes to remove the CO_2 from the flue gas and yield a CO_2-rich permeate gas stream for subsequent condensation and storage. One US Department of Energy performance target for CO_2 capture is 90% CO_2 removal from the flue gas. Flue gas contains between 10 and 15 mol% CO_2. At such a high degree of removal, at that low a feed concentration, and with the general CO_2/N_2 permselectivities offered by current membranes (Fig. 2.7), the permeate composition from the membrane unit will be too low for efficient conversion of the CO_2-rich permeate gas into a liquid. For example, assuming a CO_2/N_2 selectivity of 50, a feed gas containing 10 mol% CO_2, and an unlimited feed to permeate pressure ratio, the permeate from the first increment of membrane area will contain 84.7 mol% CO_2. By adding enough membrane to the system to remove 90% of the CO_2, the average permeate composition would drop to 67 mol% CO_2. With a finite feed to permeate pressure ratio, this concentration would be even lower.

The first approach to this permeate purity problem is to split the membrane system into two steps (or more) with a permeate stream for each step, depicted in Fig. 2.8 as "Option 1." The permeate stream from the first step, the richest in CO_2, would be processed to make it a pipeline-ready liquid. The permeate from the second step, enriched relative to the feed stream but not high enough for economical processing for pipeline transport (i.e., >95% CO_2), is recompressed as a gas and returned to the feed stream of the first membrane step in order to retain the CO_2 in the system. An alternative, system-wide, process concept, integrating the membrane separation system with the larger power plant, "Option 2" in Fig. 2.8, was proposed by Merkel et al. from Membrane Technology and Research, Inc. [62]. Instead of the membrane system operating independently of the power plant, this alternative utilizes the combustion air as a countercurrent permeate sweep gas in the second membrane step. This accomplishes several things. First, it replaces the vacuum compressor of the original two-step design, instead purging the permeate zone of the second membrane step with air already destined for the burner, saving on both energy and capital costs. Since the sweep air has minimal CO_2 and is operated in a countercurrent mode, it yields a higher driving force for mass transfer than did the vacuum, thereby greatly reducing the membrane area required in the second step. The CO_2 in the air sweep stream passes through the burner and is returned to the first membrane step. Process simulations by Merkel et al. predicted that the air sweep option ("Option 2" in Fig. 2.8) would require 57% less membrane

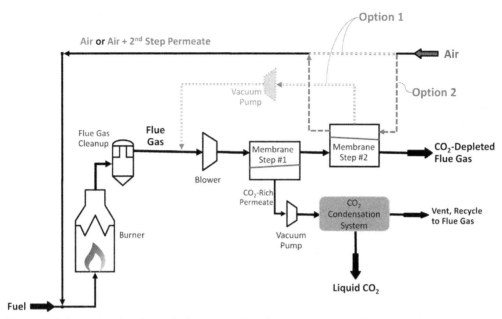

Figure 2.8 Example of carbon dioxide separation from power plant flue gas using a two-step membrane process with two options for managing the permeate from the second membrane step. In Option 1 (*purple double-dotted lines*), air is used directly in the burner while a vacuum pump creates partial pressure driving force in the second membrane step with return of the second step permeate to front of membrane process. In Option 2 (*blue dashed lines*), the combustion air is used as a countercurrent permeate sweep gas in the second membrane step. *Adapted from Figs. 11 and 12 in Merkel TC, Lin H, Wei X, Baker R. Power plant post-combustion carbon dioxide capture: an opportunity for membranes. J Membr Sci 2010;359(1–2):126–139.*

area, 33% less power, and have a 41% lower cost of capture than the original two-step design with vacuum on the second step ("Option 1") [62]. Another interesting facet of the analysis is that a membrane selectivity greater than 50 is predicted to yield minimal cost savings. The largest cost reduction would result from increasing the CO_2 permeance. This has been noted for other membrane separations: above a threshold selectivity, overall process cost is relatively constant. As a result, improvements in selectivity are not always worth pursuing, particularly considering the selectivity—permeability tradeoff noted earlier.

Example 3: Solvent/Water Separation

As highlighted by the ORNL and NRC reports from 2005 cited in the first section [1,2], alternatives to distillation are needed for the recovery of organic compounds from aqueous solutions, particularly those forming azeotropes and in dilute aqueous solutions. The dilute nature of the feed means the minimum work to perform the separation is high. Worse, the Second Law efficiency of distillation in these cases is low. Even worse, the presence of an azeotrope requires introduction of an extra separation process or agent to accomplish the full separation. This opens the door to alternatives that save energy or are not limited by an azeotrope—or both.

In this example a hybrid vapor stripping-vapor permeation process studied by the author will be discussed as an alternative to conventional distillation for alcohol/water separation. In the United States the standard process for recovering ethanol from a corn starch fermentation broth and drying it to meet fuel specifications is a process combining distillation with molecular sieve adsorption, as illustrated in Fig. 2.9. A stripping column removes the ethanol from the broth, achieving high levels of ethanol recovery (i.e., low residual ethanol in the bottoms stream). The ethanol-enriched overhead vapor from the stripping column is then sent to a second column, a rectification column wherein the ethanol is enriched to near the ethanol/water azeotrope of 95.6 wt% ethanol. The near-azeotropic overhead product from the rectification column is then dried in an adsorption step using beds of water-selective zeolite beads operated in loading/regenerating cycles. The Second Law efficiency of the conventional approach was estimated to be only 5—9% for a fermentation broth containing 10 wt% ethanol [10]. Clearly there is room for improvement. For a discussion of alternatives for alcohol recovery and drying, a review article by this author is suggested [40].

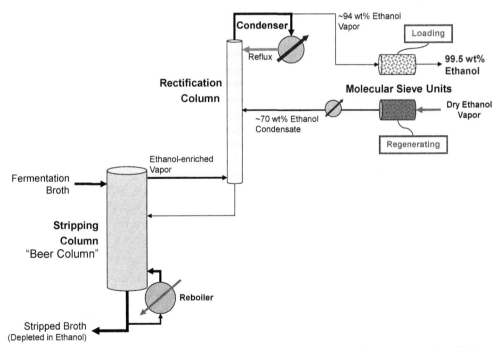

Figure 2.9 Schematic of traditional ethanol recovery and drying process for a corn-to-ethanol facility. Feed/effluent heat recovery exchanger on stripping column not shown.

It should be noted that the current "conventional" approach for ethanol/water separation supplanted heterogeneous azeotropic distillation, wherein a third compound, commonly benzene, was added to break the water/ethanol azeotrope [63]. The health and environmental concerns associated with benzene and other entrainers prompted investigation of alternative ethanol drying approaches. Advances in both zeolite adsorbents and process design made molecular sieve drying a viable alternative to azeotropic distillation. Molecular sieve drying of fuel ethanol had become a demonstrated commercialized technology not long before the corn-to-ethanol boom in the United States, allowing this separation technology to become the standard in that market.

As in the previous examples of applying membranes in desalination and carbon dioxide capture, membrane processes are promising options for alcohol/water separations [40,64]. For the recovery of the alcohol from dilute fermentation broths, hydrophobic pervaporation membranes could be applied instead of distillation. For the final drying step, pervaporation or vapor permeation with water-selective membranes could compete with molecular sieve adsorption and even rectification. However, as has been noted previously, membrane systems lose efficiency when high recoveries are required—as would be necessary to compete with the high ethanol recoveries of the stripping column of the distillation scheme in Fig. 2.9. Thus a hybrid of distillation with a membrane would seem appropriate when high recoveries are needed. One such hybrid has been under development by the author's group, in collaboration with Membrane Technology and Research, Inc., the same company previously mentioned in the CO_2 capture example [65–72]. The schematic of such a process, termed "membrane-assisted vapor stripping" (MAVS), is shown in Fig. 2.10.

The stripping column in the MAVS process serves the same function as in the traditional distillation/adsorption scheme; it provides high alcohol recovery and a low effluent concentration. The main difference is that the thermal energy to drive the stripping column is provided by recovering the latent and sensible heat from the overhead vapor leaving the stripping column. Little or no additional reboiler heat is required. First, the overhead vapor is compressed so that it can be fractionally condensed in a dephlegmator—essentially a high surface area heat exchanger with multiple VLE stages—at a temperature sufficient to transfer the heat of condensation to the reboiler of the stripping column. The water-rich condensate formed in the dephlegmator is returned to the top of the stripping column. Just a few stages of VLE are needed in the dephlegmator to substantially enrich the vapor, although the most efficient and/or cost-effective design may result in significant concentrations of water remaining in the alcohol-enriched overhead vapor. This dephlegmator overhead vapor is further compressed prior to directing the stream to the vapor permeation membrane steps for further water removal. The higher total pressure provides a higher partial pressure driving force for solution—diffusion mass transfer through the membrane. Water-selective membranes are utilized in both membrane steps. The water-rich permeate stream from the first

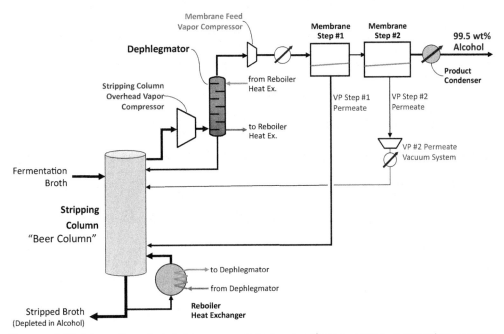

Figure 2.10 Schematic of hybrid vapor stripping-membrane vapor permeation process ("membrane-assisted vapor stripping") process for recovering and drying alcohols from water. Feed/effluent heat recovery exchanger on stripping column not shown.

membrane step is returned directly to the stripping column to form a portion of the stripping vapor. Depending on the desired water concentration in the final alcohol product, the second membrane step may require a lower permeate pressure to operate efficiently. This lower permeate pressure can be achieved by vapor compression (as in Fig. 2.10) or by condensation. Either way, the second step permeate stream is returned to the stripping column or dephlegmator, although at a higher stage than the first step permeate because of a higher alcohol concentration.

The MAVS process largely replaces the thermal energy requirement of the traditional distillation/adsorption process with the shaft work of the vapor compressors. Recalling the discussion of MVR heat pumps earlier in this chapter, there are certainly parallels between MVR and the vapor compression—fractional condensation—heat recovery portion of the MAVS process. Therefore the logical question is: "Does this swap actually save primary energy usage?" The answer is presented in Fig. 2.11. In this figure the primary energy required per kg of 99.5 wt% ethanol product for a conventional distillation/adsorption process, an MAVS process, and the theoretical minimum work scenario are presented as a function of the concentration of ethanol in the feed stream to the separation process. As described earlier, the efficiency of converting primary energy to heat and of converting primary energy to electrical energy was assumed to be 85% and

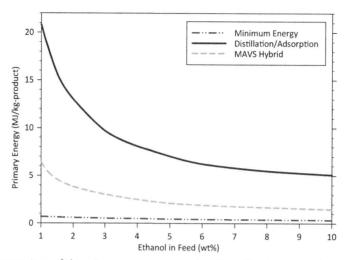

Figure 2.11 Comparison of the primary energy usage for ethanol/water separation using traditional distillation/adsorption process (Fig. 2.9) and hybrid membrane-assisted vapor stripping (MAVS; Fig. 2.10) process. Minimum energy (from minimum work calculation) shown as reference. Assumptions: 37% and 85% efficient conversion of primary energy to electrical energy and thermal energy, respectively, 0.02 wt% ethanol in stripping column bottoms, and 99.5 wt% ethanol product (0.5 wt% water).

37%, respectively. For the MAVS process the efficiency of converting electrical energy into compression work was assumed to be 75%. The curve for distillation/adsorption was established from literature references [40,73–77]. The thermal energy added by the molecular sieve section of the distillation/adsorption process was assumed to be 1.5 MJ/kg-product, the midpoint of the range reported. The MAVS process was simulated using ChemCAD 6.5.5 process simulation software with spreadsheet calculation modules for the membrane units as described in previous publications from the author's group [66,78]. The minimum work was calculated according to Eq. [2.2] assuming the bottoms stream from the stripper and the ethanol product are pure water and pure ethanol, respectively, and calculating activity coefficients for the feed stream using the NRTL thermodynamic model in ChemCAD. Minimum primary energy was then calculated from the minimum work using a 37% conversion efficiency.

The low Second Law efficiency of distillation/adsorption is apparent from the difference between the primary energy usage curves in Fig. 2.11. According to the values in the figure, for a 10 wt% feed, the traditional process has a Second Law efficiency of only 7.0%. That value is exactly in the middle of the range cited earlier for distillation/adsorption for the same separation [10]. That efficiency falls to 6.4% and 3.3% as the feed concentration is reduced to 5 wt% and 1 wt%, respectively. By comparison, the Second Law efficiency of the MAVS process is over three times that of distillation/adsorption, 24.3% efficient for a 10 wt% ethanol feed, falling to 21.4% and 10.8% for 5 wt% and

1 wt% feeds, respectively. The lower heating value (LHV) of ethanol is 27 MJ/kg. Thus for a 1 wt% ethanol feed, the amount of energy required by traditional distillation/adsorption for the same feed approaches the LHV content of the produced ethanol. On the other hand, the MAVS process uses only 24% of the LHV of ethanol to perform the separation, despite a Second Law efficiency of 10.8% at such a low feed concentration, demonstrating the potential for an alternative scheme to improve efficiency and expand the range where alcohol recovery is, at least, a net energy winner.

CONCLUDING THOUGHTS

While there is a clear need for low-energy alternatives to thermally driven separation processes, advancing low-energy separation technologies depends on the availability of MSAs that achieve selective separations and cost-effective process options. As noted in the 2005 NRC report [2]: "While membrane separations, adsorption, and extractions tend to be less energy intensive [than distillation], significant technical challenges must be overcome in the development of these alternatives in order to realize any significant reductions in energy intensity of the chemical process industry." The ORNL report echoed that sentiment and identified the following needs for materials/equipment of low-energy processes in order for them to be implemented [1]:

1. provide the selectivity required;
2. provide the throughput required;
3. provide adequate throughput for long periods of time;
4. be sufficiently durable to maintain optimum performance under harsh environments;
5. provide sufficient economies-of-scale incentive for large-volume processes.

One reminder for those of us who think we have built a better mousetrap: traditional technologies are not static. A quick look at the advances in distillation technology over the years will prove it [79]. The same economic, energy, and environmental pressures that motivate searches for alternative technologies also stimulate advances in the conventional technologies, as does the competition posed by the nascent alternatives. However, as shown in the desalination example, it is possible to develop and commercialize an alternative with a high thermodynamic efficiency. Conversely, the alcohol/water separation example indicates that there is still room for improvement. One such opportunity is combining a lower energy alternative, like a membrane process, with renewable energy [80]. Caution is needed because "renewable" does not necessarily mean "sustainable" or "economical," just as "waste heat" is not free—there is always a cost to capture and move that heat.

On a final note, this chapter has been about separating things, but the reverse process, the controlled mixing of streams, is being considered as a means to take advantage of Mother Nature's tendency to mix things up. For example, both pressure retarded osmosis (PRO) and reverse electrodialysis (RED) use permselective membranes to recover energy from the controlled mixing of seawater with a low-salinity stream, such as river water or a wastewater treatment effluent [81]. In PRO the two streams are separated by a membrane that selectively allows water to flow from the dilute stream to the high-salinity stream, creating a hydraulic head on the salty side that can then drive a turbine [82,83]. In RED, alternating cation and anion exchange membranes separate the saline and low-salinity streams allowing ions to flow from high to low salinity creating an electrical current [84]. Here the lunch may not be completely free, but at least it is a "downhill" task.

DISCLAIMER

The views expressed in this chapter are those of the author and do not necessarily reflect the views or policies of the US Environmental Protection Agency.

REFERENCES

[1] Materials for separation technologies: energy and emission reduction opportunities. Oak Ridge National Laboratory, U.S. Department of Energy; 2005.
[2] Sustainability in the chemical industry: grand challenges and research needs. (Washington, DC): National Research Council, The National Academies Press; 2005.
[3] Finalized Rule — definition of solid waste. 2014. EPA-HQ-RCRA-2010-0742-0019.
[4] Anastas PT, Zimmerman JB. Design through the 12 principles of green engineering. Environ Sci Technol 2003;37(5):94A—101A.
[5] Grübler A. Technology and global change. Cambridge (UK): Cambridge University Press; 1998.
[6] Dahmus JB, Gutowski TG. What gets recycled: an information theory based model for product recycling. Environ Sci Technol 2007;41(21):7543—50.
[7] Lightfoot EN, Cockrem MCM. Complex fitness diagrams: downstream processing of biologicals. Sep Sci Technol 2013;48(12):1753—7.
[8] Dwyer JL. Scaling up bio-product separation with high performance liquid chromatography. Bio/Technology November 1984:957—64.
[9] Separation & purification: critical needs and opportunities. (Washington, DC): National Academy Press; 1987.
[10] House KZ, Baclig AC, Ranjan M, van Nierop EA, Wilcox J, Herzog HJ. Economic and energetic analysis of capturing CO_2 from ambient air. Proc Natl Acad Sci 2011;108(51):20428—33.
[11] Elimelech M, Phillip WA. The future of seawater desalination: energy, technology, and the environment. Science 2011;333(6043):712—7.
[12] Brueske S, Sabouni R, Zach C, Andres H. U.S. manufacturing energy use and greenhouse gas emissions analysis. report ORNL/TM-2012/504. Prepared by Energetics Inc. for Oak Ridge National Laboratory; 2012.
[13] Grubler A, Johansson TB, Muncada L, Nakicenovic N, Pachauri S, Riahi K, et al. Energy primer. In: Global energy assessment: toward a sustainable future. Cambridge University Press and IIASA; 2012. p. 99—150 [chapter 1].
[14] Çengel YA, Boles MA. Thermodynamics: an engineering approach. 5th ed. New York: McGraw Hill; 2005.
[15] Luis P. Exergy as a tool for measuring process intensification in chemical engineering. J Chem Technol Biotechnol 2013;88(11):1951—8.
[16] Luis P, Van der Bruggen B. Exergy analysis of energy-intensive production processes: advancing towards a sustainable chemical industry. J Chem Technol Biotechnol 2014;89(9):1288—303.
[17] Demirel Y. Thermodynamic analysis of separation systems. Sep Sci Technol 2004;39:3897—942.
[18] Seader JD, Henley EJ, Roper DK. Thermodynamics of separation operations. In: Separation process principles. 3rd ed. J. Wiley; 2011 [chapter 2].

[19] Kister HZ. Distillation operation. New York: McGraw-Hill Education; 1990.

[20] Kister HZ. Distillation design. New York: McGraw-Hill Education; 1992.

[21] Kiss AA. Distillation technology — still young and full of breakthrough opportunities. J Chem Technol Biotechnol 2014;89(4):479—98.

[22] Becker H, Maréchal F, Vuillermoz A, editors. Process integration and opportunity for heat pumps in industrial processes. Proceedings of ECOS 2009, the 22nd International Conference on Efficiency, Costs, Optimization, Simulation and Environmental Impact of Energy Systems; 2009. (Parana, Brazil).

[23] Benstead R, Sharman FW. Heat pumps and pinch technology. Heat Recovery Syst CHP 1990;10:387—98.

[24] McMullan A. Industrial heat pumps for steam and fuel savings. U.S. Department of Energy; 2003. DOE/GO-102003—1735.

[25] How much electricity is lost in transmission and distribution in the United States? U.S. Energy Information Administration; 2015. Available from: http://www.eia.gov/tools/faqs/faq.cfm?id=105&t=3 [22.02.16].

[26] Hussy C, Klaassen E, Koornneef J, Wigand F. International comparison of fossil power efficiency and CO_2 intensity — update 2014. Final Report. ECOFYS Netherlands B.V.; 2014.

[27] van de Bor DM, Infante Ferreira CA. Quick selection of industrial heat pump types including the impact of thermodynamic losses. Energy 2013;53:312—22.

[28] Collura MA, Luyben WL. Energy-saving distillation designs in ethanol production. Ind Eng Chem Res 1988;27:1686—96.

[29] Soave G, Feliu JA. Saving energy in distillation towers by feed splitting. Appl Therm Eng 2002;22(8):889—96.

[30] Seader JD, Siirola JJ, Barnicki SD. Distillation. In: Perry RH, Green DW, editors. Perry's chemical engineers' handbook. 7th ed. New York: McGraw Hill; 1997.

[31] Kiss AA, Flores Landaeta SJ, Infante Ferreira CA. Towards energy efficient distillation technologies — making the right choice. Energy 2012;47(1):531—42.

[32] Shenvi AA, Herron DM, Agrawal R. Energy efficiency limitations of the conventional heat integrated distillation column (HIDiC) configuration for binary distillation. Ind Eng Chem Res 2010;50(1):119—30.

[33] Olujic Z, Fakhri F, de Rijke A, de Graauw J, Jansens PJ. Internal heat integration — the key to an energy-conserving distillation column. J Chem Tech Biotechnol 2003;78:241—8.

[34] Eldridge RB, Seibert AF, Robinson S. Hybrid separations/distillation technology research opportunities for energy and emissions reduction. 2005.

[35] Kiss AA, Suszwalak DJPC. Enhanced bioethanol dehydration by extractive and azeotropic distillation in dividing-wall columns. Sep Purif Technol 2012;86(0):70—8.

[36] Gudena K, Rangaiah GP, Samavedham L. HiGee stripper-membrane system for decentralized bioethanol recovery and purification. Ind Eng Chem Res 2013;52:4572—85.

[37] Offeman RD, Stephenson SK, Robertson GH, Orts WJ. Solvent extraction of ethanol from aqueous solutions. I. Screening methodology for solvents. Ind Eng Chem Res 2005;44:6789—96.

[38] Munson CL, King CJ. Factors influencing solvent selection for extraction of ethanol from aqueous solutions. Ind Eng Chem Process Des Dev 1984;23(1):109—15.

[39] Dadgar AM, Foutch GL. Evaluation of solvents for the recovery of Clostridium fermentation products by liquid—liquid extraction. Biotechnol Bioeng Symp 1985;15:611—20.

[40] Vane LM. Separation technologies for the recovery and dehydration of alcohols from fermentation broths. Biofuels Bioprod Biorefin 2008;2:553—88.

[41] Koros WJ. Evolving beyond the thermal age of separation processes: membranes can lead the way. AIChE J 2004;50:2326—34.

[42] Rajagopalan K. Membrane desalination. In: Ray C, Jain R, editors. Drinking water treatment: focusing on appropriate technology and sustainability. Springer; 2011 [chapter 4].

[43] Koros WJ, Lively RP. Water and beyond: expanding the spectrum of large-scale energy efficient separation processes. AIChE J 2012;58(9):2624—33.

[44] Koros WJ. Materials & materials processing opportunities to enable future membranes development. Cincinnati (OH, USA): National Meeting of the American Institute of Chemical Engineers; 2005.

[45] Sanders DF, Smith ZP, Guo R, Robeson LM, McGrath JE, Paul DR, et al. Energy efficient polymeric gas separation membranes for a sustainable future: a review. Polymer 2013;54:4729−61.

[46] Robeson LM. Correlation of separation factor versus permeability for polymeric membranes. J Membr Sci 1991;62(2):165−85.

[47] Robeson LM. The upper bound revisited. J Membr Sci 2008;320(1−2):390−400.

[48] Cussler EL. Non-selective membranes for separations. J Chem Technol Biotechnol 2003;78(2−3):98−102.

[49] Zhang G, Cussler EL. Distillation in hollow fibers. AIChE J 2003;49(9):2344−51.

[50] Yang D, Barbero RS, Devlin DJ, Cussler EL, Colling CW, Carrera ME. Hollow fibers as structured packing for olefin/paraffin separations. J Membr Sci 2006;279(1−2):61−9.

[51] Zhang G, Cussler EL. Hollow fibers as structured distillation packing. J Membr Sci 2003;215(1−2):185−93.

[52] Bhandari DA, Bessho N, Koros WJ. Hollow fiber sorbents for desulfurization of natural gas. Ind Eng Chem Res 2010;49(23):12038−50.

[53] Lively RP, Chance RR, Kelley BT, Deckman HW, Drese JH, Jones CW, et al. Hollow fiber adsorbents for CO_2 removal from flue gas. Ind Eng Chem Res 2009;48(15):7314−24.

[54] Agrawal R, Noble RD. Separations research needs for the 21st century. Ind Eng Chem Res 2005;44:2887−92.

[55] Namboodiri V, Rajagopalan N. 2.6-Desalination. In: Ahuja S, editor. Comprehensive water quality and purification. Waltham: Elsevier; 2014. p. 98−119.

[56] Robinson RA, Stokes RH. Electrolyte solutions. The measurement and interpretation of conductance, chemical potential and diffusion in solutions of simple electrolytes. 2nd ed. Revised. London: Butterworths & Co. LTD; 1959.

[57] Cerci Y, Cengel Y, Wood B, Kahraman N, Karakas ES. Improving the thermodynamic and economic efficiencies of desalination plants: minimum work required for desalination and case studies of four working plants. Report prepared by the University of Nevada at Reno for the U.S. Department of the Interior, Bureau of Reclamation; 2003.

[58] Elimelech M. Seawater desalination. Presented at 2012. In: National Water Research Institute Clarke Prize Conference, "Research and innovations in urban water sustainability". Newport Beach (CA); 2012.

[59] Cost and performance baseline for fossil energy plants. Volume 1: Bituminous coal and natural gas to electricity (Rev. 2a). National Energy Technology Laboratory, U.S. Department of Energy; 2013. Report DOE/NETL-2010/1397.

[60] DOE/NETL advanced carbon dioxide capture R&D program: technology update. National Energy Technology Laboratory, U.S. Department of Energy; 2013.

[61] Gerdes K. NETL studies on the economic feasibility of CO_2 capture retrofits for the U.S. power plant fleet. National Energy Technology Laboratory, U.S. Department of Energy; 2014.

[62] Merkel TC, Lin H, Wei X, Baker R. Power plant post-combustion carbon dioxide capture: an opportunity for membranes. J Membr Sci 2010;359(1−2):126−39.

[63] Swain RLB. Molecular sieve dehydrators: why they became the industry standard and how they work. In: Ingledew WM, Kelsall DR, Austin GD, Kluhspies C, editors. The alcohol textbook. 5th ed. Nottingham, UK: Nottingham University Press; 2009.

[64] Vane LM. A review of pervaporation for product recovery from biomass fermentation processes. J Chem Tech Biotechnol 2005;80:603−29.

[65] Vane LM, Alvarez FR, Rosenblum L, Govindaswamy S. Hybrid vapor stripping-vapor permeation process for recovery and dehydration of 1-butanol and acetone/butanol/ethanol from dilute aqueous solutions. Part 2. Experimental validation with simple mixtures and actual fermentation broth. J Chem Technol Biotechnol 2013;88:1448−58.

[66] Vane LM, Alvarez FR. Hybrid vapor stripping-vapor permeation process for recovery and dehydration of 1-butanol and acetone/butanol/ethanol from dilute aqueous solutions. Part 1. Process simulations. J Chem Technol Biotechnol 2013;88:1436−47.

[67] Vane LM, Alvarez FR, Rosenblum L, Govindaswamy S. Efficient ethanol recovery from yeast fermentation broth with integrated distillation-membrane process. Ind Eng Chem Res 2013;52:1033—41.

[68] Vane LM, Alvarez FR, Huang Y, Baker RW, Inventors, Membrane Technology & Research, Inc., U.S Environmental Protection Agency, Assignee. Membrane-augmented distillation with compression to separate solvents from water. US Patent 8114255. February 14, 2012.

[69] Vane LM, Alvarez FR, Huang Y, Baker RW. Experimental validation of hybrid distillation-vapor permeation process for energy efficient ethanol-water separation. J Chem Tech Biotechnol 2010; 85:502—11.

[70] Huang Y, Baker RW, Vane LM, Alvarez FR. Low-energy distillation-membrane separation process. Ind Eng Chem Res 2010;49:3760—8.

[71] Huang Y, Baker RW, Daniels R, Aldajani T, Ly JH, Alvarez FR, et al. Inventors, Membrane Technology & Research, Inc., U.S Environmental Protection Agency, Assignee. Membrane augmented distillation to separate solvents from water. US Patent 8263815. September 11, 2012.

[72] Vane LM, Alvarez FR. Membrane-assisted vapor stripping: energy efficient hybrid distillation-vapor permeation process for alcohol-water separation. J Chem Tech Biotechnol 2008;83:1275—87.

[73] Zacchi G, Axelsson A. Economic evaluation of preconcentration in production of ethanol from dilute sugar solutions. Biotechnol Bioeng 1989;34:223—33.

[74] Galbe M, Zacchi G. A review of the production of ethanol from softwood. Appl Microbiol Biotechnol 2002;59:618—28.

[75] Madson PW, Lococo DB. Recovery of volatile products from dilute high-fouling process streams. Appl Biochem Biotechnol 2000;84-86:1049—61.

[76] Madson PW, editor. Fuel ethanol feedstock challenges. The Montana symposium: energy future of the west; 2005.

[77] Madson PW. Ethanol distillation: the fundamentals. In: Ingledew WM, Kelsall DR, Austin GD, Kluhspies C, editors. The alcohol textbook. 5th ed. Nottingham, UK: Nottingham University Press; 2009.

[78] Vane LM, Alvarez FR. Effect of membrane and process characteristics on cost and energy usage for separating alcohol-water mixtures using hybrid vapor stripping-vapor permeation process. J Chem Technol Biotechnol 2015;90:1380—90.

[79] Kockmann N. 200 years in innovation of continuous distillation. Chem Bio Eng Rev 2014;1(1): 40—9.

[80] Schäfer AI, Hughes G, Richards BS. Renewable energy powered membrane technology: a leapfrog approach to rural water treatment in developing countries? Renew Sustain Energy Rev 2014;40:542—56.

[81] Logan BE, Elimelech M. Membrane-based processes for sustainable power generation using water. Nature 2012;488(7411):313—9.

[82] Helfer F, Lemckert C, Anissimov YG. Osmotic power with pressure retarded osmosis: theory, performance and trends — a review. J Membr Sci 2014;453:337—58.

[83] Banchik LD, Sharqawy MH, Lienhard VJH. Limits of power production due to finite membrane area in pressure retarded osmosis. J Membr Sci 2014;468:81—9.

[84] Vermaas DA, Veerman J, Yip NY, Elimelech M, Saakes M, Nijmeijer K. High efficiency in energy generation from salinity gradients with reverse electrodialysis. ACS Sustain Chem Eng 2013;1(10):1295—302.

Conceptual Chemical Process Design for Sustainability

R.L. Smith
US Environmental Protection Agency, Cincinnati, OH, United States

CONCEPTUAL CHEMICAL PROCESS DESIGN

Of the various approaches to developing sustainable chemical processes, the most important step may be the conceptual design. Assuming that a product is needed and will be produced, the manner in which it is generated determines many aspects of its sustainability, and the structure of the process is often set during the conceptual design. Regarding its sustainability, every resource used in a process will connect it to an upstream supply chain, and each part of that supply chain needs a network of other supply chains with potentially a multitude of products and related processes. Each of these processes has the potential to create environmental impacts. Perhaps as important as the raw material resources used in these supply chains are the numerous energy resources used, with each process leading to more energy use and more environmental impacts.

Processes generate solid waste, wastewater discharges, and air emissions. Regulations limit the effects of these releases as process operations implement pollution controls and develop methods to manufacture products with fewer releases. Various US Environmental Protection Agency (EPA) and other international regulations limit or control toxicity, smog, acidification, ozone-layer depletion, hazardous solid waste, liquid discharges, etc. By enforcing regulations in these areas the effects on the health of people and on the environment are mitigated.

As processes control pollution and use cleaner environmental manufacturing methods, they may not only meet regulations but go beyond compliance toward sustainability. Sustainability has many definitions, normally geared toward three pillars: economics, environment, and social aspects, with an understanding that sustainability is achieved over time. The Brundtland Commission [1] proposed means by which countries and international organizations could lead toward sustainable "development that meets the needs of the present without compromising the ability of future generations to meet their own needs." The National Research Council [2] emphasized a transition to sustainability that integrates knowledge and action with instruments for affecting change including tools and indicators. A more provocative statement asserts that

Sustainability in the Design, Synthesis and Analysis of Chemical Engineering Processes   2016 Published by Elsevier Inc.

67

the state of sustainability will be realized when the average actions of all of the world's people improve the planet [3]. Until that time, actions can move toward being more sustainable. In chemical process terms, more sustainable processes can increase economic value, decrease effects on the environment, and improve the (social) quality of life.

Developing more sustainable processes can include myriad competing intentions in the form of economic, environmental, and quality-of-life objectives. For example, a conflict can occur when lower releases create tradeoffs with higher costs. Decision makers are faced with an incremental tradeoff of how much money they will exchange for lower environmental effects [4]. The solutions that offer the best tradeoffs between various objectives are known collectively as a Pareto front [5]. To break through this Pareto front of solutions, one needs to innovate a modified process. Breakthroughs in research and development often maintain the product performance, increase environmental friendliness, and decrease capital and operating costs [6].

Another method for making a conceptual design breakthrough is computer-aided process design. A conceptual design method can use optimization, developing from a superstructure that can consider as many design attributes as possible. Normally it is difficult to develop a superstructure that is robust enough to consider the many possible design attributes, with some final process designs that uniquely combine operating units and/or their connections to other unit operations that require too much data to model. However, once the basic conceptual design is laid out, computer-aided methods can be (and are) used extensively to complete and optimize the design (e.g., [7]).

A review of older literature on process design shows that it is usually necessary to start with a conceptual design that includes some hierarchical-based decisions and short-cut methods. Peters and Timmerhaus [8] provided accounting methods for process designs and pointed out the need for "quick estimate designs." Ludwig [9] published a three-volume set that describes very practically oriented design information for equipment and unit processes. Another early design book by Rudd and Watson [10] emphasized simplifications that are possible for acyclic (nonrecycling) processes. Many aspects of plant design, from scoping and site selection to process design, layout, cost estimation, and pollution abatement, with many equipment heuristics, are presented in a book by Baasel [11].

In developing computer-aided methods, Siirola et al. [12] described the necessary and sufficient conditions for a process to exist upon solving the associated design equations, but as they point out the design equations are not complete during conceptual design. Mahalec and Motard [13] investigated designs that did not assume one set of raw materials, and they put forward that various process design methods use optimization differently and each method uses heuristics. A methodology to quickly generate base-case designs was described by Douglas [14] to provide a step-by-step approach to conceptual designs, where the method is full of hierarchical-based decisions and short-cut methods.

There are several recent books on process design. Robin Smith [15] uses an onion model of process design, with the core of the onion (i.e., procedure) being the reactor, followed outward by separation and recycle, heat recovery, utilities, and treatment. Seider et al. [16] provide methods for process and product design, where the latter incorporates process design using various methods. In addition to common design subjects, Turton et al. [17] include environmental, health and safety, and green engineering in their book. While writing for an academic audience, Towler and Sinnott [18] present process design and unit operations with tools and methods used in industry.

The need for conceptual design arises from new processes. When a process exists, or even a similar process exists, one naturally builds off of these designs [18]. An existing plant might need debottlenecking to increase the production rate, whereas retrofit projects incorporate some new material or technology, necessitating a design change. These methods are fine for incremental changes from an existing process; however, building off an existing design (or even a flowsheet that someone suggests) can hamper imaginative solutions because the afterimage of seeing the design will be imprinted in the mind of the designer. Such an imprint can make creative design (of flowsheets) more difficult. Although when an existing flowsheet has been viewed, the results of retrofit designs will not become stuck in a form too close to the existing design if the design procedure includes a rich enough evolutionary approach.

Depending on how large a retrofit project is, the economic sustainability of it may benefit from conceptual design. Fisher et al. [19] described methods for conceptual design of retrofits. Part of the strategy describes a method where a new flowsheet is developed for completely replacing the existing process. The issue with large retrofit projects is that the capital costs can become large and unsustainable. Unless profitability is significantly improved and quick, most changes are desired to be drop-in replacements or minor retrofits. The common short-term objective is to minimize operating costs and to spend as little capital as possible in the process.

One approach for conceptual design generation is to consider the transformations necessary in making raw materials into a product. Siirola [20] describes a hierarchy of property differences: molecular identity, quantity, concentration and/or purity, phase, temperature, pressure, and form. Normally these property differences are transformed in the order of the hierarchy presented, with reaction changing molecular identity first, mixing or splitting changing quantity, blending or separations changing concentration and/or purity, etc. However, this order is not necessarily always used, for instance, when a tighter purity constraint improves a reaction, then separation operations can precede reactors. Reaction is not necessarily done by itself, as some processes use reactor/ separator operations, or intermittent heat integration, etc. Similar breaks from the hierarchy can be made by changing phases, temperatures, and/or pressures throughout a flowsheet so that other operations can be executed. Finally, recycle streams interrupt the straight-path architecture of raw materials leading to products. Recycle streams create loops that can change these properties, which can lead to different unit operations or the same type of unit operation with a different design and/or operating conditions. Also, recycle streams can blur the organization of a flowsheet as to which operations are occurring first, especially if various raw materials enter the process at substantially different locations along a recycle loop.

Flowsheets for processes are sometimes generated without following the hierarchy of properties described previously. As an example, Siirola [20] proposed a reactive-distillation solution to make methyl acetate. Unit operations that combine the property differences present abrupt departures from common methodologies. With the advent of various pieces of equipment, such as differential side-stream feed reactors (i.e., semicontinuously fed batch reactors), continuous evaporator-reactors (e.g., wiped-film evaporators), and reactive distillation columns, one can consider these unit operations in the development of conceptual designs. As an example, Doherty and Malone [21] have presented systematic methods for reactive distillation design.

Ideas for conceptual designs of processes can be generated from experience or from computer programs. For the use of computers, Powers [22] described AIDES as a program that uses these property differences and technical options (i.e., unit operations) for closing the differences. The choice of unit operations, order of placement, and connections between them create many potential process flowsheets. In addition, the unit operations may require the addition of new materials, like solvents for absorption, reactants for neutralization, etc. Opening the process flowsheet to include these additional unknown chemicals raises the potential complexity of computer-aided design algorithms tremendously.

A methodology to quickly generate base-case designs was described by Douglas [14] to provide a step-by-step approach to conceptual designs. The procedure, further developed by Douglas [23], uses computers to evaluate flowsheets as necessary, but relies on a hierarchical approach using paper, heuristics, back of the envelope calculations, and short-cut methods as applicable. The intent is for the base case to be a starting point for the evolution of better and finally optimized process designs. The power of the method is in eliminating uneconomical processes as early as possible, as less than 1% of process designs become commercialized. The result of this synthesis—analysis approach is a reasonable base-case design and a series of alternatives for consideration.

SUSTAINABILITY APPROACH FOR CHEMICAL PROCESSES

Chemical processes have been the subject of sustainability research in gate-to-gate and life cycle perspectives. A life cycle perspective including processes has been put forward by Azapagic [24], reviewing process selection and design as well as decision-making and optimization. Bakshi and coworkers considered exergy analysis that accounts for ecosystem services [25] and developed a tool to consider the ecologically based-life cycle assessment aspects [26]. Reviews have focused on various footprint, energy, and life cycle methods [27,28]. An optimization connecting processes to the supply chain or life cycle perspective can show tradeoffs between economic, environmental, and social objectives (e.g., [29]).

Other research has focused more on gate-to-gate analyses [30]. Douglas [31] discusses his hierarchical design procedure with a view toward developing alternatives that eliminate or minimize pollution streams. Manousiouthakis and Allen [32] divided process synthesis into several steps, including material synthesis and substitution, reaction path synthesis, reactor network synthesis, etc., with each step offering interesting waste minimization opportunities. For material synthesis and substitution, the ability to predict physical and chemical properties is vitally important [33], allowing new or different materials to be used with lower environmental impact properties [34,35]. Material selection is an important aspect of mass exchange networks, which is one method for approaching the improvement of chemical processes [36]. Another property focused approach reverses the methodology through property clusters aimed at designing appropriate molecules [37].

Further advances that considered gate-to-gate processes have focused on design objectives. Shonnard and Hiew [38] demonstrated the environmental fate and risk assessment tool as an assessment methodology. Optimization has been the focus of a number of reports that emphasize environmental impacts [39,40]. In designs including structural controllability, applying fuzzy logic methods makes the control problem tractable for waste minimization [41]. Uncertainties can be considered in optimization problems with the use of Hammersley sequence sampling techniques [42]. Tools to analyze design alternatives provide a connection between methodologies and practice [43,44].

To extend the approach in this work a sustainability-based hierarchical method will be developed. The social aspects of sustainability are (initially) assumed to exist by the quality of life advanced through the positive uses of the produced chemicals and the employment of those making the products. Economic sustainability will be described with analyses along the lines prescribed by Douglas [23], and environmental sustainability will be advanced through the biodegradation-based analysis of toxic emissions. While these aspects of sustainability are analyzed, additional attributes are important to sustainability and could be included. For example, the GREENSCOPE tool developed by the US EPA offers nearly 140 indicators to describe the sustainability of processes [45]. More sustainable processes improve attributes without creating significant tradeoffs.

To evaluate attributes for a process, the waste reduction (WAR) algorithm was developed initially as a pollution balance [46] and later as a potential impact balance [47]. The potential impact balance is a relative of toxicity indexes (e.g., [48]), which divide pollution by an LD_{50} (lethal dose that kills 50% of a test population) or threshold-limiting value. However, the balance incorporates an input—output calculation of system analysis that defines a generation term for a process. In this work it is assumed that the input materials to the processes are similar in type and quantity, and thus they would subtract out in comparison calculations. The potential environmental impact (PEI) of the pollutants,

$$\mathrm{PEI}_j = \sum_i M_i \psi_{ij} \qquad [3.1]$$

is determined by the mass flows, M_i, and characterization factors, ψ_{ij}, which are ratios of potential impacts per mass that are specified for the components i and the impact category or method j. Values of ψ_{ij} can come from the WAR algorithm or other method, for instance one based on the US EPA's reference doses (RfDs), as described later.

The assumption here is that unexamined aspects are neutrally affected by changes in the process described, but if tradeoffs were created a decision maker would have to decide the relative importance of process attributes [4]. An example to show the sustainability-based hierarchical approach will be presented next, using chlor-alkali production with a focus on economics and human toxicity potential.

 ## EXAMPLE: CHLOR-ALKALI PRODUCTION WITH HUMAN TOXICITY POTENTIAL ANALYSIS

The example presented here is one based on the chlor-alkali industry, where chlorine and sodium hydroxide are produced as coproducts. This example will be presented for two processes, known as the mercury process and the diaphragm process. Going through the steps of the design hierarchy presented by Douglas [23], the design will not focus on a description of these steps, which Douglas has presented very well. This work will emphasize the economic and environmental aspects of sustainability, and then the results will be analyzed for one particular environmental impact of interest, namely, human toxicity potential by ingestion. In analyzing this impact category the intention is not to discount the other categories. Rather, in the past the release of toxic substances has been analyzed by specific methods in chemical process design and evaluation, for example, the WAR algorithm [49], or USEtox [50], without adequately addressing pollutant biodegradation. Thus biodegradation will be considered here in the context of a design evaluation, marking an important extension to sustainable process design methods. Also the societal aspects of sustainability, assumed to be satisfied earlier through increased quality of life because of use of the products and the employment of those who make them, will be a topic of interest in the later discussion.

Analysis of Process Economics

The first step in the economic design hierarchy of Douglas is the choice of batch versus continuous operation. With a production rate of 365,000 tons of chlorine per year, this

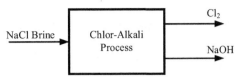

Figure 3.1 Input—output flowsheet structure of the chlor-alkali process.

design falls into the continuous category, where the advantages of scale can be realized. The second step of the hierarchy is an input—output flowsheet structure. The flows into and out of the overall process are described, and in this example shown by the flowsheet in Fig. 3.1.

With the known production rate of chlorine, one can also determine the coproduct and feedstock flows. Having known reactor technologies to consider for the two processes, the mercury process and diaphragm process, one can apply the stoichiometry from the appropriate reactions to calculate input and output flow rates.

The diaphragm process uses the electrolysis of sodium chloride brine with a diaphragm between the anode and cathode compartments in what is called a reactor cell [51]. The brine is fed into the anolyte compartment, and at the anode chlorine ions react to form chlorine gas (Cl_2), which subsequently exits up out of the reactor cell. The brine passes across the diaphragm barrier. At the cathode, water reacts with electrons to form hydrogen gas and hydroxide ions. (In this analysis it is assumed that the hydrogen is used locally in the process, although it could be sold.) The remaining hydroxide ions and sodium from the disassociation of salt together form sodium hydroxide, NaOH. The diaphragm keeps the chlorine from contacting the hydrogen and NaOH, which would react to form (in this case, undesirable) byproducts of HCl, sodium hypochlorite, and sodium chlorate. The overall reaction is,

$$2NaCl + 2H_2O \rightarrow 2NaOH + Cl_2 + H_2$$

Based on a stoichiometric number of moles of chlorine and NaOH produced, the quantity of NaOH product is 412,000 tons per year. The salt used, assuming perfect stoichiometric reaction, is 602,000 tons per year. Reasonable costs for these compounds were found [52]: Cl_2 at \$0.975/kg, NaOH at \$0.260/kg, and NaCl at \$0.100/kg. Using Douglas' definition of economic potential, for the input—output level, the value of products minus raw materials gives a potential profit of \$403 million per year.

The mercury process has a similar anode reaction producing chlorine gas, but the cathode reaction occurs on mercury flowing through the reactor cell [51]. Sodium and mercury form an amalgam that flows out of the reactor to a decomposer reactor. There the amalgam reacts with water to form NaOH, hydrogen, and mercury for recycling back to the electrolysis cell.

$$2NaHg + 2H_2O \rightarrow 2NaOH + 2Hg + H_2$$

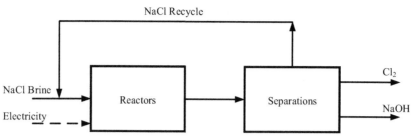

Figure 3.2 Recycle–reactor system structure flowsheet for the chlor-alkali process. The energy flow, electricity, is shown with a *dashed arrow*. This also represents the mercury process separation system structure flowsheet, as it does not need to further concentrate NaOH as the diaphragm process does.

The NaOH from the mercury process is concentrated to 50%, which is the same as the final concentration of the diaphragm process NaOH. Thus both processes obtain the same product, NaOH (50%). The resulting input–output economic potential for the mercury process is also $403 million per year.

The next hierarchical design level is the recycle–reactor system structure flowsheet. This flowsheet can be seen in Fig. 3.2, with the recycle of NaCl shown. (Details of these processes, including auxiliary parts of the process not considered in the designs of this work, can be found in Bommaraju et al. [51] and Schmittinger et al. [53].) The reactor cells and second decomposer reactor for the mercury process are grouped in Fig. 3.2 in the reactors block. Had these elements been developed for the first time, they would be fleshed out at this point. The anachronism of studying already existing processes means the designs are only discussed here rather than developed.

At this design level, losses of NaCl are considered. Bommaraju et al. [51] report that, instead of the stoichiometric amount, 621,000 tons of NaCl are used per year. This wasted salt reduces the economic potentials reported above by $2 million per year.

Reactor system design costs are included in the economic potential at this level. Bommaraju et al. [51] report a diaphragm process cell cost, which for our purposes is scaled to produce 1000 tons of Cl_2 per day, totaling $137 million per year (Douglas' method uses a capital charge factor to put costs on an annualized basis). All costs in this work have been adjusted to 2014 dollars with the Chemical Engineering Plant Cost Index [54]. The mercury process reactor system cost is estimated at 50% more than the diaphragm cell, thus reactor costs total $205 million per year. Schmittinger et al. [53] report an electricity usage for the mercury process electrolysis at 3300 kWh/ton of Cl_2, and for the diaphragm process, 15% less, i.e., 2800 kWh/ton. The electricity costs, using a wholesale value for electricity of $0.0437/kWh [55], are approximately $53 million and $45 million per year, respectively. The recycle–reactor system structure economic potentials for the diaphragm process and mercury process are now $219 million and $143 million per year, respectively. Thus at this design level the higher reactor system and electricity costs of the mercury process are dominating the differences between the processes.

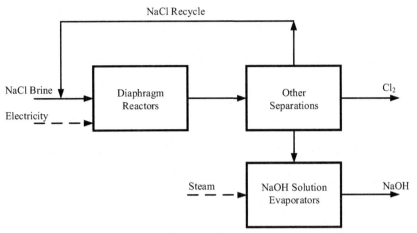

Figure 3.3 Separation system structure flowsheet for the chlor-alkali diaphragm process. Energy flows, electricity and steam, are marked with *dashed arrows*.

The final design level considered here, separation system structure flowsheet, is shown in Fig. 3.3. The only separation considered here is the evaporation of water for the diaphragm process, which needs to concentrate the NaOH from 11% up to 50% for sale. This matches the NaOH from the mercury process, which produces NaOH at 50% concentration from the decomposer reactor.

To concentrate the NaOH, four evaporators are designed by the methods described in Towler and Sinnott [18]. The evaporators annualized cost is $3 million per year, and the steam cost is approximated at $10 million per year, using steam tables and Douglas [23] for the cost of steam. For this example the final economic potential is $206 million for the diaphragm process. The mercury process remains at the same value as the last level, $143 million per year. Thus this analysis indicates that the diaphragm process is more economically sustainable, although for both processes the other parts of the full processes would have to be designed. See Bommaraju et al. [51] for block diagrams of these full processes.

Analysis of Human Toxicity Potential

In analyzing the toxicity potential of chlor-alkali processes, five species were chosen for analysis, representing environmental releases with relatively high toxicities, although other releases would have to be included for this analysis to be considered complete. The five species considered in this example include mercury, benzene, toluene, formaldehyde, and hexane. Among these, an obvious concern for a mercury process is the release of mercury. Concerns are so high that studies have been done to measure how encapsulation technologies perform for mercury solid wastes [56]. Additional concerns arise in foods like high-fructose corn syrup, where some samples have been found to have mercury, perhaps polluted by food industry ingredients produced from mercury process chlor-alkali plants [57]. A final EPA rule [58] also points to a mass balance issue at mercury processes, stating that "significant mercury remains unaccounted" in chlor-alkali processes. The harm that mercury can cause to humans is outlined according to its numerous effects by the US EPA [59]; various pharmacokinetics, toxicity, etc. aspects are reviewed by Bernhoft [60]; and personal health issues and interventions are described by Cutler [61].

The analysis now explores environmental aspects, in particular the human toxicity potential by ingestion, although many other environmental (and other) indicators are available in the US EPA's GREENSCOPE tool [62]. Two methods were used to determine emissions from the diaphragm process and mercury process, including 2011 national emissions inventory (NEI) process data [63] and emission factors for electricity and steam generation. The NEI emission data were set to a ratio by specific production rates (e.g., [64]) using information found in Linak and Inui [65]. The resulting average emission factors are presented in Table 3.1.

The five emissions considered for this analysis include mercury, benzene, toluene, formaldehyde, and hexane. The amounts of these emissions are of interest, and an analysis of Table 3.1 presents a scenario where it appears that mercury emissions for the mercury process will be traded off versus no mercury emissions and higher other toxic emissions for the diaphragm process. To compare the pollutants, the analysis was performed as if all pollutants for 1 year were released at the same time and the toxicity potentials for the pollutants can be added together as PEI, as introduced by the WAR algorithm. Guidelines for considering when it is appropriate to consider adding chemical doses and responses are discussed in documents by the US EPA [66,67].

An emphasis in this work is placed on the biodegradation of the emissions. An analysis was done using the tool PBT Profiler [68], which determined relative releases to water, soil, sediment, and air. An assumption was made that the releases ended up (immediately and forever) in the water compartment, where for all compounds (except

Table 3.1 Process Emission and Toxicity Characterization Factors for Five Pollutants of the Mercury Process and Diaphragm Process

	Process Emission Factors (kg/MT)		Toxicity Characterization Factors, ψ_{ij}	
	Mercury Process	**Diaphragm Process**	**WAR Algorithm (HTPI)**	**RfD Method**
Mercury	3.60×10^{-4}	0	0.4726	17,000
Benzene	5.90×10^{-7}	1.92×10^{-5}	0.1136	250
Toluene	9.55×10^{-7}	7.10×10^{-5}	0.0751	13
Formaldehyde	2.11×10^{-5}	6.95×10^{-5}	0.4696	5
Hexane	4.55×10^{-4}	1.68×10^{-3}	0.0131	33

Emission factor units are per metric ton of chlorine product. *HTPI*, Human toxicity potential by ingestion; *RfD*, reference dose.

mercury) the PBT Profiler reported 39%+ as an initial distribution to water. A more complex and realistic fate and transport model, for instance Mackay [69], could be used to describe the distribution of pollutants. In this work, the half-lives taken from the PBT Profiler in water for benzene, toluene, formaldehyde, and hexane are 38, 15, 15, and 8.7 days, respectively. Using first-order biodegradation the mass of each component emitted from the mercury and diaphragm processes is shown over 180 days in Fig. 3.4. Note that mercury, being an element, does not biodegrade, and its mass remains constant at 119 kg for the year. In order to show biodegradation starting from initial conditions, the yearly emissions have been displayed in Fig. 3.4 as if they all occurred on day 0. These emissions are actually spread over the year.

An analysis was accomplished using toxicity characterization factors for human toxicity potential by ingestion, $\psi_{i,HTPI}$, from the WAR algorithm [49] as shown in Table 3.1. Based on reciprocal LD_{50} values divided by an average value for the category, the characterization factors for mercury, benzene, toluene, formaldehyde, and hexane were determined. These PEI do not represent end effects, but only relative scores of potential effects among the set of pollutants. Using these scores in Eq. [3.1], a series of human toxicity potential profiles can be shown in Fig. 3.5. The results show that the toxicity potentials are dominated by mercury from the mercury process and the mercury toxicity potential remains high because mercury as an element does not biodegrade. It is important to realize in using a single characterization factor for mercury that mercury exists in many reacted forms, so it will not actually remain constant over time. However,

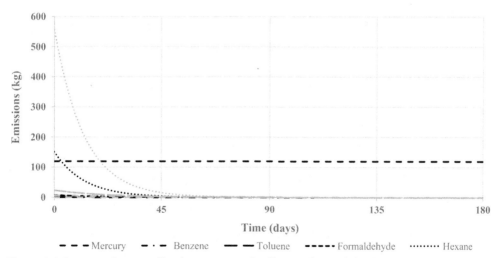

Figure 3.4 Process release profiles for amounts of pollutants (kg) with biodegradation described by half-life in the water compartment. *Bold lines* represent mercury process pollutants; *light gray lines* represent diaphragm process pollutants. For analysis purposes all pollution for a year is assumed to be emitted at once.

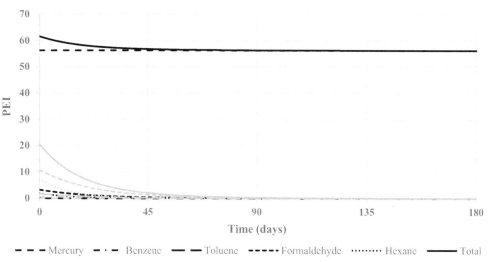

Figure 3.5 Potential environmental impact (PEI) profiles for amounts of pollutants with biodegradation described by half-life in the water compartment. *Bold lines* represent mercury process pollutants; *light gray lines* represent diaphragm process pollutants. PEI have been calculated using the waste reduction algorithm to present human toxicity potential by ingestion.

even with fluctuations because of the cycling of mercury through its various forms, the mercury toxicity potential will exist, as mercury remains in the biosphere. Driscoll et al. [70] have found that mercury settles into deep ocean waters and recalcitrant soil pools after hundreds of years.

Another method for analyzing the toxicity of emissions is to relate them to the US EPA's RfD for assessing the risk of health effects of systemic toxicity (i.e., toxicity other than cancer and gene mutations). RfD is an approximation of the daily exposure for which no effect would be expected over a lifetime [71]. Instead of using LD_{50} values from the WAR algorithm, which are based on various animal studies, RfD applies directly to human health. The RfD values for mercury, benzene, toluene, formaldehyde, and hexane are 6×10^{-5}, 4.0×10^{-3}, 0.08, 0.2, and 0.03 mg/kg/day, respectively. The inverse of these RfD are $\psi_{i,RfD}$ characterization factors (see Table 3.1), which can be implemented in Eq. [3.1]. PEI can be calculated using these RfD characterization factors and the releases and degradation rates presented in Fig. 3.4. A graph showing the results of these calculations as PEI would appear as two horizontal lines, one for the mercury process at 2.08×10^{6} and one for the diaphragm process on the order of 10^{4}, i.e., in a relative sense, near zero. This mirrors the results shown in Fig. 3.5 for the WAR algorithm calculations, where the mercury process had much higher potential impacts. This is an important result of the WAR algorithm and RfD method example calculations presented, that while the magnitudes of results are different, the trends for the results are consistent. The WAR algorithm characterization factors present a reasonable (although imperfect, see formaldehyde $\psi_{i,j}$ in Table 3.1) alternative to the RfD method, especially when RfD values are not available. (Note that the RfD values were determined as they would be for risk assessments. The mercury RfD was *estimated* as the maximum contaminant level, 0.002 mg/L [72], multiplied by 2 L of water a day, and divided by 70 kg mass per person. The hexane RfD was determined by using the subchronic Provisional Peer-Reviewed Toxicity Value for Superfund [73] and dividing by 10 to obtain a chronic value. The other three values came directly from the US EPA's IRIS database list of RfD values [74].)

A further analysis of the diaphragm process emissions using the RfD method presents interesting results, shown in Fig. 3.6. In this figure one can see that the total PEI follows hexane for about 50 days, and then the total is dominated by benzene at longer times. This behavior occurs because of the large amount of hexane emitted (as shown in Fig. 3.4) and the much longer half-life of benzene (38 days compared to 8.7 days for hexane). There is no mercury involved in the diaphragm process and so all of the emissions biodegrade over time. In this case the PEI is dominated by compounds with larger emissions, larger characterization factors (i.e., lower RfD values), and longer half-lives.

Additional analysis is necessary when one considers energy-related emissions. First, a description of the energy-related emission factors will be presented. Assuming that coal is burned for electricity generation [75], and using an EPA report [76] for the amount of mercury in coal and emitted after pollution controls (assuming 90% capture), the controlled emission rate is 5.58×10^{-9} kg mercury/kWh. For the benzene, toluene, formaldehyde, and hexane emissions from coal-fired electricity generation, EPA and

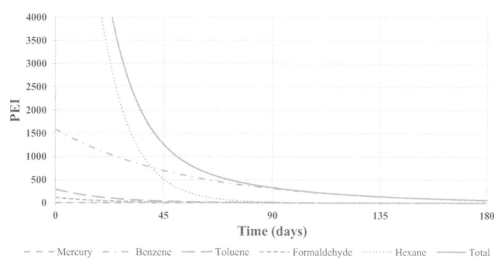

Figure 3.6 Diaphragm process potential environmental impacts (PEI) using the reference dose (RfD) method. Hexane initially has a PEI value of 18,500, and the total is initially 20,500.

Environmental Impact Assessment references [76–78] were used to develop the following controlled emission rates: 3.21×10^{-7}, 5.93×10^{-8}, 5.93×10^{-8}, and 1.65×10^{-8} kg/kWh, respectively. It was assumed that steam from natural gas combustion was used to heat the evaporators for the diaphragm process. Emission factors were developed from US EPA [77] for mercury and Ventura County APCD (air pollution control district) [79] for the other pollutants, with values of 1.10×10^{-13}, 7.17×10^{-13}, 3.29×10^{-12}, 1.52×10^{-12}, and 5.48×10^{-13} kg/kJ of natural gas for mercury, benzene, toluene, formaldehyde, and hexane, respectively.

Based on calculations done for the process designs described previously, the resulting emissions that include energy processes lead to PEI similar to that described earlier, with nearly horizontal lines for the mercury toxicities (not changing over time) and the total values approaching the mercury PEI values asymptotically (for the mercury process at 2.21×10^{6}, and for the diaphragm process at 1.2×10^{5}). Compared to the values reported earlier (without energy-related emissions) each of the PEI values has increased by about 10^{5}. This represents a very large percentage increase for the diaphragm process, which without energy process emissions had a PEI on the order of 10^{4}. Thus energy-related emissions can dominate the results if the process emissions are relatively low. A detailed graph shows the PEI using the RfD method for only the diaphragm process in Fig. 3.7. As compared to Fig. 3.6, which considers only process emissions, the addition of energy-related emissions causes an order of magnitude increase in the toxicity potentials. Here mercury and benzene have the largest toxicity potentials, followed distantly by hexane and the other pollutants. The reduced prominence for hexane can be

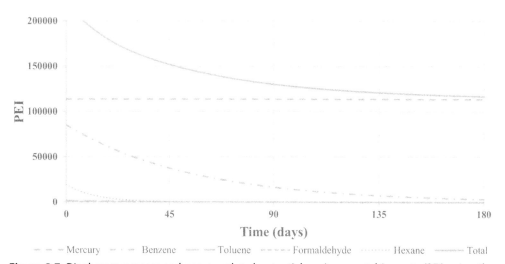

Figure 3.7 Diaphragm process and energy-related potential environmental impacts (PEI) using the reference dose (RfD) method. The total PEI is initially 220,000, an order of magnitude larger than for the exclusively process-related PEI (shown in Fig. 3.6).

attributed to its lower emission factor for coal-fired electricity generation. Both mercury and benzene have a combination of emission factors and characterization factors (i.e., high $\psi_{i,\text{RfD}}$ values) that lead to these two pollutants dominating the diaphragm process and energy-related initial toxicity potentials. Over time, mercury dominates as benzene biodegrades.

DISCUSSION

Processes with releases of elements and slowly degrading compounds distribute hazardous pollution into the environment with the prospect of constantly cycling effects as the pollution continuously transfers through environmental media. As the chlor-alkali example shows, a toxic element can dominate PEI when the effects are calculated over a long time period. What the example did not show is the effect of a second year of releases on the results, or for that matter 10, 20, or 50 years of nondegrading pollution. Without doing further calculations, it is clear that mercury impacts will stack on top of each other year after year, while compounds that (rapidly) degrade have effects that remain relatively low.

Some processes are evaluated with the tacit assumption that compounds recycled internally stay within the process. Almost always a makeup stream of the recycling compound is introduced (although perhaps intermittently). A makeup stream should be a clear indication that the recycling compound is leaving the process. If the recycling compound has a high enough PEI, then analyses must take it into account in order to be complete.

The analysis shown for the chlor-alkali process did not consider the input of mercury or its upstream supply chain. Mercury was not considered as an input in the chlor-alkali example, but it should have been had the analysis been able to consider all inputs. In addition, any time a toxic element, like mercury, is used in a process it must be generated in upstream processes. Based on a life cycle perspective of this supply chain, it is likely that mercury would dominate the long-term toxicity evaluated over decades, or for that matter over much shorter time periods.

Perhaps unexpectedly, mercury can dominate the long-term toxicity of processes that *do not* use mercury as an input but *do* use energy with associated mercury releases. This was shown in the chlor-alkali example, where the diaphragm process PEI showed mercury overshadowing the other releases when energy use was included. It may be that *every* process environmental analysis developed in the literature should be (re)evaluated for energy use and the effects of nondegrading releases.

Finally, to completely analyze a process and its effects one should return to the social impacts that this study has so far assumed to be satisfied. If mercury, or any compounds, are so long lived that they are cycling through communities, then they are potentially diminishing people's quality of life with a high (but unrecognized) social cost [80]. A truly sustainable design will consider economic, environmental, and social aspects, and the results shown here suggest the need for a new complete-analysis paradigm that includes this societal aspect of some releases in evaluating and designing sustainable processes.

CONCLUSIONS

A method for addressing human toxicity potential by ingestion was presented along with hierarchical designs for chlor-alkali processes. This method can be extended to other impact categories, for instance, exposure by inhalation could be estimated using reference concentration values. In addition to presenting the potential toxicity impacts, the method also applied biodegradation to the releases. Mercury, being an element, does not biodegrade, and thus the importance of mercury to toxicity impacts over time is evident. Other nonbiodegradable or slowly biodegraded pollutants, like benzene, could have similar environmental and over the long term social effects, and these effects should be evaluated and acted upon in assessing toxicity impacts (and other impacts where applicable) for process evaluation and design.

DISCLAIMER

The views expressed in this chapter are those of the author and do not necessarily represent the views or policies of the US Environmental Protection Agency.

REFERENCES

[1] World Commission on Environment and Development. Our common future. Oxford: Oxford University Press; 1987.
[2] National Research Council. Our common journey: a transition toward sustainability. Washington (DC): National Academy Press; 1999.
[3] Dernbach JC. Agenda for a sustainable America. Washington (DC): ELI Press; 2009. p. 10.
[4] Smith RL, Ruiz-Mercado GJ. A method for decision making using sustainability indicators. Clean Technol Environ Policy 2014;16:749—55.
[5] Pareto V. Cours d'Economie Politique, vol. 1. Lausanne, Switzerland: Rouge; 1896.
[6] Smith RL. Sustainability and life cycle assessments. In: Presentation 28c, AIChE annual meeting, Nashville, TN; 2009.
[7] Biegler L, Grossmann I, Westerberg A. Systematic methods of chemical process design. Upper Saddle River (NJ): Prentice Hall; 1997.
[8] Peters MS, Timmerhaus KD. Plant design and economics for chemical engineers. New York: McGraw-Hill; 1958.
[9] Ludwig EE. Applied process design for chemical and petrochemical plants. Houston (TX): Gulf Publishing Company; 1964.
[10] Rudd DF, Watson CC. Strategy of process engineering. New York: John Wiley and Sons; 1968.
[11] Baasel WD. Preliminary chemical engineering plant design. New York: Elsevier; 1976.
[12] Siirola JJ, Powers GJ, Rudd DF. Synthesis of system designs: III. Toward a process concept generator. AIChE J 1971;17(3):677—82.
[13] Mahalec V, Motard RL. Procedures for the initial design of chemical processing systems. Comput Chem Eng 1977;1:57—68.
[14] Douglas JM. A hierarchical decision procedure for process synthesis. AIChE J 1985;31(3):353—62.

[15] Smith R. Chemical process design and integration. West Sussex, England: John Wiley & Sons; 2005.

[16] Seider WD, Seader JD, Lewin DR, Widagdo S. Product and process design principles: synthesis, analysis and design. Hoboken (NJ): John Wiley & Sons; 2008.

[17] Turton R, Bailie RC, Whiting WB, Shaeiwitz JA, Bhattacharyya D. Analysis, synthesis, and design of chemical processes. Upper Saddle River (NJ): Prentice Hall; 2012.

[18] Towler GP, Sinnott RK. Chemical engineering design: principles, practice and economics of plant and process design. Oxford: Elsevier; 2013.

[19] Fisher WR, Doherty MF, Douglas JM. Screening of process retrofit alternatives. Ind Eng Chem Res 1987;26(11):2195−204.

[20] Siirola JJ. An industrial perspective on process synthesis. In: Biegler LT, Doherty MF, editors. Fourth international conference on foundations of computer-aided process design. AIChE symposium series 304; 1995. p. 222−33.

[21] Doherty MF, Malone MF. Conceptual design of distillation systems. Boston: McGraw-Hill; 2001.

[22] Powers GJ. Heuristic synthesis in process development. Chem Eng Prog 1972;68(8):88−95.

[23] Douglas JM. Conceptual design of chemical processes. Boston (MA): McGraw-Hill; 1988.

[24] Azapagic A. Life cycle assessment and its application to process selection, design and optimisation. Chem Eng J 1999;73:1−21.

[25] Hau JL, Bakshi BR. Expanding exergy analysis to account for ecosystem products and services. Environ Sci Technol 2004;38:3768−77.

[26] Zhang Y, Baral A, Bakshi BR. Accounting for ecosystem services in life cycle assessment, part II: toward an ecologically based LCA. Environ Sci Technol 2010;44:2624−31.

[27] Zhang Y, Singh S, Bakshi BR. Accounting for ecosystem services in life cycle assessment, part I: a critical review. Environ Sci Technol 2010;44:2232−42.

[28] Ingwersen W, Cabezas H, Weisbrod AV, Eason T, Demeke B, Ma X, et al. Integrated metrics for improving the life cycle approach to assessing product system sustainability. Sustainability 2014;6(3):1386−413.

[29] You F, Tao L, Graziano DJ, Snyder SW. Optimal design of sustainable cellulosic biofuel supply chains: multiobjective optimization coupled with life cycle assessment and input−output analysis. AIChE J 2012;58(4):1157−80.

[30] Cano-Ruiz JA, McRae GJ. Environmentally conscious chemical process design. Annu Rev Energy Environ 1998;23:499−536.

[31] Douglas JM. Process synthesis for waste minimization. Ind Eng Chem Res 1992;31:238−43.

[32] Manousiouthakis V, Allen D. Process synthesis for waste minimization. In: Fourth international conference on foundations of computer-aided process design. Snowmass (CO): FOCAPD; 1994. p. 72−86.

[33] Joback KG. Knowledge bases for computerized physical property estimation. Fluid Phase Equilibr 2001;185:45−52.

[34] Martin TM, Young DM. Prediction of the acute toxicity (96-h LC50) of organic compounds to the fathead minnow (Pimephales promelas) using a group contribution method. Chem Res Toxicol 2001;14:1378−85.

[35] US EPA. Quantitative structure activity relationship. http://www.epa.gov/nrmrl/std/qsar/qsar.html [accessed on 26.08.15].

[36] El-Halwagi MM. Pollution prevention through process integration. San Diego (CA): Academic Press; 1997.

[37] Eden MR, Jorgensen SB, Gani R, El-Halwagi MM. Property integration − a new approach for simultaneous solution of process and molecular design problems. Comput Aided Chem Eng 2002;10:79−84.

[38] Shonnard DR, Hiew DS. Comparative environmental assessments of VOC recovery and recycle design alternatives for a gaseous waste stream. Environ Sci Technol 2000;34:5222−8.

[39] Pistikopoulos EN, Stefanis SK, Livingston AG. A methodology for minimum environmental impact analysis. In: El-Halwagi MM, Petrides DP, editors. Pollution prevention via process and product modifications. AIChE symposium series, vol. 303(90); 1994. p. 139−50.

[40] Smith RL, Ruiz-Mercado GJ, Gonzalez MA. Using GREENSCOPE indicators for sustainable computer-aided process evaluation and design. Comput Chem Eng 2015;81:272–7.

[41] Edgar TF, Huang YL. Artificial intelligence approach to synthesis of a process for waste minimization. In: Tedder DW, Pohland FG, editors. Emerging technologies in hazardous waste management IV. ACS symposium series, 554. Washington (DC): American Chemical Society; 1994. p. 96–113.

[42] Fu Y, Diwekar UM, Young D, Cabezas H. Process design for the environment: a multi-objective framework under uncertainty. Clean Prod Process 2000;2:92–107.

[43] Gonzalez MA, Smith RL. A methodology to evaluate process sustainability. Environ Prog 2003;22(4):269–76.

[44] Carvalho A, Matos HA, Gani R. SustainPro—a tool for systematic process analysis, generation and evaluation of sustainable design alternatives. Comput Chem Eng 2013;50:8–27.

[45] Ruiz-Mercado GJ, Smith RL, Gonzalez MA. Sustainability indicators for chemical processes: I. Taxonomy. Ind Eng Chem Res 2012;51:2309–28.

[46] Hilaly AK, Sikdar SK. Pollution balance: a new methodology for minimizing waste production in manufacturing processes. J Air Waste Manag Assoc 1994;44:1303–8.

[47] Mallick SK, Cabezas H, Bare JC, Sikdar SK. A pollution reduction methodology for chemical process simulators. Ind Eng Chem Res 1996;35:4128–38.

[48] Grossmann IE, Drabbant R, Jain RK. Incorporating toxicology in the synthesis of industrial chemical complexes. Chem Eng Commun 1982;17:151–70.

[49] Young DM, Scharp R, Cabezas H. The waste reduction (WAR) algorithm: environmental impacts, energy consumption, and engineering economics. Waste Manag 2000;20:605–15.

[50] Rosenbaum RK, Bachmann TM, Gold LS, Huijbregts MAJ, Jolliet O, Juraske R, et al. USEtox — the UNEP-SETAC toxicity model: recommended characterisation factors for human toxicity and freshwater ecotoxicity in life cycle impact assessment. Int J LCA 2008;13(7):532–46.

[51] Bommaraju TV, Luke B, O'Brien TF, Blackburn MC. Chlorine. In: Seidel A, editor. Kirk-othmer encyclopedia of chemical technology. John Wiley and Sons; 2002.

[52] Alibaba.com. Product prices. Chlorine, http://www.alibaba.com/product-detail/chlorine-liquid-gas_1783235183.html?s=p; 50% NaOH solution, http://www.alibaba.com/product-detail/sodium-hydroxide-solution-50-Factory-supply_503639302.html; NaCl, http://www.alibaba.com/product-detail/sodium-chloride-price_60139459886.html [accessed on 19.05.15].

[53] Schmittinger P, Florkiewicz T, Curlin LC, Luke B, Scannell R, Navin T, et al. Chlorine. In: Ullmann's encyclopedia of industrial chemistry. Weinheim: Wiley-VCH; 2011.

[54] Vatavuk WM. Updating the CE plant cost index. Chem Eng 2002;109(1):62–70.

[55] Smith RL, Sengupta D, Takkellapati S, Lee CC. An industrial ecology approach to municipal solid waste management: II. Case studies for recovering energy from the organic fraction of MSW. Resour Conserv Recy 2015. http://dx.doi.org/10.1016/j.resconrec.2015.04.005.

[56] Randall P, Chattopadhyay S. Advances in encapsulation technologies for the management of mercury-contaminated hazardous wastes. J Hazard Mater 2004;B114:211–23.

[57] Wallinga D, Sorensen J, Mottl P, Yablon B. Not so sweet: missing mercury and high fructose corn syrup. Minneapolis (MN): Institute for Agriculture and Trade Policy; 2009.

[58] Federal Register. National emission standards for hazardous air pollutants: mercury emissions from mercury cell chlor-alkali plants, vol. 68(244); 2003. p. 70904–46.

[59] US EPA. Mercury, health effects. http://www.epa.gov/mercury/effects.htm [accessed on 29.05.15].

[60] Bernhoft RA. Mercury toxicity and treatment: a review of the literature. J Environ Public Health 2012. http://dx.doi.org/10.1155/2012/460508.

[61] Cutler AH. Amalgam illness: diagnosis and treatment. Cutler; 1999.

[62] Ruiz-Mercado GJ, Smith RL, Gonzalez MA, Smith RL. Sustainability indicators for chemical processes: III. Biodiesel case study. Ind Eng Chem Res 2013;52:6747–60.

[63] US EPA. Technology transfer network, clearinghouse for inventories & emissions factors, emission inventories. NEI Data; 2011. http://www.epa.gov/ttn/chief/eiinformation.html [accessed on 25.02./15].

[64] Sengupta D, Hawkins TR, Smith RL. Using national inventories for estimating environmental impacts of products from industrial sectors: a case study of ethanol and gasoline. Int J LCA 2015;20:597–607.

[65] Linak E, Inui Y. Chemical economics handbook, marketing research report, chlorine/sodium hydroxide. Menlo Park (CA): SRI International; 2002.

[66] US EPA. Guidelines for the health risk assessment of chemical mixtures, EPA/630/R-98/002. Washington (DC): Risk Assessment Forum; 1986.

[67] US EPA. Supplementary guidance for conducting health risk assessment of chemical mixtures, EPA/630/R-00/002. Washington (DC): Risk Assessment Forum; 2000.

[68] PBT Profiler. Persistent bioaccumulative and toxic (PBT) chemical program. http://www.epa.gov/pbt/tools/toolbox.htm [accessed on 09.04.15].

[69] Mackay D. Multimedia environmental models: the fugacity approach. 2nd ed. Boca Raton (FL): CRC Press; 2001.

[70] Driscoll CT, Mason RP, Chan HM, Jacob DJ, Pirrone N. Mercury as a global pollutant: sources, pathways, and effects. Environ Sci Technol 2013;47:4969–83.

[71] US EPA. Reference dose (RfD): description and use in health risk assessments, background document 1A. 1993. http://www.epa.gov/iris/rfd.htm [accessed on 20.05.15].

[72] US EPA. Maximum contaminant level for mercury. http://water.epa.gov/drink/contaminants/basicinformation/mercury.cfm [accessed on 20.05.15].

[73] US EPA. Provisional peer reviewed toxicity values for superfund. 2009. http://hhpprtv.ornl.gov/index.html [accessed on 20.05.15].

[74] US EPA. Integrated risk information system (IRIS). http://cfpub.epa.gov/ncea/iris/index.cfm?fuseaction=iris.showSubstanceList [accessed on 20.05.15].

[75] EIA. Electric power annual 2013, table 8.1, average operating heat rate for selected energy sources. 2015.

[76] US EPA. Control of mercury emissions from coal-fired electric utility boilers, interim report, EPA-600/R-01-109, tables 2-2, 2-5, and 6-5. 2002. http://permanent.access.gpo.gov/lps34674/P10071NU.pdf [accessed on 21.05.15].

[77] US EPA. *AP-42*, Chapter 1.1, bituminous and subbituminous coal combustion; table 1.1-14, emission factors for various organic compounds from controlled coal combustion. Chapter 1.4, natural gas combustion, table 1.4-4 used for mercury emissions. 1998.

[78] EIA. Monthly energy review; Washington, DC; table A5, approximate heat content of coal and coal coke. March 2015.

[79] Ventura County APCD. AB 2588 combustion emission factors, used >100MMBtu/h column in table of NG external equipment emission factors. 2001. In: http://www.aqmd.gov/docs/default-source/permitting/toxics-emission-factors-from-combustion-process-.pdf?sfvrsn=0 [accessed on 21.05.15].

[80] Hylander LD, Goodsite ME. Environmental costs of mercury pollution. Sci Total Environ 2006;368:352–70.

CHAPTER FOUR

Process Integration for Sustainable Design

M.M. El-Halwagi
Texas A&M University, College Station, TX, United States

INTRODUCTION

Some of the primary pillars supporting sustainable design of industrial processes are:

- Effective utilization and conservation of natural resources (mass and energy)
- Development and application of greener technologies and molecules
- Enhancement of process yield, efficiency, and profitability

These pillars can be methodically and effectively achieved through the use of a powerful framework and set of tools referred to as "process integration." A possible definition of process integration is that it is a "holistic approach to process design and operation that emphasizes the unity of the process" [1,2]. All units, streams, species, and resources are analyzed, synergized, and optimized from a big-picture perspective. There are three key categories of process integration: (1) energy, (2) mass, and (3) property integration, which offer a holistic approach to design with specific attention to various forms of energy, species and streams, and functionalities, respectively. Process integration usually involves the following steps [2,3]:

1. Task identification: to determine the goals and actionable activities
2. Targeting: to establish a benchmark of excellence for performance
3. Generation and selection of alternatives (synthesis): to generate alternatives that can reach the desired targets and screen them based on specific criteria (e.g., economic, environmental, etc.)
4. Analysis of selected alternative(s): to predict, validate, or improve the performance of the chosen implementation schemes

There are several books and review articles that provide comprehensive coverage of process integration [3–12]. This chapter offers a brief overview of the primary aspects of sustainable design through process integration. First, the three primary categories of process integration (mass, property, and energy integration) are covered. Then, applications including multiscale approaches to process integration are discussed.

Sustainability in the Design, Synthesis and Analysis of Chemical Engineering Processes
87

MASS INTEGRATION

Mass integration is a "systematic methodology that provides a fundamental understanding of the global flow of mass within the process and employs this understanding in identifying performance targets and optimizing the generation and routing of species throughout the process" [1]. It is a holistic approach to the tracking, management, and optimization of species and streams. One way of understanding the integration of species and streams is to represent the process flowsheet from the perspective of species (Fig. 4.1). The process streams are referred to as "sources." The process units are referred to as "sinks." Each sink has a number of constraints on what constitutes an acceptable feed to the unit. It is possible to segregate, split, and mix the streams to meet the requirements for the process units. This is referred to as "direct recycle." It may also be necessary to add new units to adjust the flowrates, composition, and characteristics of the sources and render them in a condition that is acceptable as feed streams to the process sinks. The collection of these new units is referred to as the "interception network." It involves the use of mass and energy separating agents. Mass integration provides a set of methodologies and systematic tools that determine the best possible performance of the process and how to separate, transform, and allocate the various streams and species throughout the process while maintaining the perspective of what is best for the whole process.

Benchmarking tools may be used to determine the targets for consumption, production, and discharge of each species involved in the process. Once these targets are determined, mass-integration tools may be used to achieve these benchmarks. Various strategies may be used for implementation. These strategies may be low, medium, or high in cost. Fig. 4.2 is a hierarchical representation of mass-integration strategies. No-/low-cost strategies include modest modifications of the process structure (e.g., segregation, blending, and reuse/recycle of streams) or minor changes in operating

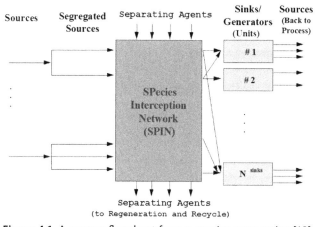

Figure 4.1 A process flowsheet from a species perspective [13].

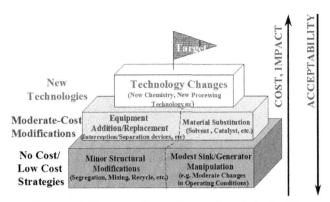

Figure 4.2 Hierarchy of mass-integration strategies [2].

conditions (e.g., temperature and pressure). Typically, these are the "low hanging fruits" that are associated with a very attractive payback period (almost instantaneous), a modest impact, and a high level of acceptability (likelihood of commercial application). Next, as new process units are added and molecules are substituted, the cost and impact increase but the acceptability decreases. Finally, with substantial technological changes, the ultimate process targets are achieved, but typically the cost is highest and the acceptability is lowest. All of these strategies are important and must be seriously considered. Depending on the company's sustainability objectives and management culture and constraints, the implementation strategies are selected.

As an illustration of a class of problems in mass integration, let us consider the direct recycle problem. The objective of this problem is to determine the optimal allocation of sources to sinks without the addition of new equipment. Each source is characterized by a flowrate and composition. For each sink there are constraints on the acceptable feed, which are given in terms of lower and upper bounds on flowrate and composition. One approach to solving this problem is through the use of the material recovery/recycle pinch diagram developed by El-Halwagi et al. [14]. Fig. 4.3 is a representation of this graphical tool. First, flowrate and composition data are collected for all the recyclable streams (referred to as sources) and units (referred to as sinks) that can accept the recycle to reduce the consumption of the fresh resources. The flowrate and composition of impurities are used to calculate the load of impurities in each source as follows:

Load of impurities = Flowrate of the source

$$* \text{ Composition of impurities in the source} \qquad [4.1]$$

Each source is represented as a straight line on a load versus flowrate diagram. The slope of each line is the composition of impurities in that stream. Using superposition, a source composite curve is constructed to represent the cumulative load of impurities versus the flowrate of the streams. The streams are represented in ascending order of the

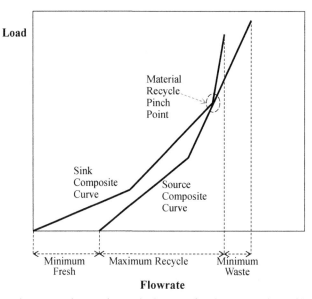

Figure 4.3 The materials recycle pinch diagram for direct recycle problems [14].

composition of impurities. The same approach is adopted to create a sink composite curve representing all the units in the process. The load of each sink is the maximum allowable load of impurities that may be fed to the unit. To insure feasibility the sources composite curve must lie below the sink composite curve. The two composites are moved until they touch with the sources composite lying completely below the sink composite. The point where the two composites meet is referred to as the "material recycle pinch point." Three important targets are identified from this diagram: minimum consumption of the fresh resources, maximum recycle of the process resources, and minimum discharge of the waste streams. In addition to these rigorous targets, the diagram also offers three design rules that may be used in synthesizing new recycle networks or critiquing existing systems. These rules are [14]:

- No flowrate should be passed through the pinch (i.e., the two composites must touch)
- No waste should be discharged from sources below the pinch
- No fresh feed should be used in any sink above the pinch

An important category of the resource conservation and recycle networks is the minimization of fresh water usage and wastewater discharge [15]. Another important class of direct recycle problems is hydrogen management [16]. It involves the recycle and rerouting of hydrogen-rich sources to hydrogen-using units to reduce the usage of fresh hydrogen. The recycled streams must meet the purity levels required by the units. In addition to the hydrogen purity constraints, one must also account for the additional consideration of pressure. Compressors and throttling devices may be used to adjust the pressure to required levels.

While direct recycle problems deal with the segregation, mixing, and allocation of streams to sinks, some mass-integration tasks may require separation (interception) to render the sources acceptable for recycle to the sinks. In this context an important class of mass integration problems is the synthesis of mass-exchange networks (MENs) introduced by El-Halwagi and Manousiouthakis [17]. The problem involves the transfer of a number of components from streams that are rich in these components to streams that are lean in the same components (mass-separating agents [MSAs]). One possible objective for the synthesis of an MEN is to minimize the usage of external separating agents and to maximize the use of process (internal) resources as separating agents. Practical feasibility of mass exchanges is ensured by using thermodynamic equilibrium and a minimum mass-transfer driving force. For instance, consider a mass exchanger for which the equilibrium relation for the transfer of a certain species from a rich stream to the jth lean stream (MSA) is given by the following linear equation:

$$y^* = m_j x_j^* + b_j \qquad\qquad [4.2]$$

where y^* and x^* are the compositions of the transferrable component in the rich and the lean phases, respectively. The equilibrium coefficients m_j and b_j represent the slope and intercept of the equilibrium function when the composition of the rich phase is plotted versus the composition of the lean phase. By employing a minimum allowable composition difference of ε_j, the maximum practically attainable composition in the lean phase is calculated through:

$$x_j^{\text{max,practical}} = x_j^* - \varepsilon_j \qquad\qquad [4.3]$$

A mass–exchanged load versus composition plot is developed. First, each rich stream is represented as a straight line extending from the supply (inlet) to target (outlet) composition of the rich stream. The slope of this line is the flowrate of the stream. Using superposition the rich streams are collectively represented with a rich composite stream. Next, Eq. [4.3] is used to create a correspondence among the rich and lean streams for which mass transfer is practically feasible. Each process lean stream is represented as an arrow from supply composition to target composition with a slope of the flowrate of the process lean stream. Again, superposition is used to construct a lean composite stream representing all the process MSAs in the plant. Mass can be feasibly transferred from the rich streams to the lean streams lying to their left (because this satisfies the thermodynamic constraints and the minimum mass-transfer driving force). This diagram is referred to as the mass-exchange pinch diagram and can be used to determine minimum usage of external separating agents and maximum utilization of process resources for mass exchange as shown by Fig. 4.4.

There are different variants of the MEN problem and tools such as reactive MENs, heat induced MENs, and batch MENs [9,10,18–22].

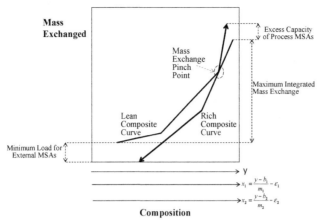

Figure 4.4 Mass-exchange pinch diagram [17]. *MSAs*, Mass-separating agents.

PROPERTY INTEGRATION

In mass integration, the focus is on species and streams. There are important sustainable design problems that are dependent on properties and functionalities (e.g., toxicity, chemical oxygen demand (COD), biological oxygen demand (BOD), color, pH, density, viscosity, etc.) and not necessarily chemical species. For such problems, property integration offers a powerful framework for design. El-Halwagi and coworkers [23,24] introduced framework of property integration, which is a "functionality-based, holistic approach to the allocation and manipulation of streams and processing units, which is based on the tracking, adjustment, assignment, and matching of functionalities throughout the process" [24].

Much research has been carried out in the area of sustainable design through property integration [25–38]. The following sections present some sample tools and applications. Let us first start with the problem of the property-based material recycle pinch diagram. It is analogous to the previously discussed direct recycle via the material recycle pinch diagram. The key difference is that instead of using chemical composition to characterize the sources and the constraints on the sinks, properties are used. Since properties are not conserved, there is a need to track properties via mixing operators. A commonly used form is:

$$\overline{F} * \psi(\overline{p}) = \sum_i F_i * \psi(p_i)$$ [4.4]

where $\psi(p_i)$ is the property-mixing operator and $\overline{F}$ is the total flowrate of the mixture, which is given by:

$$\overline{F} = \sum_i F_i$$ [4.5]

An example of a mixing rule that may be derived from first principles is the following expression for estimating the density resulting from the blending of several incompressible fluids:

$$\frac{\overline{F}}{\overline{\rho}} = \sum_i \frac{F_i}{\rho_i} \qquad [4.6]$$

In the aforementioned expression, the density operator is defined as:

$$\psi(\rho_i) = \frac{1}{\rho_i} \qquad [4.7]$$

In other cases, the property-mixing rules may be developed from the regression of experimental data over a limited range of application [23,24]. In more complex cases the mixing operator depends on multiple interdependent properties and must be created through the accounting of such interdependence [25].

If the property load is defined as the product of the flowrate times the property operators, then it is possible to develop composite representations for the sources and sinks and to obtain benchmarks for direct recycle. Kazantzi and El-Halwagi [26] developed the property-based material-recycle pinch diagram as shown by Fig. 4.5. Similar to the benchmarks described earlier for the composition-based material recycle pinch diagram, the property-based analog can be used to identify three rigorous targets: minimum usage of the fresh resources, maximum recycle of the process streams, and minimum discharge of waste.

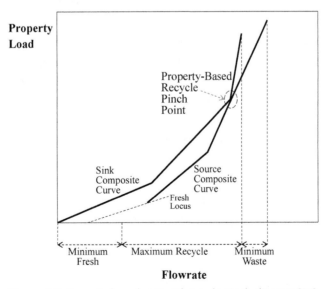

Figure 4.5 Property-based material recycle pinch diagram [26].

When multiple properties are to be considered simultaneously, other approaches may be used. For example, the cluster ternary diagram may be used as a graphical tool for the design of property-based recycle networks. Shelley and El-Halwagi [23] introduced the concept of property clusters as the basis for property-based component-free design. Based on the property mixing expressions (e.g., Eq. [4.1]), the property operators may be normalized via the division by a reference value:

$$\Omega_{r,i} = \frac{\psi_r(p_{r,i})}{\psi_r^{\text{ref}}} \qquad [4.8]$$

When these normalized values for each property are added for all the properties, one obtains an augmented property (AUP) index for each stream, i, which is given by:

$$\text{AUP}_i = \sum_{r=1}^{Np} \Omega_{r,i} \quad i = 1, 2, ..., N_{\text{Sources}} \qquad [4.9]$$

The cluster for property r in stream i, $C_{r,i}$, is defined through the following ratio between the normalized operator and the augmented index:

$$C_{r,i} = \frac{\Omega_{r,i}}{\text{AUP}_i} \qquad [4.10]$$

It can be shown that these clusters enjoy intrastream and interstream conservation [23]. As such they may be used as surrogate quantities to properties with the added advantage of being conserved. Hence, lever arm rules on a ternary diagram work well for clusters and may be used to benchmark performance and to synthesize recycle strategies. For instance, if a designer would like to minimize the usage of a fresh resource (F) by maximizing the recycle of a process or a waste stream (W), the ternary cluster diagram may be used to determine the optimal mixing ratio as shown by Fig. 4.6. Minimizing the fresh arm corresponds to minimizing the fresh usage. Also, interstream conservation of clusters indicates that all possible mixtures of F and W will lie on the straight line connecting them. Therefore, while mixtures such as the one denoted by points a, b, and c are feasible, mixture a corresponds to the minimum fresh arm and, consequently, minimum usage of the fresh resource. An example of this level-arm rule is shown by Fig. 4.7 for a metal degreasing plant, which uses a solvent based on its density, sulfur content, and Reid vapor pressure (RVP). A gaseous waste may be condensed to yield a recyclable condensate that may be mixed with the fresh solvent to reduce both the fresh usage of the solvent and the hydrocarbon discharge in the gaseous waste. The circles in Fig. 4.7 represent the recyclable hydrocarbon stream at different condensation temperatures. The dotted line represents the boundaries of the feasibility region of the degreaser as dictated by the constraints on the properties of the solvent. If the condensate at 240 K is to be mixed with a fresh solvent, the

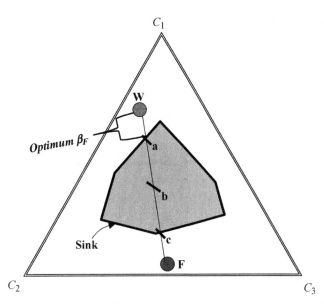

Figure 4.6 Minimizing the usage of fresh resources using the ternary cluster diagram [25].

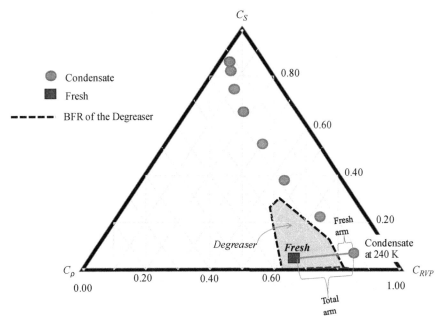

Figure 4.7 Determining the fresh usage via the ternary property—cluster diagram for the degreaser [3,24,25]. *BFR*, Boundaries of the feasibility region.

lever-arm rule is used to determine the reduced usage of the fresh solvent and the recyclable flowrate of the condensate.

For more than three properties, algebraic [37,38] and optimization [31–34] approaches may be used.

ENERGY INTEGRATION

Energy integration is another holistic approach to design and operation, which focuses on the exchange, transformation, and allocation of energy in its different forms. One of the principal problems in energy integration is the synthesis of heat exchange networks (HENs), which is intended to transfer heat from process hot streams to process cold streams. A particularly useful tool is the HEN thermal pinch diagram [39,40]. It involves the construction of a hot composite curve and a cold composite curve to represent all the process hot and cold streams. By ensuring thermodynamically feasible transfer of heat from the hot composite to the cold composite, the designer can determine three rigorous targets: minimum external heating utility, minimum external cooling utility, and maximum process heat exchange. In order to screen and optimize the selection of utilities, another tool (referred to as the grand composite curve [GCC] may be used [8,41]. The GCC represents the net heat residuals (difference between heat supplied by the hot and heat gained by the cold) versus temperature (average temperature of the hot and the cold streams). This representation is the GCC. The pinch point corresponds to the zero-residual point. Fig. 4.8 is a representation of a typical GCC.

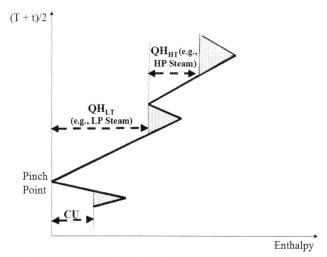

Figure 4.8 The grand composite curve for identifying and selecting minimum heating and cooling utilities. *HP*, High pressure; *LP*, low pressure; QH_{LT}, heating utility at low temperature; QH_{HT}, heating utility at high temperature; *CU*, cooling utility.

When the residual increases, it implies that the heat available from the hot streams exceeds the heat removed by the cold streams. This excess heat may be passed to a lower temperature to provide the deficiency in heat. This type of heat integration is shown by the shaded pockets in Fig. 4.8. When the residual heat moves from left to right, this indicates a surplus heat that may be transferred to a lower temperature. On the other hand, when the residual heat moves from right to left, this indicates that there is heat deficiency. Hence, a surplus from a higher temperature may be passed to a deficiency at a lower temperature and the "pocket" is closed through heat integration. Next, the utility requirements are determined starting with the cheapest utility (e.g., low-pressure steam) followed by high-pressure steam. Below the pinch, the minimum cooling utility requirement is determined to close the heat balance. The same targets may also be obtained using mathematical programming approaches [42,43]. Extensive reviews are available for the synthesis of HENs [3—12].

Although heat is a major form of energy, there are other forms of energy that are commonly used in the process industries. Of particular interest is the integration of work (or power, which is work per unit time) with heat. This is referred to as cogeneration or combined heat and power (CHP). When refrigeration is also included, the problem is referred to as trigeneration. There are various approaches to the design of CHP systems. For details the reader is referred to the literature [5,6,8,44—48]. Graphical, algebraic, and optimization approaches have been developed for CHP systems. As a sample of the graphical approaches is the benchmarking approach of El-Halwagi et al. [48]. The basic idea is to carry out energy balances around the steam headers at different pressure and temperature levels starting with the process resources for heating. As a result, some headers will offer surplus energy and other headers will show deficiency in energy. The extractable power for each header is defined as the product of the surplus (or deficiency) flowrate of steam times the enthalpy of the steam at the header conditions times the turbine efficiency. The headers are arranged in ascending order of pressure. Next, a composite curve is developed using superposition for the surplus headers and another composite is developed for the deficit headers. By combining both composites and accounting for the amount of energy that can be extracted through steam turbines, targets can be determined for minimum cogeneration potential and excess steam (Fig. 4.9).

An example of mathematical programming approaches for CHP is the procedure proposed by Gutiérrez-Arriage et al. [47], which involves the use of organic Rankine cycles (ORCs) along with excess process heat. Fig. 4.10A,B shows a typical ORC and its thermodynamic performance on a temperature—entropy diagram. Because of the complexity in modeling the different steps of the process and the integration with the process thermal loads, an optimization approach is needed. One approach is to use genetic algorithms (GAs) in a structured procedure as the one shown by Fig. 4.11

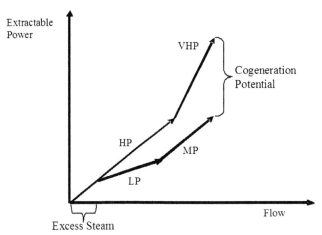

Figure 4.9 Benchmarking cogeneration targets using a pinch diagram [48]. *LP*, Low pressure; *HP*, high pressure; *MP*, medium pressure; *VHP*, very high pressure.

When applied to the case study shown by Fig. 4.12, the results displayed by Fig. 4.13 illustrate that the excess process heat is efficiently used to reduce the cooling duty of the plant, to drive the ORC, to generate electric power, and to produce a refrigeration duty.

It is also worth noting that various forms of renewable energy may be used in CHP systems. Abdelhady et al. [49] addressed the problem of incorporating solar energy into a CHP system for a processing facility. Fig. 4.14 is a schematic representation of the problem involving the integration of excess process heat, fossil fuel, and solar energy to design a cogeneration system. Thermal storage is used to handle the diurnal variation of solar energy. The heating and cooling demands are collected for the process. Heat integration is carried out to minimize the external heating and cooling utilities for the process. A cogeneration system is then considered with the objectives of utilizing excess process heat, external solar fuels, and solar energy. Because of the diurnal changes in solar energy collection, two strategies are adopted: the use of thermal storage and dispatch and the application of fossil fuels to supplement the usage of the solar energy.

To solve the above-mentioned problem, a hierarchical design approach (Fig. 4.15) may be used. It includes a combination of modeling, heat integration, multiperiod mixed integer nonlinear programming, and technoeconomic analysis to determine the optimal distribution of energy mix and the design of the system while accounting for the time-based variability of the solar energy. Fig. 4.16 shows the optimal hourly contribution of fossil fuels, directly used solar energy, and stored/dispatched solar energy for the month of January for a petrochemical plant in the city of Jeddah, Saudi Arabia.

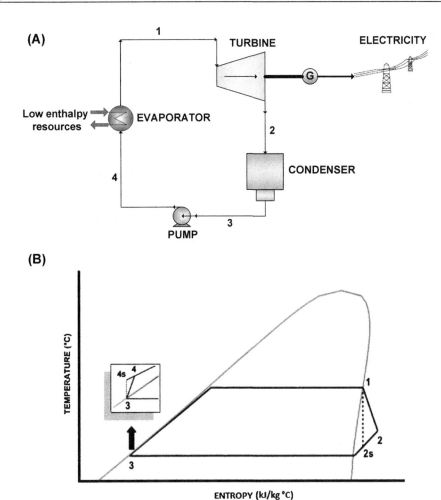

Figure 4.10 Organic Rankine cycle. (A) Schematic diagram; (B) temperature versus entropy diagram [47]. *ORC*, Organic Rankine cycle.

MULTISCALE APPROACHES

A particularly attractive feature of process integration tools is that they naturally extend to cover multiscale applications of sustainable design. Fig. 4.17 is a representation of the multiscale examples that can be consistently tackled using process integration tools.

In this section, three examples of multiscale application are presented:
- Integration of process and molecular design
- Integration of the process with surrounding environment
- Eco-industrial parks

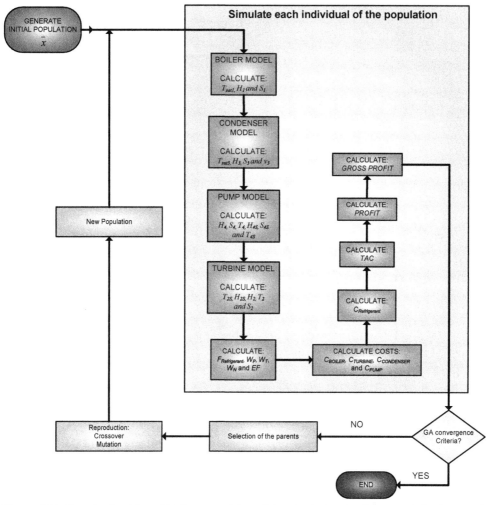

Figure 4.11 Genetic algorithms (GAs) sequential modular simulation for the organic Rankine cycle (ORC) [47].

Integration of Process and Molecular Design

As an illustration of the consistent basis for tackling process and molecular design problems, let us consider the coupling between property integration and molecular design. The molecular design problem requires information on the desired set of properties and generates, in turn, desirable molecular structures. Basic thermodynamic and thermophysical properties may be estimated using group contribution methods. Hence, the desired properties may be related to the molecular design problem. On the other hand, property integration can be used to identify targets for

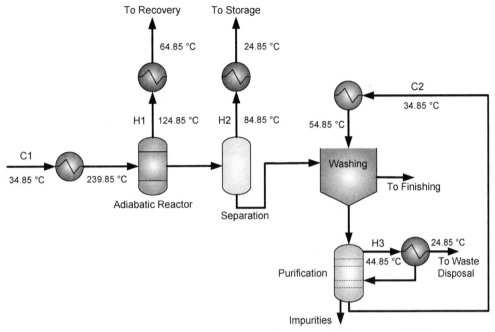

Figure 4.12 Flowsheet for the organic Rankine cycle (ORC) case study [47].

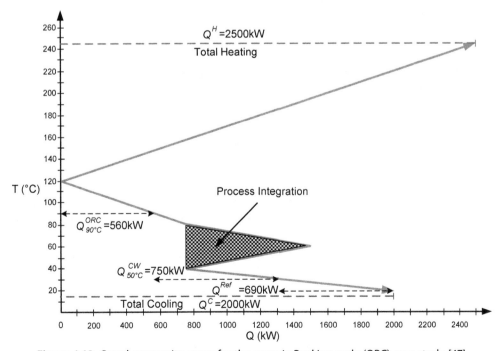

Figure 4.13 Grand composite curve for the organic Rankine cycle (ORC) case study [47].

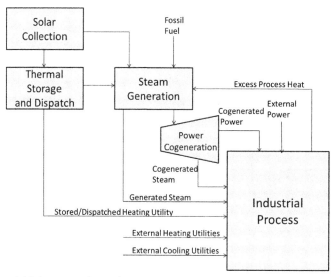

Figure 4.14 Incorporating solar energy in a process cogeneration system [49].

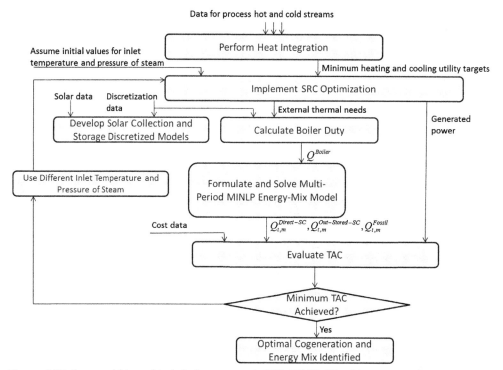

Figure 4.15 Proposed hierarchical design approach [49]. *MINLP*, Mixed integer nonlinear programming; *TAC*, total annualized cost; *SRC*, steam Rankine cycle.

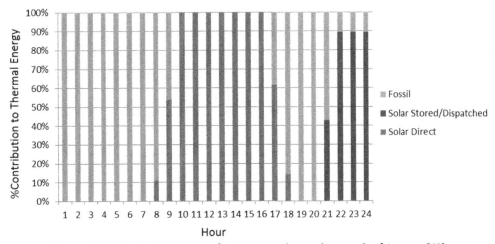

Figure 4.16 Hourly distribution of energy mix during the month of January [49].

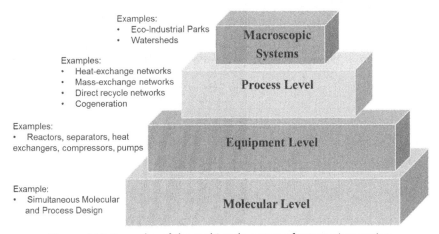

Figure 4.17 Examples of the multi-scale nature of process integration.

the desired properties needed for the molecules to be used in the process. Eljack et al. [30] introduced a framework for integrating property integration with molecular design. Using targeting techniques of property integration, a reverse-problem formulation is solved to determine the optimal set of properties needed for the molecules to be used in the process (Fig. 4.18). For instance, the property-based material recycle pinch diagram previously shown by Fig. 4.5 can be used to determine the bounds on the property targets for the molecules to be used in recycle/reuse networks [26,27], as shown by Fig. 4.19. These targets are then used to synthesize the set of desired molecules.

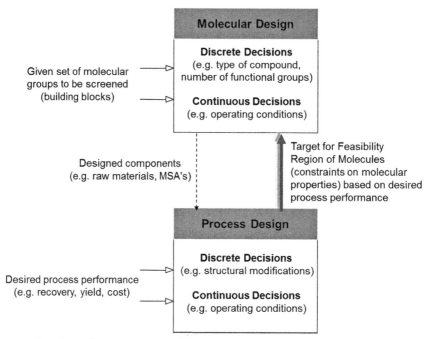

Figure 4.18 Coupling of property integration and molecular design [30]. *MSAs,* Mass-separating agents.

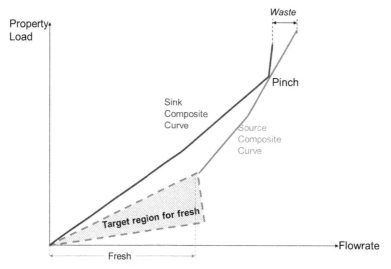

Figure 4.19 Using the property-based material recycle pinch diagram to target the properties of the molecules.

Integration of the Process With Surrounding Environment

In addition to developing an integrated approach to the processing facilities, it is important to account for the interaction of these facilities with the surrounding environments. For instance, when an industrial facility impacts a surrounding watershed, it becomes necessary to optimize the process design not just to meet the process objectives but to also address the environmental impact on the surrounding environment. A material flow analysis (MFA) may be used to track the physical, chemical, and biological implications of process discharges on the watershed. Examples of MFA include water quality models and atmospheric dispersion models. In addition to process discharges, other inputs within the watershed (e.g., residential, agricultural) are also accounted for as shown by Fig. 4.20. Coupling of process integration with the MFA (Fig. 4.19) provides a framework for macroscopic system integration with the processing facility. The watershed receives and dispatches water from and to various sectors including the industrial processes, rural areas, residential area, farms, and natural precipitation. Hence, water usage and discharge of an industrial facility must be balanced with the other sectors that interact via the watershed. Furthermore, the physical, chemical, biological, and biochemical processes taking place within the watershed must be incorporated in the analysis of the watershed and the synthesis of cost-effective solutions. An example of such phenomena is the nitrogen cycle shown by Fig. 4.21.

Models such as the MFA interact well with a mass-integration framework. The MFA model provides the data needed for mass integration (e.g., flows, concentrations of sources and sinks, impact of a change in input on the output characteristics, etc.).

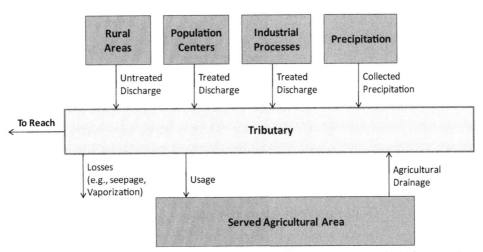

Figure 4.20 A material flow analysis (MFA) representation for integrating an industrial process into a watershed [50].

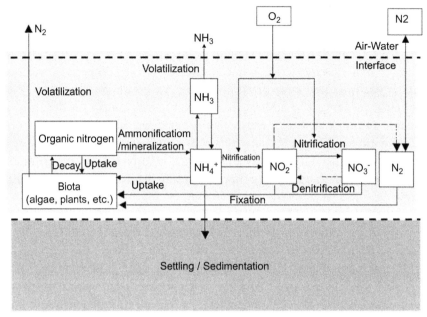

Figure 4.21 Accounting for the nitrogen cycle in a watershed [50].

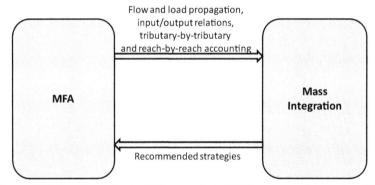

Figure 4.22 Coupling of material flow analysis (MFA) and mass integration [51].

In return, mass integration is used to devise solution strategies for certain objectives of the watershed. The interaction is shown by Fig. 4.22.

The application of mass–integration tools to the watersheds establishes tradeoffs between several objectives such as cost of managing water versus environmental impact [52]. For instance, in the case study given by Lira–Barragán et al. [52] for the Balsas watershed system located in Mexico, a new chemical plant is to be installed. There are 20 candidate sites as shown by Fig. 4.23. Mass integration coupled with MFA can be used to evolve water-management strategies and to aid in the site selection for the chemical

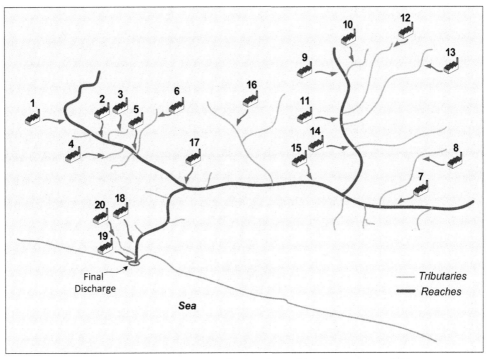

Figure 4.23 The Balsas watershed and possible plant locations [52].

plant. The Pareto (tradeoff) curve for the cost versus the concentration of pollutants at the outfall of the watershed is shown by Fig. 4.24.

Eco-Industrial Parks

An eco-industrial park (EIP) may be defined [53]: "as a cluster of processing and other facilities together with supporting infrastructure which create the circular economy and promote sustainable development. Such a park will be developed and managed as a real estate development, seeking high environmental (e.g., reduction of pollutant discharge), economic (e.g., reduction of cost or enhancement of profitability), and social (e.g., creation of high-caliber jobs) benefits as well as business excellence. It will implement conditions for participation and will recruit members who in total achieve the desired objective. This park will be bottom line driven and the tenants selected in such a way that economic performance increases as the level of integration and, therefore, efficiency increases. Participation will provide clear economic and other advantages over the current stand-alone processing model." Spriggs et al. [53] introduced an integrated representation for designing an EIP and for coordinating the exchange and treatment of streams and species (Fig. 4.25). Lovelady and El-Halwagi [54] developed a

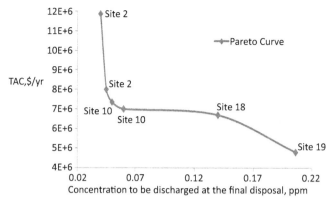

Figure 4.24 The Pareto curve showing the tradeoff between total annualized cost (TAC) and the pollutant concentration at the final discharge location [52].

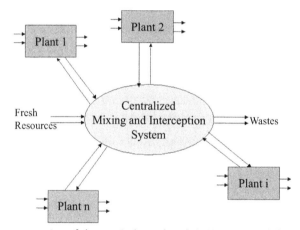

Figure 4.25 Representation of the eco-industrial park (EIP) integrated-design problem [53].

mass–integration approach to the optimal design of EIPs. The representation (Fig. 4.26) and the associated mathematical programming formulation are analogous to those used in integrating mass within the process.

In addition to optimizing the EIP based on streams and chemical components, the same approach can be extended to the atomic level. Noureldin and El-Halwagi [55] introduced the concept of a carbon, hydrogen, oxygen symbiosis network (CHO–SYN) to integrate multiple processing facilities by identifying opportunities based on atomic integration of carbon, hydrogen, and oxygen. The overall representation is illustrated by Fig. 4.27 and a typical solution is shown by Fig. 4.28. The atomic insights are coordinated with the process and multiprocess approaches to yield an optimal integrative framework for the EIP based on a consistent basis at all scales.

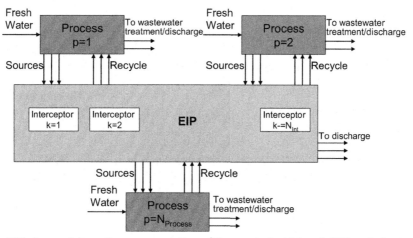

Figure 4.26 A mass-integration representation of the eco-industrial park (EIP) optimization [54].

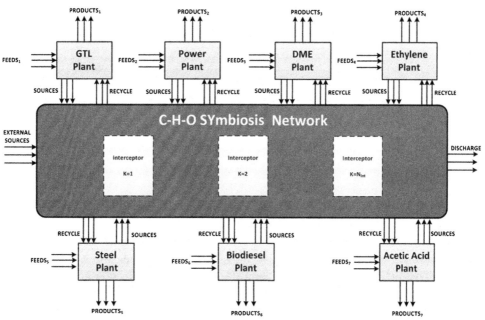

Figure 4.27 A superstructure representation of a carbon, hydrogen, oxygen symbiosis network (CHOSYN) network [55].

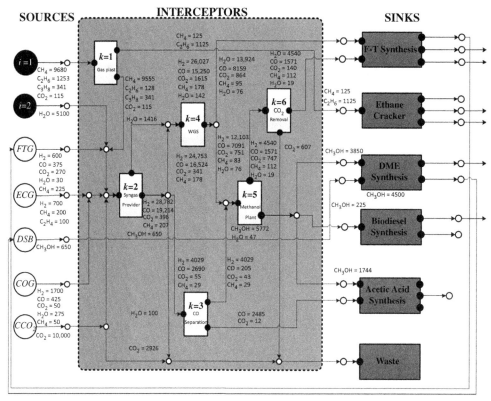

Figure 4.28 A superstructure solution of a carbon, hydrogen, oxygen symbiosis network (CHOSYN) network [55].

CONCLUSIONS

This chapter has presented an overview of key tools and applications of process integration for sustainable design. Mass, property, and energy integration frameworks have been discussed along with the enabling tools needed for implementation. Process integration provides a powerful approach to sustainable design because it systematically addresses principal objectives such as the conservation of mass and energy resources, the optimization of process performance, and the reconciliation of multiple criteria. The key theme in all these approaches is to embrace a holistic approach, determine performance benchmarks, then zoom in on the details for implementation. Another important observation is that process integration tools work equally well at multiple scales ranging from the atomic level all the way to a multiplant or a regional system. Examples were given for direct recycle systems, MENs, heat integration, cogeneration, molecular design, macroscopic systems, and eco-industrial parks. Graphical, algebraic, and optimization approaches can be used as implementation tools. Process integration approaches systematize numerous decisions associated with sustainable design and offer the designer a methodical and efficient platform for navigating through the complexity of sustainable design and for trading off the various objectives.

REFERENCES

[1] El-Halwagi MM. Pollution prevention through process integration: systematic design tools. San Diego: Academic Press; 1997.

[2] El-Halwagi MM. Process integration. Amsterdam: Elsevier; 2006.

[3] El-Halwagi MM. Sustainable design through process integration: fundamentals and applications to industrial pollution prevention, resource conservation, and profitability enhancement. Butterworth—Heinemann/Elsevier; 2012.

[4] El-Halwagi MM, Foo DCY. Process synthesis and integration, in Kirk-Othmer encyclopedia of chemical technology. Wiley; 2014. http://dx.doi.org/10.1002/0471238961.1618150308011212. a01.pub2.

[5] Klemeš J. Handbook of process integration: minimisation of energy and water use. Woodhead Publishing Limited; 2013.

[6] Klemeš J, Varbanov PS, Wan Alwi SR, Manan ZA. Process integration and intensification: saving energy, water, and resources. Berlin: De Gruyter; 2014.

[7] Foo DCY. Process integration for resource conservation. Boca Raton (Florida): CRC Press; 2012.

[8] Kemp I. Pinch analysis and process integration — a user guide on process integration for the efficient use of energy. 2nd ed. Butterworth—Heinemann; 2009.

[9] Majozi T. Batch chemical process integration: analysis, synthesis, and optimization. Heidelberg: Springer; 2010.

[10] Majozi T, Seid ER, Lee J-Y. Synthesis, design, and resource optimization in batch chemical plants. Boca Raton: CRC Press; 2015.

[11] Noureldin MB. Pinch technology and beyond: new vistas on energy efficiency optimization. Nova Science Publishers; 2011.

[12] Smith R. Chemical process design and integration. New York: Wiley; 2005.

[13] El-Halwagi MM, Hamad AA, Garrison GW. Synthesis of waste interception and allocation networks. AIChE J 1996;42(11):3087—101.

[14] El-Halwagi MM, Gabriel F, Harell D. Rigorous graphical targeting for resource conservation via material recycle/reuse networks. Ind Eng Chem Res 2003;42:4319—28.

[15] Wang YP, Smith R. Wastewater minimisation. Chem Eng Sci 1994;49(7):981—1006.

[16] Alves JJ, Towler GP. Analysis of refinery hydrogen distribution systems. Ind Eng Chem Res 2002;41:5759—69, 1994.

[17] El-Halwagi MM, Manousiouthakis V. Synthesis of mass exchange networks. AIChE J 1989;35(8):1233—44.

[18] Gadalla MA. A new graphical-based approach for mass integration and exchange network design. Chem Eng Sci 2015;127:239—52.

[19] Azeez OS, Isafiade AJ, Fraser DM. Supply-based superstructure synthesis of heat and mass exchange networks. Comput Chem Eng 2013;56:184—201.

[20] Hallale N, Fraser DM. Capital and total cost targets for mass exchange networks: part 1: simple capital cost models. Comput Chem Eng 2000;23(11—12):1661—79.

[21] Papalexandri KP, Pistikopoulos EN. A multiperiod MINLP model for the synthesis of heat and mass exchange networks. Comput Chem Eng 1994;18(12):1125—39.

[22] Srinivas BK, El-Halwagi MM. Synthesis of reactive mass-exchange networks with general nonlinear equilibrium functions. AIChE J 1994;40(3):463—72.

[23] Shelley MD, El-Halwagi MM. Component-less design of recovery and allocation systems: a functionality-based clustering approach. Comput Chem Eng 2000;24:2081—91.

[24] El-Halwagi MM, Glasgow IM, Eden MR, Qin X. Property integration: componentless design techniques and visualization tools. AIChE J 2004;50(8):1854–69.

[25] Sandate-Trejo MDC, Jiménez-Gutiérrez A, El-Halwagi MM. Property integration models with interdependence mixing operators. Chem Eng Res Des 2014;92(12):3038–45.

[26] Kazantzi V, El-Halwagi MM. Targeting material reuse via property integration. Chem Eng Prog 2005;101(8):28–37.

[27] Kazantzi V, Qin X, El-Halwagi M, Eljack F, Eden M. Simultaneous process and molecular design through property clustering – a visualization tool. Ind Eng Chem Res 2007;46:3400–9.

[28] Hortua AC, Ng D, Foo DCY, El-Halwagi MM. An integrated approach for simultaneous mass and property integration for resource conservation. ACS Sustain Chem Eng 2013;1:29–38.

[29] Lira-Barragán F, Ponce-Ortega JM, Nápoles-Rivera F, Serna-González M, El-Halwagi MM. Incorporating property-based water networks and surrounding watersheds in site selection of industrial facilities. Ind Eng Chem Res 2013;52:91–107.

[30] Eljack F, Eden M, Kazantzi V, Qin X, El-Halwagi MM. Simultaneous process and molecular design – a property based approach. AIChE J 2007;35(5):1232–9.

[31] Jiménez-Gutiérrez A, Sandate-Trejo MDC, El-Halwagi MM. An MINLP model that includes the effect of temperature and composition on property balances for mass integration networks. Processes 2014;2(3):675–93.

[32] Kheireddine H, Dadmohammadi Y, Deng C, Feng X, El-Halwagi MM. Optimization of direct recycle networks with the simultaneous consideration of property, mass, and thermal effects. Ind Eng Chem Res 2014;50(7):3754–62.

[33] Ponce-Ortega JM, Hortua AC, El-Halwagi MM, Jiménez-Gutiérrez A. A property-based optimization of direct-recycle networks and wastewater treatment processes. AIChE J 2009;55(9):2329–44.

[34] Nápoles-Rivera F, Ponce-Ortega JM, El-Halwagi MM, Jiménez-Gutiérrez A. Global optimization of mass and property integration networks with in-plant property interceptors. Chem Eng Sci 2010;65(15):4363–77.

[35] Tula AK, Eden MR, Gani R. Process synthesis, design and analysis using process-group contribution method. Comput Aided Chem Eng 2014;34:453–8.

[36] Sotelo-Pichardo C, Bamufleh H, Ponce-Ortega JM, El-Halwagi MM. Optimal synthesis of property-based water networks considering growing demand projections. Ind Eng Chem Res 2014;53(47):18260–72.

[37] Qin X, Gabriel F, Harell D, El-Halwagi M. Algebraic techniques for property integration via componentless design. Ind Eng Chem 2004;43:3792–8.

[38] Ng DKS, Foo DCY, Tan RR, El-Halwagi MM. Automated targeting techniques for concentration- and property-based total resource conservation networks. Comp Chem Eng 2010;34(5):825–45.

[39] Hohmann EC. Optimum networks for heat exchange [Ph.D. thesis]. Los Angeles: University of Southern California; 1971.

[40] Linnhoff B, Flower JR. Synthesis of heat exchanger networks: I. Systematic generation of energy optimal networks. AIChE J 1978;24(4):633–42.

[41] Linnhoff B, Townsend DW, Boland D, Hewitt GF, Thomas BEA, Guy AR, et al. A user guide on process integration for the efficient use of energy. IChemE UK; 1982.

[42] Papoulias SA, Grossmann IE. A structural optimization approach in process synthesis. II. Heat recovery networks. Comput Chem Eng 1983;7(6):707–21.

[43] Floudas CA, Ciric AR, Grossmann IE. Automatic synthesis of optimum heat exchange network configurations. AIChE J 1968;32(2):276–90.

[44] Al-Azri N, Al-Thubaiti M, El-Halwagi MM. An algorithmic approach to the optimization of process cogeneration. J Clean Tech Env Policy 2009;11(3):329–38.

[45] Mavromatis SP, Kokossis AC. Conceptual optimization of utility networks for operational variation-I targets and level optimization. Chem Eng Sci 1998;53:1585.

[46] Papadopoulos AI, Stijepovic M, Linke P, Seferlis P, Voutetakis S. Multi-level design and selection of optimum working fluids and ORC systems for power and heat cogeneration from low enthalpy renewable sources. Comp Chem Eng 2012;30:66–70.

[47] Gutiérrez-Arriage CG, Abdelhady F, Bamufleh H, Serna-González M, El-Halwagi MM, Ponce-Ortega JM. Industrial waste heat recovery and cogeneration involving organic rankine cycles. Clean Technol Environ Policy 2015;17(3):767—79.

[48] El-Halwagi MM, Harell D, Spriggs HD. Targeting cogeneration and waste utilization through process integration. Appl Energy 2009;86(6):880—7.

[49] Abdelhady F, Bamufleh H, El-Halwagi NM, Ponce-Ortega JM. Optimal design and integration of solar thermal collection, storage, and dispatch with process cogeneration systems. Chem Eng Sci 2015;136(2):158—67.

[50] El-Baz AA, Ewida KT, Shouman MA, El-Halwagi MM. Material flow analysis and integration of watersheds and drainage systems: I. Simulation and application to ammonium management in Bahr El-Baqar drainage system. Clean Techn Environ Policy 2005;7:51—61.

[51] El-Baz AA, Ewida KT, Shouman MA, El-Halwagi MM. Material flow analysis and integration of watersheds and drainage systems: II. Integration and solution strategy with application to ammonium management in Bahr El-Baqar drainage system. Clean Techn Environ Policy 2005;7:78—86.

[52] Lira- Barragán F, Ponce-Ortega JM, Serna-González M, El-Halwagi MM. An MINLP model for the optimal location of a new industrial plant with simultaneous consideration of economic and environmental criteria. Ind Eng Chem Res 2011;50(2):953—64.

[53] Spriggs HD, Lowe EA, Watz J, Lovelady EM, El-Halwagi MM. Design and development of eco-industrial Parks. Paper #109a. AIChE Spring meeting, New Orleans, April 2004.

[54] Lovelady EM, El-Halwagi MM. Design and integration of eco-industrial parks. Environ Prog Sustain Energy 2009;28(2):265—72.

[55] Noureldin MMB, El-Halwagi MM. Synthesis of C-H-O symbiosis networks. AIChE J 2005;64(4):1242—62.

Modeling and Advanced Control for Sustainable Process Systems

F.V. Lima, S. Li, G.V. Mirlekar
West Virginia University, Morgantown, WV, United States

L.N. Sridhar
University of Puerto Rico, Mayaguez, PR, United States

G. Ruiz-Mercado
US Environmental Protection Agency, Cincinnati, OH, United States

INTRODUCTION TO SUSTAINABLE PROCESS SYSTEMS

Process industries utilize a wide range of feedstocks to manufacture chemical, petroleum, gas, petrochemical, pharmaceutical, biofuel, food, microelectronics, metal, textile, and forestry products. Chemical processes exert some of the most profound impacts on the environment especially because of their ubiquitous nature and the importance of chemical products to our modern society (e.g., antibiotics, varieties of foods, energy, fuels, etc.). In recent years, environmental consciousness has been growing and has become a critical factor in the decision-making step of industrial processes. This awareness has led to the formulation of the concept of sustainability and sustainable development in the past century [1,2].

In particular, sustainable development has been defined in terms of economic, environmental, and social factors, such as the development of economic prosperity and the establishment of a more equitable society without depriving the future generations' standard of living [3]. The definition of sustainable development provides a guideline for working toward a new and better society. From the perspective of engineers, sustainability is typically considered to have environmental, economic, and social dimensions that should be balanced and jointly optimized. The following engineering definition has been proposed for sustainability [4,5]: "for a man-made system, sustainable development is a continual improvement in one or more of the three domains of sustainability, i.e., economic, environmental and societal without causing degradation in any of the rest, either now or in the future, when compared, with quantifiable metrics, to a similar system that it is intended to replace." In terms of engineering decision making, the

115

following definition [6] is also important: "a sustainable process is one that constrains resource consumption and waste generation to an acceptable level, making a positive contribution to the satisfaction of human needs, and providing enduring economic value to the business enterprise."

Despite these available definitions, there is still an ongoing debate on the characterization of sustainable development and the assessment of a process design and operation in terms of sustainability. For example, in process systems engineering, what should be the methodologies and approaches employed to obtain a sustainable process operation through design, optimization, and control? In the past, there were several contributions and methods developed that focus on the minimization of the environmental impact of chemical industries. Specifically, three basic principles for green engineering, end-of-pipe technologies, pollution prevention, and water minimization, were introduced [7]. A sustainable process design was proposed through the integration of energy, materials, and processing tasks, minimizing the amounts of energy and materials needed and size of equipment required to produce a given quantity of product per unit of time [8,9]. Environmental constraints/concerns were incorporated when formulating a multiobjective optimization framework to arrive at cost optimal designs [10]. Particularly, minimum environmental risks/impacts related to nonroutine and routine releases were considered. This problem was solved by employing the ε-constraint method, assuming that the environmental impact targets were within ranges imposed by a selected amount defined by ε. Also using the same solution method, life cycle assessment (LCA) principles have been taken into account in the formulation of superstructure multiobjective optimization problems for the design of sustainable chemical process flowsheets [11]. Along the same line a framework that combines a multiobjective optimization approach, LCA, and principal component analysis was introduced and applied to the synthesis of biological processes with economic and environmental concerns [12]. However, LCA methodologies need a large amount of data within a wide boundary, which are usually difficult to obtain, especially in the beginning of a process design. In addition, it is not clear how certain are the life cycle inventory data obtained from commercial packages for a particular process under study and LCA does not provide boundary values for process-related aspects at any stage and scale of a chemical process design. In an effort to develop methods for the sustainable design of chemical processes, the GREENSCOPE tool (gauging reaction effectiveness for the environmental sustainability of chemistries with a multi-objective process evaluator) was proposed as a systematic methodology for the evaluation of process performance and sustainability [13]. A set of environmental, material, energy, and economic indicators was employed to describe process aspects in terms of a quantitative sustainability measurement scale [14,15]. Thus GREENSCOPE quantitatively characterizes how sustainable the process utilizes energy, material goods, and services to generate a valuable product, while maximizing its social and economic benefits, and minimizing or eliminating negative environmental impacts. Besides, GREENSCOPE can provide clear quantified limits for comparing process performances and designs between themselves and determining if some sustainability improvements or better operating conditions have been achieved. In addition, GREENSCOPE worst and best case scenarios can be adapted according to particular design needs, decision makers, and product specifications. The desired GREENSCOPE indicators could also be employed to optimize the process performance if incorporated into multiobjective optimization problems. A recent contribution detailed the implementation and use of this sustainability evaluation and design tool for the manufacturing of biodiesel [16]. It also demonstrated the capability of this tool to identify the major process aspects or conditions favorable for the attainment of sustainable improvements. This is a key approach with reduced complexity and adoption cost for proposing and performing process changes that are potentially effective at influencing its sustainability. Finally, a review [17] of multiobjective optimization associated with economic and environmental objectives from the process synthesis and supply chain management point of view was presented. Also an overview [1] of the issue, methodologies, and needs for integrating environmental concerns into process design has been introduced.

Therefore the reported systems literature studies address design methodologies focused on three components: decision-making sequence, sustainable design framework, and design assessment tools. Thus there is a lack of studies on sustainable process operations employing process control. The typical objective functions other than profit were only motivated but not implemented for control purposes in the literature, such as meet customer specifications consistently, minimize product variability, meet safety and regulatory requirements, maximize asset utilization, operate the plant flexibly, improve operating range and reliability [18,19]. This chapter describes the first attempt to directly integrate sustainability assessment tools with advanced control strategies to simultaneously optimize and assess chemical processes during operation. Through a case study of a fermentation process for bioethanol production, a set of steady-state alternatives is generated, the implementation of an advanced biomimetic control strategy is discussed, and the obtained process operating points are evaluated in terms of sustainability employing the GREENSCOPE tool. The outline of the rest of this chapter is as follows: first, the proposed modeling, advanced process control, and sustainability assessment approaches are presented. These approaches provide a general framework for optimizing and controlling chemical processes in terms of sustainability. Then, the proposed approach is implemented for the case study of a fermentation process involving *Zymomonas mobilis*. In addition, a discussion on sustainability evaluation and control for process optimization is described. The chapter is then closed with conclusions and future directions.

PROPOSED APPROACH: MODELING, ADVANCED CONTROL, AND SUSTAINABILITY ASSESSMENT

Fermentation Process Model

Ethanol derived from renewable sources such as corn, sugar cane, and beet is an attractive clean fuel to control and decrease air pollution from internal combustion engines and reduce the dependence on fossil fuels. In this section we consider a process model for a homogeneous, perfectly mixed continuous culture fermentor equipped with an ethanol-selective removal membrane. This membrane can stabilize the fermentation process by removing ethanol from the fermentor, thus reducing the effect of product inhibition. Fig. 5.1 shows a schematic diagram of the fermentor with the in situ ethanol-removal membrane. For the *Z. mobilis* fermentation process addressed here, repeated oscillations that affect the process dynamics have been widely reported [20–23]. The oscillations in fermentation processes correspond to changes in product, substrate, and biomass concentrations that are caused by substrate and product inhibition. To mathematically describe and analyze these oscillations, Jöbses et al. [24,25] proposed the indirect inhibition structural model, in which the concentration of a key component (which includes RNA and proteins in biomass) is affected by substrate and product concentration. Also the formation of this key component is inhibited by ethanol concentration (C_P) and is a function of substrate concentration (C_S). The rate of formation expression for the key component is given by:

$$r_e = f(C_S)f(C_P)(C_e)$$

A Monod-type equation is taken for $f(C_S)$:

$$f(C_S) = \left(\frac{C_S}{K_S + C_S}\right)$$

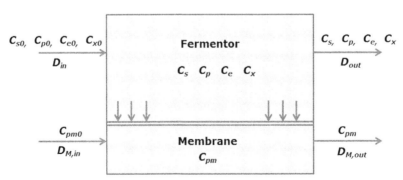

Figure 5.1 Schematic diagram of the fermentor unit.

The function $f(C_P)$ is described by a second-order polynomial of the following form:

$$f(C_P) = k_1 - k_2 C_P + k_3 C_P^2$$

The dynamic models for the key component, biomass, and substrate are expressed as:

$$\frac{dC_e}{dt} = \left[k_1 - k_2 C_P + k_3 C_P^2 \right] \left(\frac{C_S C_e}{K_S + C_S} \right) + D_{in} C_{e_0} - D_{out} C_e$$

$$\frac{dC_X}{dt} = P \left(\frac{C_S C_e}{K_S + C_S} \right) + D_{in} C_{X_0} - D_{out} C_X$$

$$\frac{dC_S}{dt} = P \left(\frac{-1}{Y_{SX}} \right) \left(\frac{C_S C_e}{K_S + C_S} \right) - m_S C_X + D_{in} C_{S_0} - D_{out} C_S$$

As the product flows out of the system through the fermentor and the membrane, mass balances are also derived for both compartments that are given by:

$$\frac{dC_P}{dt} = P \left(\frac{-1}{Y_{PX}} \right) \left(\frac{C_S C_e}{K_S + C_S} \right) + m_P C_X + D_{in} C_{P_0} - D_{out} C_P - \left(\frac{a}{V_F} \right) (C_P - C_{PM})$$

$$\frac{dC_{PM}}{dt} = \left(\frac{a}{V_F} \right) (C_P - C_{PM}) + D_{M,in} C_{PM_0} - D_{M,out} C_{PM}$$

in which

$$a = A_M P_M$$

Using the overall mass balances for the membrane and fermentor, the outlet dilution rate for both compartments are respectively defined as:

$$D_{M,out} = D_{M,in} + \frac{a(C_P - C_{PM})}{V_M(\rho)}$$

$$D_{out} = D_{in} - \frac{a(C_P - C_{PM})}{V_F(\rho)}$$

Thus the fermentation model addressed is represented by a set of five differential equations and two algebraic equations. Table 5.1 provides the base values of the parameters used in this *Z. mobilis* fermentation problem (see all variables' definitions and units in the Nomenclature section).

Table 5.1 Base Set of Parameters Used for the Fermentation Process [21]

Parameter	Value	Units
k_1	16.0	h^{-1}
k_2	0.497	$m^3/kg \cdot h$
k_3	0.00383	$m^6/kg^2 \cdot h$
m_S	2.16	$kg/kg \cdot h$
m_P	1.1	$kg/kg \cdot h$
Y_{SX}	0.02444498	kg/kg
Y_{PX}	0.0526315	kg/kg
K_S	0.5	kg/m^3
P_M	0.1283	m/h
$D_{M,in}$	0.5	h^{-1}
C_{S_0}	150.3	kg/m^3
C_{X_0}	0	kg/m^3
C_{P_0}	0	kg/m^3
C_{e_0}	0.02	kg/m^3
V_F	0.003	m^3
V_M	0.0003	m^3
ρ	789	kg/m^3
P	1.0	h^{-1}

Advanced Control Approach

One of the major challenges in the fermentation process model described is that its process dynamics exhibit oscillatory behavior, making it difficult to produce ethanol smoothly, thus affecting process efficiency (material and energy) and its sustainability performance. Here an advanced biomimetic control approach to address this challenge in the fermentation process is briefly described. This approach consists of a biologically inspired multiagent-based algorithm that combines the ants' rule of pursuit idea [26] with optimal control concepts for the calculation of optimal trajectories of individual agents. The ants' rule of pursuit, which is shown schematically in Fig. 5.2, is an excellent

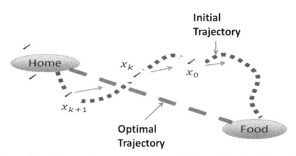

Figure 5.2 Schematic representation of the ant's rule of pursuit.

example of how biological systems can efficiently solve problems encountered in nature by cooperative behavior. According to this ants' rule, the first ant is supposed to find food by walking around at random. This pioneer ant would then trace a wiggly path back to the nest and start "group recruitment." The subsequent ants (or agents) would one after the other straighten the trail a little starting from the original path until the agents' paths converge to a line connecting the nest and the food source, despite the individual ant's lack of sense of geometry. Thus, by cooperating in large numbers, ants (or agents) accomplish tasks that would be difficult to achieve individually. The following is an outline of the developed algorithm in Fig. 5.3 and also a summary of this algorithm for the advanced control approach (see Ref. [27] for details regarding the control method).

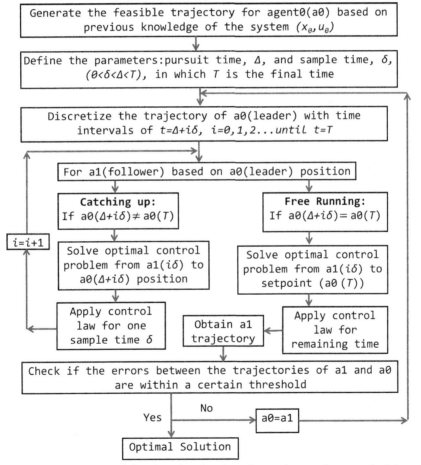

Figure 5.3 General structure of the algorithm for the advanced control approach. Adapted from Hristu-Varsakelis D, Shao C. A bio-inspired pursuit strategy for optimal control with partially constrained final state. Automatica 2007;43(7):1265—1273.

In summary, starting from an initially feasible trajectory (which satisfies the system of equations that represent the fermentation process model) for the leader agent, which can be obtained from the previous knowledge about the process, each follower agent improves its path toward the optimal solution by employing optimal control laws. As the number of agents progresses, the trajectories converge to an optimal solution. In general, for each follower, the optimal control trajectory to its leader is computed numerically by calling *dynopt* [28] (an optimal control toolbox in MATLAB), and the obtained control law is applied for δ time units, before repeating the procedure outlined in Fig. 5.3.

In this case, the objective of the implementation performed for the fermentation process is the setpoint tracking of the product concentration. The objective function is thus given by $\min(C_P - \text{setpoint})^2$. Then, to generate the initial feasible trajectory, the first agent moved optimally by solving the problem using *dynopt* all at once with the given time horizon and setpoint. The subsequent agents applied the proposed strategy (with $\delta = 1$ and $\Delta = 14$) in order to find the overall optimal trajectory. In particular, for the catching-up part of the algorithm, a dynamic objective function is used such that the setpoint changes for each intermediate optimal control problem according to the C_P profile obtained from the leader's trajectory. For the free-running portion of the algorithm, the optimal control problem is solved for each agent in a single segment until it reaches the setpoint.

As mentioned earlier, the developed algorithm employs *dynopt* to solve the intermediate problems associated with the local interaction of the agents. Specifically, *dynopt* is a set of MATLAB functions that use the orthogonal collocation on finite elements method for the determination of optimal control trajectories. As inputs, this toolbox requires the dynamic process model, the objective function to be minimized, and the set of equality and inequality constraints. The dynamic model here is described by the set of ordinary differential equations and differential algebraic equations that represent the fermentation process model. For the purpose of optimization, the MATLAB Optimization Toolbox, particularly the constrained nonlinear minimization routine *fmincon* [29], is employed.

Sustainability Assessment

Assessing and measuring a concept as critical as sustainability across a system is a challenging task. Although several sustainability assessment methodologies and tools as mentioned earlier are available, it is difficult to arrive at the same conclusion on the level of sustainability performance for a given process. In an effort to develop an all-inclusive methodology for evaluating a new or existing process, researchers of the US Environmental Protection Agency (EPA) proposed in 2003 a framework for sustainability assessment named GREENSCOPE [13]. This methodology was later deployed as a software tool based on Microsoft Excel. In this section, the GREENSCOPE assessment tool (Microsoft Excel version) and some of its indicators are briefly described for their subsequent employment to assess a set of process operating points associated with the fermentation process.

The GREENSCOPE assessment tool includes about 140 indicators in the following aspects: Efficiency (26 indicators), Energy (14 indicators), Economics (33 indicators), and Environment (66 indicators) [30]. This sustainability assessment tool was designed for quantifying the sets of indicators for several processes with the same product that employ different raw materials or technologies, or for evaluating the improvements on sustainability performance after making process modifications. In GREENSCOPE, each indicator has a score on a scale of 0–100% between two reference states representing the best (100% sustainability) and the worst (0% sustainability) case scenarios. After the identification and selection of the upper and lower bounds for each indicator, GREENSCOPE defines a dimensionless score for the actual indicator values using the following equation:

$$\text{Indicator Score} = \frac{|\text{Actual} - \text{Worst}|}{|\text{Best} - \text{Worst}|} \times 100\%$$

The obtained dimensionless indicator results can then help the decision maker or process engineer to evaluate whether the implementation of new technologies or process modifications is moving toward a more sustainable scenario. The selection of suitable sustainability indicators depends on the desired level of assessment for a particular case, data availability, particular sustainability areas of interest from the decision makers (e.g., water consumption, solid generation, release physical state, human health exposure, etc.), and process characteristics (e.g., economic sector, manufacturing scale, etc.). In this work, given the current limitations in data availability for evaluating some indicators and the process characteristics (isothermal, production rate, process design complexity, equipment inventory), we have decided to employ nine indicators in three categories (Economic, Efficiency, and Environmental) to assess the sustainability performance of the fermentation process. In addition, small variations in energy indicator results are expected for different process conditions because of the current small energy demand from this isothermal–isobaric process. However, this hypothetical scenario of low sensitivity to energy variations might change if pollution control and mitigation strategies were implemented and the rise of product demand required a process scaling-up by adding more equipment and complexity to the current design. The selected indicators as well as their reference values are presented in Table 5.2. Also the definitions of these indicators are given in the next paragraphs.

Efficiency Indicators

The efficiency indicators describe the process sustainability performance in terms of the amount of material and services required to generate the desired product or complete a specific process task. Some indicators in this category provide valuable information on

Table 5.2 Selected GREENSCOPE Indicators and Their Reference Values

Category	Indicator	Unit	Reference Value	
			Best Case	Worst Case
Efficiency	Reaction yield (RY)	kg/kg	1.0	0
	Renewability-material index (RI_M)	kg/kg	1.0	0
	Water intensity (WI)	m³/$	0	1.0
Environmental	Environmental quotient (EQ)	m³/kg	0	5.0
	Environmental hazard, water hazard, (EH_{water})	m³/kg	0	100.0
	Global warming potential (GWP)	kg/kg	0	Any waste released has a potency factor at least equal to 1
Economic	Economic potential (EP)	$/kg	0.5	0
	Specific raw material cost (C_{SRM})	$/kg	0	1.0
	Total water cost ($C_{water\ tot.}$)	$/year	1.5	0

the sustainability improvement that can be achieved by changing the process chemical reaction. Here also three indicators (Reaction yield, Renewability-material index, Water intensity) are chosen to assess the studied fermentation process.

1. **Reaction yield, RY**: Defined as the ratio between the actual mass of the produced valuable product and the theoretical mass of product that could be generated by assuming that all of the limiting reagent is consumed. This yield is calculated by the ratio of the real yield (i.e., mass of desired product over the mass of the limiting reagent) over the theoretical yield (i.e., theoretical product mass flow over the mass of the limiting reagent). The computation of this indicator is given by:

$$RY = \frac{\dot{m}_{product}}{\frac{\dot{m}_{limit,reagent}^{in}}{MW_{limit,reagent}} \times \frac{\beta_{product}}{\alpha_{limit,reagent}} \times MW_{product}}$$

2. **Renewability-material index, RI_M**: This indicator consists of the fraction of the total consumed feedstocks that are categorized as renewable resources. Data entries for its calculation include the renewable materials list and the process input mass flows. This calculation is as follows:

$$RI_M = \frac{\sum_{i=1}^{N} \left(\dot{m}_i^{in}\right)_{renewable}}{\sum_{i=1}^{N} \dot{m}_i^{in}}$$

3. **Water intensity, WI**: This is the amount of water consumed by the manufacturing process per sales revenue of valuable product. The following equation describes the calculation of this indicator:

$$WI = \frac{\dot{Q}_{H_2O}}{S_{product}}$$

Environmental Indicators

The indicators in this category demonstrate how a new reaction/process, operating conditions, releases, and feedstocks affect the environment by providing quantification of increase/decrease level of environmental sustainability. Once again, three indicators (Environmental quotient, Environmental hazard for water, Global warming potential) are selected to evaluate the analyzed fermentation process.

4. **Environmental quotient, EQ**: This represents the characterization of the environmental unfriendliness of the produced waste. This indicator is computed as shown in the following equation by multiplying the E factor (an Efficiency indicator) by the quotient Q. This quotient is an assigned coefficient that can be 1 if the waste is identified as innocuous, while for toxic materials such as heavy metals, Q could be a scalar between 100 and 1000.

$$EQ = \frac{\sum_{i=1}^{I} \dot{m}_{waste\ \ i \neq H_2O}^{out} \times Q_i}{\dot{m}_{product}}$$

5. **Environmental hazard, water hazard, EH$_{water}$**: This evaluates the process toxicity to the aquatic environment. This indicator corresponds to a hypothetical volume of water polluted per unit mass of desired product. GREENSCOPE employs a set of equations for the calculation of this indicator. A definition of this indicator is given by (refer to Ref. [7] for details on this calculation):

$$EH_{water} = \frac{V_{t,water\ polluted}}{\dot{m}_{product}}$$

6. **Global warming potential, GWP**: This is the total mass of carbon dioxide emitted per unit mass of valuable product(s). This indicator can include CO_2 emitted as a byproduct from the treatment of waste streams and from the burning of fuel needed to generate energy for the process. The CO_2 emissions resulting from the generation of electricity and steam can also be included in this indicator, even when energy

utilities (electricity, steam, and cooling water) are purchased rather than generated onsite. The following equation describes the calculation of this indicator:

$$GWP = \frac{\sum_{i=1}^{I} \dot{m}_i^{out} \times PF_{CO_2,i}}{\dot{m}_{product}}$$

Economic Indicators

Economic performance is the most important attribute for process industries when evaluating a process design in a commercial setting. Although the emergence of the concept design for sustainability has occurred, economic feasibility is still one of the most fundamental indicators for the stakeholders to determine the existence and continuation of a product development. Therefore a new or current process must be economically sustainable [13]. Here three indicators will be used to describe the economic performance of the fermentation process operation.

7. **Economic potential, EP**: The EP indicator is a good approximated cost model that fits into the conceptual process design framework and only requires data about product sales revenue, feedstock costs, and utility costs. The following equation describes the calculation of this indicator:

$$EP = \frac{PWF_{cf,m}\left(S_m - C_{RM,m} - C_{UT,m}\right) - FCI_{L,m}}{\sum_{i=1}^{I} \dot{m}_{m,product\ i}}$$

8. **Specific raw material cost (C_{SRM})**: C_{SRM} can be used as an economic indicator at the basic conceptual process design stage. This indicator assumes 100% reaction yield, i.e., the minimum raw material cost is computed. Some process design routes can be discarded when C_{SRM} exceeds the targeted product value. The following equation shows how to obtain this indicator:

$$C_{SRM,m} = \frac{\sum_{i=1}^{I}\left(m_{m,i}^{in} \times C_{m,i}\right)}{\sum_{i=1}^{I} \dot{m}_{m,product\ i}}$$

9. **Total water cost, $C_{water\ tot.}$**: These are the costs related to the water demand during the day-to-day operation of a manufacturing plant: drinking water, process use water, boiler feed water, deionized water, etc. The following equation expresses the computation of this indicator:

$$C_{water\ tot.} = \sum_{i=1}^{W} \dot{m}_{H_2O,i}^{in} \times C_{H_2O,i}$$

This selected list of indicators is used next to assess the fermentation process model in focus. It is worth mentioning that this work aims to compare the sustainability performance of chemical processes under different operating conditions using the foregoing selected indicators that implicitly take into account environmental, process, and government regulation limitations by their user-defined reference values. Therefore, if the maximum environmental limitation for a specific indicator could be accurately specified by the government, society, researchers, or any other stakeholder, then these limitations could be explicitly incorporated into the worst and best case reference values for the indicator score to reflect how close the operating condition would be to the environmental limit.

CASE STUDY: FERMENTATION FOR BIOETHANOL PRODUCTION SYSTEM

As mentioned previously the studied fermentation process to produce ethanol presents challenges in its process dynamics that exhibit oscillatory behavior, which affect process productivity and sustainability. To address these challenges, this section introduces a new sustainable process control framework that combines the biomimetic control strategy detailed earlier and the GREENSCOPE sustainability assessment tool. In this case study, the controlled variable is the concentration of product, C_P (see objective function defined earlier), and the dilution rate, D_{in}, is chosen as the manipulated variable. GREENSCOPE is employed to evaluate the sustainability performance of the system in open-loop and closed-loop operations. The obtained GREENSCOPE indicator scores provide information on whether the implementation of the biomimetic controller for the fermentation process enables a more efficient and sustainable process operation.

To analyze the effect of D_{in} in the entire process, a set of open-loop simulations was completed in which D_{in} varied from 0.05 to 0.5 h^{-1}. For these simulations, the system was integrated using *ode15s* solver in MATLAB for the given differential and algebraic equations that were solved simultaneously. In agreement with the literature [31], oscillatory behavior (periodic attractors as well as multiplicity of steady states), which arises from the interaction between cell growth, substrate consumption, and the ethanol production, occurred. Figs 5.4 and 5.5 show the concentration profiles of the key component, biomass, substrate, and product in the fermentor side for the open-loop simulations with different dilution rates. Specifically, the concentrations of substrate, biomass, and ethanol exhibit such oscillations until they reach steady state after about 50 h of operation. It is also important to note in Fig. 5.5 that the steady-state concentration of product, C_P, varies significantly with D_{in}. Thus it is critical to obtain for this system the best possible steady-state operating conditions. For fermentation processes, yield of the desired product (ethanol in this case study) and productivity of the reactor

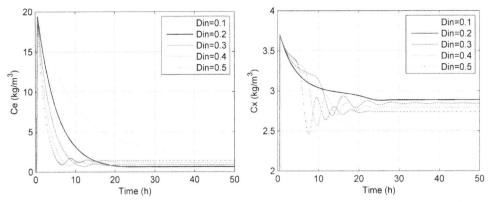

Figure 5.4 Open-loop simulations: concentration profiles of key component and biomass for different dilution rates.

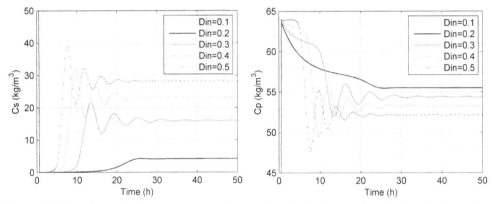

Figure 5.5 Open-loop simulations: concentration profiles of substrate and product in the fermentor side for different dilution rates.

with the given volume are both important performance considerations from a stake-holders' commercial point of view. Therefore, according to the characteristics of this system's open-loop dynamics and performance analyses carried out in MATLAB, two cases are first selected for the implementation of the advanced biomimetic control strategy: (1) the highest yield that occurs when D_{in} is at $0.2\ \text{h}^{-1}$ and (2) the highest productivity in terms of utility of the reactor that takes place when D_{in} is at $0.5\ \text{h}^{-1}$. These two cases consider the setpoint control at two different steady-state concentrations of product. For such cases the closed-loop and open-loop performances are compared for the corresponding C_P steady state. Also it is worth noting from the process dynamics in Fig. 5.5, which exhibit underdamped oscillations, that there might exist other steady states with higher concentrations of product that may be achievable in the presence of the

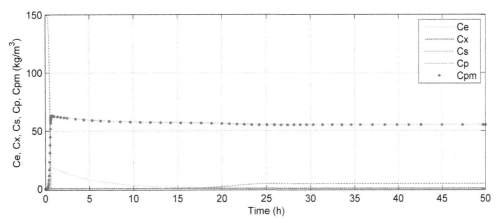

Figure 5.6 Open-loop simulation (case 1): concentration profiles of key component, biomass, substrate, product in fermentor, and membrane sides at D_{in} of 0.2 h^{-1}.

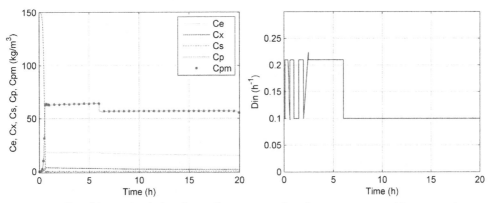

Figure 5.7 Closed-loop simulation (case 1): concentration (key component, biomass, substrate, product in fermentor, and membrane sides) and input profiles.

controller. Thus two other cases are also considered, in which a steady state D_{in} is assumed for each case and the controller tries to push the process performance to the highest possible steady state C_P using the specific input. The closed-loop cases with the associated open-loop scenarios are also compared in these cases.

In case 1 a setpoint of 55.52 kg/m^3, which corresponds to the steady state C_P for the open-loop simulation at D_{in} of 0.2 h^{-1}, is used for the closed-loop simulation. The sustainability indicators of the open-loop and closed-loop simulations with $C_P = 55.52$ kg/m^3 are then compared. Fig. 5.6 depicts the concentration profiles for the open-loop simulation when D_{in} equals 0.2 h^{-1}, while Fig. 5.7 shows the concentrations of key component, biomass, substrate, and product as well as the D_{in} profiles for the closed-loop simulation. With the implementation of the control strategy, note in Fig. 5.7

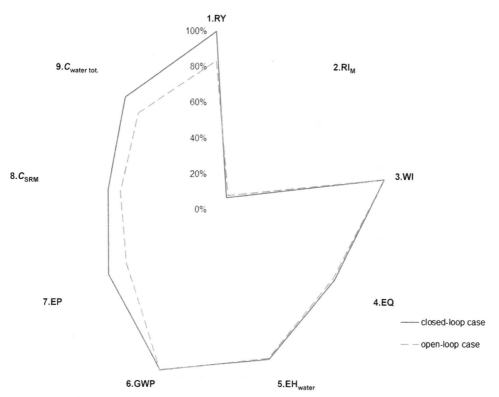

Figure 5.8 Radar plot with GREENSCOPE indicators for the closed-loop and open-loop simulations (case 1).

that all the substrate is converted into product. The radar plot in Fig. 5.8 shows that the controller implementation also improves some GREENSCOPE indicators in three categories (Efficiency, Environmental, and Economic), such as RY, EP, C_{SRM}, and $C_{water,tot.}$, toward a more sustainable process operation.

In case 2, we select another steady-state C_P of the open-loop simulation at the upper bound D_{in} value of 0.5 h^{-1} as the setpoint for the closed-loop simulation. As shown in Fig. 5.9, the steady-state concentration of product for this open-loop scenario after 50 h is 52 kg/m^3. Fig. 5.9 also depicts the concentration profiles of key component, biomass, substrate, and product for the open-loop simulation at D_{in} of 0.5 h^{-1}, while Fig. 5.10 shows the concentration and input profiles for the closed-loop case. From Figs. 5.9 and 5.10, note that the original oscillations of large amplitude in the open-loop dynamics are eliminated after the implementation of the biomimetic control strategy. Additionally, some GREENSCOPE indicators in Fig. 5.11, such as RY, EQ, EH$_{water}$, EP, C_{SRM}, and $C_{water,tot.}$, demonstrate the higher degree of sustainability for the closed-loop scenario.

As briefly discussed earlier, the dynamic analysis in Fig. 5.5 showed the potential of the fermentation process to achieve higher product concentrations, and thus higher

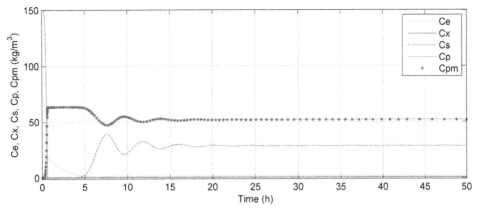

Figure 5.9 Open-loop simulation (case 2): concentration profiles of key component, biomass, substrate, product in fermentor, and membrane sides at D_{in} of 0.5 h^{-1}.

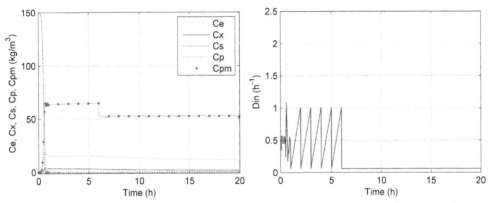

Figure 5.10 Closed-loop simulation (case 2): concentration (key component, biomass, substrate, product in fermentor, and membrane sides) and input profiles.

efficiency, than the values in cases 1 and 2. In case 3, we define a value of $C_P = 63$ kg/m^3 as the setpoint for the closed-loop scenario as a tentative to make the fermentation process more efficient. Fig. 5.12 shows the concentrations of key component, biomass, substrate, and product as well as the D_{in} profiles for the closed-loop simulation in this case. Note that the closed-loop case arrives at the steady state when $D_{in} = 0.18$ h^{-1}. When compared to the open-loop case with the same D_{in} (shown in Fig. 5.13), the closed-loop scenario shows some improvement of the fermentation process after the implementation of the biomimetic control strategy, by keeping the concentration of product at the higher steady state and converting all the substrate into product. The GREENSCOPE indicators in Fig. 5.14 also demonstrate the aforementioned improvements in efficiency (RY) and economic aspects (EP, C_{SRM}).

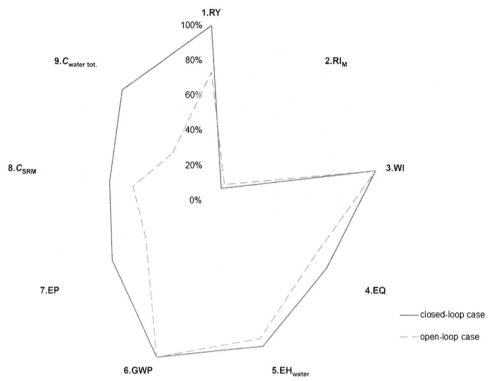

Figure 5.11 Radar plot with GREENSCOPE indicators for the closed-loop and open-loop simulations (case 2).

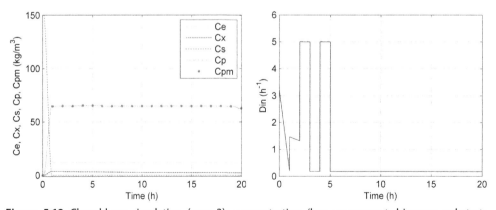

Figure 5.12 Closed-loop simulation (case 3): concentration (key component, biomass, substrate, product in fermentor, and membrane sides) and input profiles.

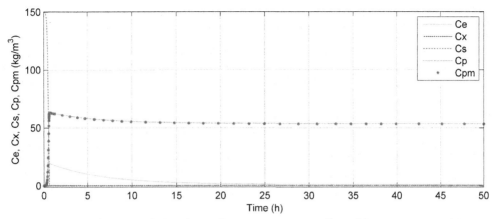

Figure 5.13 Open-loop simulation (case 3): concentration profiles of key component, biomass, substrate, product in fermentor, and membrane sides at D_{in} of 0.18 h^{-1}.

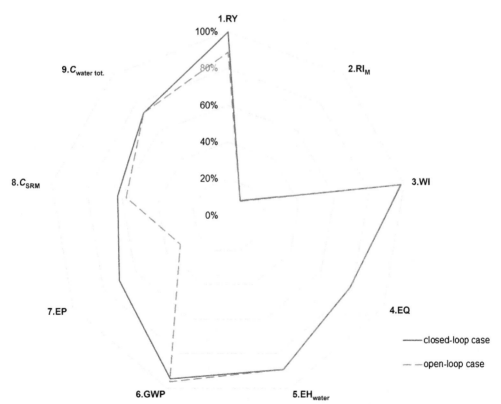

Figure 5.14 Radar plot with GREENSCOPE indicators for the closed-loop and open-loop simulations (case 3).

In the last case we select $C_P = 68$ kg/m^3 as the setpoint for the closed-loop scenario, which is slightly higher than the setpoint of 63 kg/m^3 in case 3. The concentration and D_{in} profiles for the closed-loop case are shown in Fig. 5.15, in which D_{in} takes the steady-state value of 0.15 h^{-1}, a lower value than the steady state of 0.18 h^{-1} in case 3. Fig. 5.16 depicts the concentrations of key component, biomass, substrate, and product for the open-loop case with the same input value. When comparing to the open-loop case, the closed-loop scenario shows once again improvement in concentration of product. Finally, Fig. 5.17 shows that in this case the sustainability indicators in environmental and economic aspects have achieved improvement after the advanced control implementation.

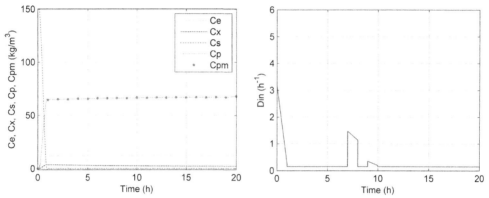

Figure 5.15 Closed-loop simulation (case 4): concentration (key component, biomass, substrate, product in fermentor, and membrane sides) and input profiles.

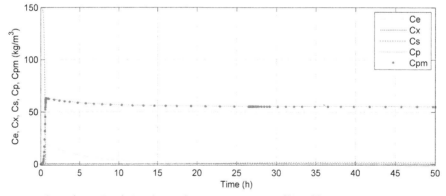

Figure 5.16 Open-loop simulation (case 4): concentration profiles of key component, biomass, substrate, product in fermentor, and membrane sides at D_{in} of 0.15 h^{-1}.

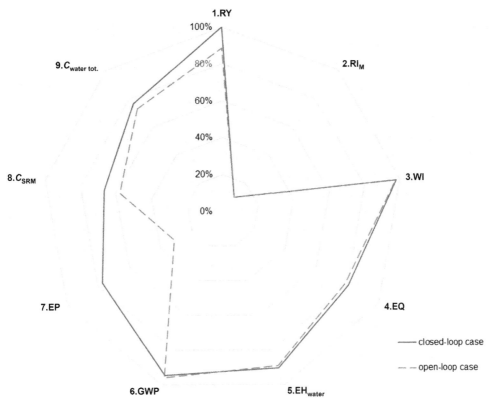

Figure 5.17 Radar plot with GREENSCOPE indicators for the closed-loop and open-loop simulations (case 4).

SUSTAINABILITY ASSESSMENT AND PROCESS CONTROL

As described in the previous section, sustainability evaluation was performed during open- and closed-loop control simulation cases at each of the four different initial setpoints. Sustainability improvements were found in most of the cases when comparing open and closed-loop control scenarios. Particularly, score increments were achieved for three economic indicators (EP, $C_{water\ tot.}$, C_{SRM}) and one efficiency indicator (RY). When comparing all four case conditions, case 2 described the major sustainability improvements characterized by the highest productivity steady state in terms of utility of the reactor for the open-loop condition. In addition, this case was the only one describing improvements in the environmental indicator scores (EH_{water} and EQ) for the closed-loop scenario. Therefore this evidenced a high influence of the dilution rate manipulation to achieve a more sustainable steady-state process condition. In addition, note that higher sustainability scores for the RY indicator were achieved for all four cases after the implementation of the controller. This might suggest that a maximum product yield is found before the dynamic system reaches a final steady state. However, this extra residence time would be needed by the system in order to overcome its dynamic oscillatory behavior before a desired steady state is attained. To confirm this fact, as a future direction, a dynamic sustainability analysis would have to be performed. In contrast, some other indicators did not show substantial changes by variations in the dilution rate. This indicates that other manipulated and controlled variables would be needed in order to show changes and improvements in other process sustainability aspects after implementing the proposed biomimetic process control strategy.

CONCLUSIONS AND FUTURE DIRECTIONS

In this chapter we presented a novel approach for process systems to achieve sustainable operation through the application of an advanced control strategy combined with sustainability assessment tools. The effectiveness of the developed framework was highlighted via a case study of a bioethanol production process. In this case study, open-loop and closed-loop scenarios were compared considering two cases with fixed product concentration setpoints and two cases with fixed steady-state input values. This comparison was performed through indicators of the GREENSCOPE sustainability assessment tool in three categories (Efficiency, Environmental, and Economic). The obtained results showed that the implementation of the advanced biomimetic controller improved the system's sustainability performance in general. As future work, global optimization approaches could be taken into account in the formulated framework. These approaches may improve the optimal solution obtained by the controller given the nonlinearity of the model employed in this chapter. For this purpose the rearrangement of the process model could be investigated [32]. In addition, note that the implemented control strategy was based on an efficiency objective function associated with the product concentration. In the future, objective functions based on desired GREENSCOPE indicators will be explored. Because of the conflicting nature of some of the indicators, multiobjective approaches may be considered in the control problem. Another potential research direction consists of the placement of additional process sensors to collect data so that the sustainability performance of the process can be optimized based on the additional collected information [19].

NOMENCLATURE

List of Symbols
A_M Area of membrane (m^2)
C_i Concentration of component i (kg/m^3)

$C_{RM,m}$ Cost of raw materials in year m ($)
$C_{UT,m}$ Cost of utilities in year m ($)
C_m Price of raw material in year m ($/kg)
C_{H_2O} Price of water ($/kg)
D_{in} Inlet fermentor dilution rate (h^{-1})
D_{out} Outlet fermentor dilution rate (h^{-1})
$D_{M,in}$ Inlet membrane dilution rate (h^{-1})
$D_{M,out}$ Outlet membrane dilution rate (h^{-1})
$FCI_{L,m}$ Fixed capital cost investment without land cost in year m ($)
K_S Monod constant (kg/m^3)
k_1 Empirical constant (h^{-1})
k_2 Empirical constant ($m^3/kg \cdot h$)
k_3 Empirical constant ($m^6/kg^2 \cdot h$)
m_S Maintenance factor based on substrate ($kg/kg \cdot h$)
m_P Maintenance factor based on product ($kg/kg \cdot h$)
MW Molecular weight (g/mole)
P_M Membrane permeability (m/h)
P Maximum specific growth rate (h^{-1})
$PWF_{cf,m}$ Present worth factor in year m
$PF_{CO_2,i}$ CO_2 potential factor of emissions i
Q_{H_2O} Volumetric flow of the consumed water (m^3/h)
Q_i Assigned coefficient for pollution
r_i Production rate of component i (kg/m^3)
S_m Total income from all sales in year m ($)
$S_{product}$ Sales revenue from valuable product ($/h)
V_F Fermentor volume (m^3)
V_M Membrane volume (m^3)
$V_{t,water\ polluted}$ Hypothetical volume of limit concentration water (m^3)
Y_{SX} Yield factor based on substrate (kg/kg)
Y_{PX} Yield factor based on product (kg/kg)

Greek Symbols
$\alpha_{limit,reagent}$ Limited reagent stoichiometric coefficients
$\beta_{product}$ Product stoichiometric coefficients
ρ Ethanol density (kg/m^3)

Subscripts
e Key component inside the fermentor
e_0 Inlet key component to the fermentor
P Product (ethanol) inside the fermentor
P_0 Inlet product to the fermentor
PM Product (ethanol) inside the membrane
S Substrate inside the fermentor
S_0 Inlet substrate to the fermentor
X Biomass inside the fermentor
X_0 Inlet biomass to the fermentor

ACKNOWLEDGMENTS

The authors gratefully acknowledge the financial support from West Virginia University and DOE through award # DE-FE0012451. The authors also thank Matthew Steinheimer and Zachary Chow (WVU undergraduate students) for the useful discussions on the GREENSCOPE tool and the fermentation process model.

DISCLAIMER

The views expressed in this contribution are those of the authors solely and do not necessarily reflect the views or policies of the US EPA.

REFERENCES

[1] Cano-Ruiz JA, McRae GJ. Environmentally conscious chemical process design. Annu Rev Energy Environ 1998;23(1):499−536.

[2] Zimmerman JB, Anastas PT. Peer reviewed: design through the 12 principles of GREEN engineering. Environ Sci Technol 2003;37(5):95A−101A.

[3] UUNE Program. Report of the world commission on environment and development: our common future. 1987. Available from: http://www.un-documents.net/wced-ocf.htm.

[4] Sikdar SK. Sustainable development and sustainability metrics. AIChE J 2003;49(8):1928−32.

[5] Martins AA, Mata TM, Costa CA, Sikdar SK. Framework for sustainability metrics. Ind Eng Chem Res 2007;46(10):2962−73.

[6] Bakshi BR, Fiksel J. The quest for sustainability: challenges for process systems engineering. AIChE J 2003;49(6):1350−8.

[7] Allen DT, Shonnard DR. Green engineering: environmentally conscious design of chemical processes. Pearson Education; 2001.

[8] El-Halwagi MM. Pollution prevention through process integration: systematic design tools. Academic press; 1997.

[9] El-Halwagi MM. Sustainable design through process integration: fundamentals and applications to industrial pollution prevention, resource conservation, and profitability enhancement. Elsevier; 2012.

[10] Stefanis SK, Pistikopoulos EN. Methodology for environmental risk assessment of industrial nonroutine releases. Ind Eng Chem Res 1997;36(9):3694−707.

[11] Guillén-Gosálbez G, Caballero JA, Jiménez L. Application of life cycle assessment to the structural optimization of process flowsheets. Ind Eng Chem Res 2008;47(3):777−89.

[12] Brunet R, Guillén-Gosálbez G, Jiménez L. Cleaner design of single-product biotechnological facilities through the integration of process simulation, multiobjective optimization, life cycle assessment, and principal component analysis. Ind Eng Chem Res 2012;51(1):410−24.

[13] Gonzalez MA, Smith RL. A methodology to evaluate process sustainability. Environ Prog 2003;22(4):269−76.

[14] Ruiz-Mercado GJ, Smith RL, Gonzalez MA. Sustainability indicators for chemical processes: I. Taxonomy. Ind Eng Chem Res 2012;51(5):2309−28.

[15] Ruiz-Mercado GJ, Smith RL, Gonzalez MA. Sustainability indicators for chemical processes: II. Data needs. Ind Eng Chem Res 2012;51(5):2329−53.

[16] Ruiz-Mercado GJ, Gonzalez MA, Smith RL. Sustainability indicators for chemical processes: III. Biodiesel case study. Ind Eng Chem Res 2013;52(20):6747−60.

[17] Grossmann IE, Guillén-Gosálbez G. Scope for the application of mathematical programming techniques in the synthesis and planning of sustainable processes. Comput Chem Eng 2010;34(9):1365−76.

[18] Edgar TF. Control and operations: when does controllability equal profitability? Comput Chem Eng 2004;29(1):41−9.

[19] Siirola JJ, Edgar TF. Process energy systems: control, economic, and sustainability objectives. Comput Chem Eng 2012;47:134−44.

[20] Daugulis AJ, McLellan PJ, Li J. Experimental investigation and modeling of oscillatory behavior in the continuous culture of *Zymomonas mobilis*. Biotechnol Bioeng 1997;56(1):99—105.

[21] Garhyan P, Elnashaie S. Static/dynamic bifurcation and chaotic behavior of an ethanol fermentor. Ind Eng Chem Res 2004;43(5):1260—73.

[22] Sridhar LN. Elimination of oscillations in fermentation processes. AIChE J 2011;57(9):2397—405.

[23] Mahecha-Botero A, Garhyan P, Elnashaie S. Non-linear characteristics of a membrane fermentor for ethanol production and their implications. Nonlinear Anal Real World Appl 2006;7(3):432—57.

[24] Jöbses I, Egberts G, Luyben K, Roels J. Fermentation kinetics of *Zymomonas mobilis* at high ethanol concentrations: oscillations in continuous cultures. Biotechnol Bioeng 1986;28(6):868—77.

[25] Jöbses I, Roels J. The inhibition of the maximum specific growth and fermentation rate of *Zymomonas mobilis* by ethanol. Biotechnol Bioeng 1986;28(4):554—63.

[26] Bruckstein AM. Why the ant trails look so straight and nice. Math Intell 1993;15(2):59—62.

[27] Hristu-Varsakelis D, Shao C. A bio-inspired pursuit strategy for optimal control with partially constrained final state. Automatica 2007;43(7):1265—73.

[28] Dynopt Document and Code. Available from: http://www.kirp.chtf.stuba.sk/moodle/course/view.php?id=187.

[29] Information on fmincon. Available from: http://www.mathworks.com/help/optim/ug/fmincon.html?s_tid=doc_12b.

[30] Ruiz-Mercado GJ, Smith RL, Gonzalez MA. GREENSCOPE.xlsm User's Guide. Excel Version 1.1. 2013.

[31] Wang H, Zhang N, Qiu T, Zhao J, He X, Chen B. Analysis of Hopf points for a *Zymomonas mobilis* continuous fermentation process producing ethanol. Ind Eng Chem Res 2013;52(4):1645—55.

[32] Sridhar LN. Global optimization of continuous fermentation involving *Zymomonas mobilis*. J Sustain Bioenergy Syst 2013;3:64—7.

CHAPTER SIX

Sustainable Engineering Economic and Profitability Analysis

Y. Jiang, D. Bhattacharyya
West Virginia University, Morgantown, WV, United States

INTRODUCTION

For designing chemical and energy-generating processes, screening alternative technologies, or optimizing operating conditions, the traditional approach is to perform techno-economic analysis. Such designs do not ensure that the resulting process is sustainable or that the process that performs best in terms of the techno-economic criteria is indeed superior in terms of sustainability measures. Therefore, in the future, it is imperative that sustainability measures become an integral part of the process design. The four pillars of sustainability analysis are energy, efficiency, environment, and society [1]. The techno-economic analysis automatically reflects the impact of energy and efficiency, but the environmental and social impacts still need to be evaluated. For example, consider two competing technologies, Processes A and B, for producing a certain product. Process A requires lower cooling duty, but at low temperature level it requires an ammonia vapor compression cycle, while Process B requires higher cooling duty, but air coolers are sufficient. Comparing the techno-economic analysis of these two processes, Process A may turn out to be superior, but it has stronger environmental and social impact in comparison to Process B. In addition, the refrigeration loop in Process A operates at very high pressure, low temperature, and contains toxic ammonia, exposing workers to a higher risk during normal operation as well as startup and shutdown. Comparison of appropriate sustainability measures of these two processes will be able to unearth the superiority of one process to the other.

Since the technologies that are superior in terms of sustainability measures are not necessarily superior in terms of economic measures, tax benefits and other environmental credits are being offered in a number of countries as an incentive to promote development and commercialization of sustainable processes. For example, several states in the United States offer tax credits for selling and buying of bioethanol and biodiesel [2,3]. Because of rising concern regarding the increasing concentration of greenhouse gas in the atmosphere, carbon tax levied on the carbon content of fuels is being considered as an incentive to promote technologies for a carbon-constrained world [4,5].

141

Carbon tax or similar incentives can not only help to develop renewable energy sources such as wind power, solar energy, hydropower, noncarbonaceous or low-carbon transportation fuels such as hydrogen and ethanol, but also help to develop technologies for CO_2 capture, sequestration, and utilization. In addition, owners and operators of process and energy plants would have a stronger incentive to invent and promote novel technologies that can result in lower greenhouse gas emissions as well as higher overall efficiency.

There are two main difficulties while comparing sustainability measures of multiple designs or technologies. First, while there is a strong effort to assign scores to environmental impacts under different categories, such scoring is generally difficult for societal impacts and strongly depends on user perception, existing experience, local socioeconomic and cultural background, and the categories being evaluated. Certain societal indicators can be assigned scores unambiguously. For example, if fire and explosion during process operation is one of the criteria being considered, then appropriate scores can be assigned by following a methodology such as the Dow Fire and Explosion Index (FEI) [6]. However, it is difficult to come up with the appropriate social indicators as well as their appropriate scores. In this chapter we will present some of the existing approaches that can be embedded while performing economic and profitability analysis at the design stage. One particular focus of this chapter is to point to the limitations of the existing approaches and where further research can be beneficial.

ECONOMIC SUSTAINABILITY ANALYSIS

Total capital investment for any process technology can be evaluated by considering the fixed capital investment (grassroots cost for new projects) that includes all costs associated with new construction, land, and working capital (WC). The WC includes contingency as well as costs for everything such as raw material inventories, salary, and associated costs for the personnel until the revenues from the process become available. While a common practice is to consider the WC to be a small percentage of the fixed capital investment (such as 10—20% or so) [7], careful attention must be paid for new and unproven technologies. One way of considering the risk in the technology and absence of prior experience is to consider additional contingency. Processes that operate at extremely high or low temperature and pressure might need additional contingency. In addition, if there is any specific equipment item that is critical to a process, but is relatively unproven for large-scale applications, then initial troubles with such equipment items can lead to longer startup time. Thus additional WC may be needed. Even though obvious, it must be noted that a longer startup time not only indicates higher WC, but also that the revenue will start generating at a later time. For novel processes such as many upcoming new and environment-friendly biochemical processes, it is difficult to evaluate the capital cost. For simple cases, a modular approach can be satisfactory. In such an approach, the equipment item can be decomposed into items for which capital costs are available and then some added cost is considered for incorporating the items together. However, each case must be duly considered. For example, if a reactor with an embedded heat exchanger is considered, the cost for incorporating the heat exchanger can be significantly different depending on the metallurgy of the reactor shell or if the reactor is refractory lined. Operating costs can be reasonably estimated for most processes if a satisfactory process model is available that can provide satisfactory material and energy balances. In most processes, cost of the raw materials is the main operating cost and therefore accurate estimation of raw materials consumption is absolutely critical. Overall conversion of feed and selectivity are the main two factors that affect consumption of raw materials. If the product is new to the market or the product has a small market, then the market elasticity should be considered while considering the product cost. Following the raw materials cost, usually the next main contributor to the manufacturing cost is the utility cost. If there is lack of information for heat of reaction, especially how the heat of reaction changes depending on the operating conditions, it can result in overdesign or underdesign of the reactor heating/cooling system. If a significant amount of cooling or heating is needed, then a wrong estimate can strongly affect process operation such as reduction in plant throughput in comparison to the design, and as a result negatively impact the plant economics.

For evaluating different projects, three discounted profitability criteria have been widely used: time criterion, cash criterion, and interest rate criterion [7]. The time criterion is the discounted payback period (DPBP), which is the time that is needed to recover the fixed capital investment (except land and WC) after the plant starts up. The cash criterion is known as net present value (NPV), which is the discounted cumulative cash position that can be calculated by summing all cash flows, positive and negative, as shown in Eq. [6.1]:

$$\text{NPV} = \sum_{i=0}^{k} \frac{C_k}{(1+r)^k} \qquad [6.1]$$

Here, C_k denotes nondiscounted yearly cash flows assuming that the yearly cash flows follow at the end of each year k. r denotes the discount rate. Obviously, at $i = 0$, C_k denotes capital investment. For large plants, capital investment including WC can spread out over several years. For those years, nondiscounted cash flow remains negative. The discount rate is usually the internal hurdle rate or the acceptable rate of return for any company for any given investment. For new businesses or for new products, the discount rate depends on the market assessment, risk that a company is ready to accept, forecasting, socioeconomic assessment, and several other factors. One option is to consider the interest rate of long-term public bonds [8] as the nominal interest rate i_n and then calculate r by considering the inflation f that can be obtained from the consumer price index (CPI) as shown in Eq. [6.2]:

$$r = \frac{1 + i_n}{1 + f} - 1 \qquad [6.2]$$

The interest rate criterion is known as discounted cash flow rate of return (DCFROR) or internal rate of return (IRR), which is defined as the discount rate calculated by setting NPV = 0 in Eq. [6.1] at the end of the period of the project. An IRR higher than the internal hurdle rate means that a project is worth investing from the perspective of a given company.

In the existing literature, a number of criteria have been considered for economic analysis [9]. Criteria such as total cost, profit, economic potential, DPBP, return on investment, NPV, and IRR have been used. However, while considering two or more investment alternatives, it is not necessary that all these criteria will be superior for one technology in comparison to all others. Therefore, the economic criterion that is used to select the best alternative does matter. One possible approach is to combine these criteria. For example, Othman et al. [1] have suggested combining NPV and DCFROR by assigning some weights on them. However, depending on the weights assigned, the best investment might significantly change. By performing an incremental analysis considering all possible investment alternatives, it has been shown [7] that a preferred approach is to select the alternative with the highest possible NPV. A similar suggestion has been provided by other authors as well [9].

The economic value of environmental quality is another economic measure for sustainability analysis. Several monetary measures have been developed to evaluate the impact of chemical and energy processes on utility and/or welfare, which can reflect the effects of air pollutants and other environmental stressors on unmanaged (natural) ecosystems and services they provide [10]. However, commodities that need to be evaluated for assessing the environmental quality are different to the market commodities and therefore must be valued using nonmarket-based valuation techniques [11]. One of the monetary measures that can be used for quantifying net utility changes is Hicksian consumer's surplus, which is the difference between the gross value and financial value holding utility constant. Two important Hicksian measures are compensating surplus (CS) and equivalent surplus (ES), which reflect the welfare changes attributed to restricted quantity changes. Eq. [6.3] defines the expenditure function E, which is the minimum amount of expenditures required to achieve the utility level U with fixed market good price and environmental quality level Q. CS and ES are differences between the expenditure functions evaluated at the initial and the subsequent environmental quality levels (Q^0 and Q^1) in terms of money. CS is evaluated at the initial utility level (U^0), while ES is evaluated at the subsequent utility level (U^1), as shown in Eqs. [6.4] and [6.5] [11].

Table 6.1 Economic Value of Environmental Quality

Environmental Quality Variable	Original Economic Value	Year	Dollar Value in 2015	References
Reduction in health damages caused by SO_2 emission from power plant	$11 per person per year	1976	$46	[12]
Reduction of aesthetic damages to landscape caused by coal surface mining	$55 per acre	1978	$201	[13]
Reduction of aesthetic damages to a National Forest caused by geothermal development	$2.54 per household per year	1977	$9.97	[14]

$$E = E[P, Q, U] \qquad [6.3]$$

$$CS = \left| E[P^0, Q^1, U^0] - E[P^0, Q^0, U^0] \right| \qquad [6.4]$$

$$ES = \left| E[P^0, Q^1, U^1] - E[P^0, Q^0, U^1] \right| \qquad [6.5]$$

Several studies have been conducted to estimate the economic values of environmental quality. Table 6.1 lists values of some of the environmental quality variables related to the chemical and energy processes [11]. In Table 6.1 the dollar values in 2015 have been calculated by escalating the values available for earlier years using the consumer price index (CPI) published by the Bureau of Labor Statistics.

ENVIRONMENTAL SUSTAINABILITY ANALYSIS

For evaluating environmental sustainability at the design stage, a number of approaches can be considered. Some of the popular approaches are life cycle analysis (LCA) [15–18], exergy analysis [19], emergy analysis [20], and evaluation of the sustainable process index (SPI) for ecological evaluation [21]. Choice of a particular approach depends on the scope of a project, available information, and time and resources spent on such analysis. No matter which approach is adopted, it is important to quantify the materials and energy used and the wastes and emissions released to the environment. While reasonable values of material and energy streams and large emission streams (such as CO_2 produced in a pulverized coal power plant) are available from process simulation as long as a satisfactory model is developed, most process models may not provide information about the trace species that are traditionally not important from a process economics perspective, but have strong impact on the environment. Many such species may form because of complex chemistry. A portion of these species can be destroyed elsewhere in the process and be distributed in different streams depending on the separation processes used. Formation of such species strongly depends on the operating conditions, throughput of the specific equipment items, and specifically the design of the relevant equipment items. For example, thermal NO_x formed in a typical burner strongly depends on the burner design, specifically on the maximum flame temperature in the region where O_2 is still available in reasonable quantities. These estimates can be found from the datasheet of specific vendors. Because of the complex formation, destruction, and distribution mechanisms of many trace species that impact the environment, quite often measurements are taken at the process output boundaries to get an estimate of the emission factors. For new processes where such information is not available, it is important to develop chemistry models and/or models of separations equipment. More information on the chemistry model and separations model can be found in Chapter 1, "Towards More Sustainable Chemical Engineering Process: Integrating Sustainable and Green Chemistry Into the Engineering Design Process" and Chapter 2, "Separations Versus Sustainability: There Is No Such Thing As a Free Lunch," respectively.

In LCA, the first step is to quantify the environmental burdens, as with any other methods used for environmental assessment. It is important to draw the boundary and scope at this stage. If the entire life cycle is considered, then one would be interested in a cradle-to-grave analysis. The scope can be reduced to undertake only cradle-to-gate assessment where the product life cycle is covered from manufacturing to factory gate. One popular algorithm in this area is the waste reduction (WAR) algorithm developed at the US Environmental Protection Agency [22,23]. An example of how the WAR algorithm can be used for evaluating alternative processes can be found in the example provided at the end of this chapter. The second step is to evaluate the impact of the environmental burdens quantified in the first step. The third step is to identify and evaluate strategies for improvement. More information on LCA can be found in Chapter 12, "Chemical Engineering and Biogeochemical Cycles: A Techno-Ecological Approach to Industry Sustainability" and Chapter 13, "Challenges for Model-Based Life Cycle Inventories and Impact Assessment in Early to Basic Process Design Stages" in this book and elsewhere.

The exergy of a system is defined as the maximum useful work available from a process depending on the reference state of its surroundings. It depends on the physical, chemical, and thermal state of a system [24] and can be calculated by:

$$\Xi = S(T - T_0) - V(p - p_0) + \sum_i n_i(\mu_i - \mu_{i0}) \qquad [6.6]$$

In Eq. [6.6], S, T, V, p, n, and μ denote entropy, temperature, volume, pressure, amount, and chemical potential, respectively. Subscript "0" denotes the corresponding quantities for the environment and $_i$ denotes species. Cleary when the system is at equilibrium with the surroundings, Ξ vanishes. Since chemical reactions result in change or one or more intensive quantities (such as T, p, and μ) and/or extensive quantities (such as n, S, and V), exergy dissipation takes place as seen from Eq. [6.6]. Thus exergy dissipation caused by reactions results in loss of available energy in the form of heat and loss of reactants in the form of undesired pollutants [25]. Therefore exergy analysis can be used as a tool for environmental sustainability analysis as has been suggested by a number of researchers [19,26]. If the exergy efficiency of a technology is better than the other(s), then that technology will be more preferred for sustainability. At the extreme, if the exergy efficiency approaches 100%, then the environmental impact approaches zero. A number of valuable resources exist in this area that interested readers can refer to the work of Ayres et al. and Szargut et al. [27,28].

Emergy is defined as the "total amount of exergy of one type that is directly or indirectly required for making a given product or to support a given flow" [29]. In emergy analysis, all forms of energy, human service, and resources are expressed in terms of a baseline energy that is used to produce them. The most common baseline is the solar emergy, mainly because of convenience and ease of use. The unit used for emergy is emjoule or eJ. While emergy analysis has been proposed to be useful for economic analysis of investments as well, here we will confine our discussions to its use for environmental impact analysis. Several indicators such as environmental loading ratio (ELR) [30] and emergy sustainability index (ESI) [20], which are based on emergy analysis, have been used in the literature. ELR is defined as the ratio of nonrenewable emergy to renewable emergy use. Before defining ELR, emergy yield ratio (EYR), which is used for economic evaluation of a project, needs to be defined. EYR is defined as the total emergy released to amount of invested emergy. The ESI is defined as a ratio of EYR and ELR and therefore weighs the impact of process on economy to its impact on the environment. Interested readers are referred to the work of Tilley and Swank [30] and Arbault et al. [20] for details on these indicators.

The SPI is based on the quantification of environmental impacts for goods and services based on the inventory data from LCA. It is calculated by considering the area of land needed to provide the raw materials and energy demands and to accommodate by-product flows from a process in a sustainable way. As an ecological indicator, SPI evaluates the viability of process under sustainable economic conditions [21]. For computing the SPI, material and energy streams (such as raw material extraction, emissions, and waste generation), which are exchanged with the environment along the life cycle, are identified by area [31]. For example, the raw materials fed into the system are computed in terms of the area requirement to produce raw material; utility consumptions are computed in terms of the area necessary to provide those utilities; and so on. SPI is the ratio of the area to provide one inhabitant with a certain service or product to the area theoretically available for a person to guarantee its sustainable subsistence. This is a measure of the expense of the given service in an economy oriented to sustainability.

In other words, with a lower SPI, the process is more sustainable, competitive under sustainable conditions, and environmentally compatible [31]. Readers interested in detailed methodology for computing SPI can refer to the work of Krotscheck and Narodoslawsky [21].

SOCIAL SUSTAINABILITY ANALYSIS

While Chapter 14, "Life Cycle Sustainability Assessment: A holistic evaluation of social, economic and environmental impacts" has in-depth analysis of various aspects that are considered from the perspective of social sustainability analysis, here we present some of the key aspects of social sustainability that should be incorporated at the design stage. Obviously the indicators for social sustainability analysis that can be used at the design stage should be such that enough information is available at the design stage to evaluate them. Also they should not, by and large, capture the overall statistics for a given company based on its past records. Then such analysis is not necessarily specific to the problem at hand. Some of the social indicators proposed in the open literature [32] fall under this category. For example, the employment situation in the workplace may not be reflective of a particular technology that is being designed, but by the employer in general. Similarly, data on other societal markers such as personnel and legality issues may neither be available (e.g., a startup company) nor be reflective of the technology being considered. Plus, such indicators may be the same for both technologies and therefore is not a good tool for differentiating two technologies. From this perspective, two indicators that can be considered are health and safety. Notably, health and safety are routinely considered as part of risk assessment. Some of the tools used for risk analysis are dependent on historical data and therefore cannot be used unless such history is available or if the technology is novel. For example, the Occupational Safety and Health Administration incidence rate is calculated by considering the number of illnesses and injuries for 200,000 h of exposure. Another such measure is the fatal accident rate, which is defined as the number of fatalities for 10^8 h of exposure. Obviously, such data are not available for novel technologies. If such data are available for specific equipment item or specific plant sections, then this information can be leveraged to partially circumvent the issues just mentioned. For example, in a bioprocessing plant it may be known that most fatal accident rates take place in the grinder. Then such information can be utilized if a grinder of similar severity and service is used elsewhere as an approximation. It should be noted that the worst case scenario analysis such as hazard and operability study undertaken routinely in process design is used to mitigate or eliminate risks. But there are always possibilities that certain risks may be overlooked or considered infeasible because of lack of experience or information. This is particularly applicable to novel technologies.

The health risk mainly results from exposure to toxic chemicals. Such chemicals can enter the human body by ingestion, inhalation, or through the skin. The chemical risk depends on the type of the chemical as well as the quantity (storage or concentration in the air caused by release or similar measure) of it. One of the comprehensive ways of evaluating potential chemical risk has been provided in the work of Martins et al. [33]. The authors have classified the chemicals depending on their frequency of use ("frequency class"), danger characteristics ("hazard class"), relative quantity ("quantity class"), and potential exposure ("potential exposure class") [33]. Each class is then divided into various levels. At the design stage it is possible to determine the levels under each class. For example, the "quantity class" is divided into five levels depending on the ratio of the quantity of a chemical to the quantity of the most used chemical in the process. Obviously, the level of a chemical can be determined if the required information is available from the material balance calculations. One of the issues with this approach is that it does not consider the impact of storage condition. With the simple example of water, when it is stored under atmospheric pressure and temperature, the risk is minimal, but steam under high temperature and pressure even at small quantities can cause high risk. Yet another example is a chemical that is liquid under room conditions may have minimal risk in terms of toxicity since the accidental leakage will be in the form of a liquid and exposure may be avoided. However, if that chemical is being used in a reactor under high temperature and pressure conditions, it can be released as a vapor to the atmosphere if there is an accidental leakage. Under such circumstances, it may be unavoidable to avoid the exposure, and the concentration can even be above LC_{50}, which is defined as the lethal dose required to kill 50% of a population. Thus the same chemical stored under different conditions should be weighed differently. This issue has been addressed in Dow's chemical exposure index (CEI) calculation [34]. The methodology involves both CEI and Hazards Distance calculation for airborne releases. The approach estimates the amount of material that becomes airborne following a release. In case of vapor release the airborne quantity is the highest total flow rate calculated based on the operating conditions of the vessel. For liquid release the methodology accounts for the conditions that the liquid released from a vessel or a pipe can form a pool on the ground, can partially vaporize forming a vapor cloud in addition to the liquid pool, or flash to form vapor. The total airborne quantity of a chemical is calculated by adding the airborne quantity resulting from the flash with the airborne quantity evaporating from the pool surface. Chemicals involved in all potential releases can be considered and the largest airborne release rate is used for the CEI calculation.

The main safety concern for many process plants is caused by fires and explosions. A number of indicators have been used in the open literature. Othman et al. [1] have extracted indicators such as "safety during startup and shutdown," "safety during operation," "operability of the plant," and "design should meet location specific demands" from the work of Herder and Weijnen [35] and considered them in a scale of 1 (worst case) to 10 (best case). Another approach is to use the inherent safety index proposed by Heikkilä [36]. One widely used methodology is Dow FEI [6]. The plant is divided into unit categories and for each one of them the critical item(s) is(are) identified. Typical unit categories are raw materials storage, process stream storage, reactor feed pumps, reactor(s), stripper(s), recovery vessel(s), flash drum(s), and others. The critical item(s) is(are) identified by considering a number of criteria such as substance processed, quantity stored by the unit, process conditions, design conditions, corrosion hazard, etc. After that the material factor (MF) is calculated. The MF accounts for the intrinsic potential energy released by combustion, explosion, or chemical reaction of the materials restrained by the equipment that is being studied. Following that a general process hazards factor (F1) is calculated by considering six factors—exothermic chemical reaction(s), endothermic processes, material handling and transfer, access to the area, drainage and spill control, and enclosed or indoor process units. After that the special process hazards factor (F2) is calculated by considering 12 factors—toxic material(s), subatmospheric pressure, operation in or near a flammable range, dust explosion, relief pressure, low temperature, quantity of flammable/unstable material, corrosion and erosion, leakage joint and packing, use of fire equipment, hot oil heat exchanger system, and rotating equipment. The process unit hazard factor (F3) is calculated by multiplying F1 by F2. Then FEI is calculated by multiplying F3 with MF. In addition, a loss control credit factor is calculated by considering the process control credit factor, material isolation credit factor, and fire protection credit factor. While calculating these indices, operation during startup and shutdown must be considered. These require specifics and experience and are not so easily quantified. For example, consider a distillation tower where an intermediate liquid is withdrawn. The withdrawal tray is directly connected to the pump suction via a pipe. During startup the pump can severely cavitate because of lack of priming or slugs of vapor coming through the pipe or entrapped vapor in the long suction pipes arriving at the pump suction intermittently. Cavitation for a long time can cause failure of the pump seal and cause fire hazards. During normal operation of the unit the possibility of such incident is minimal as long as the desired liquid level in the withdrawal tray is maintained. It should be noted that the process plants—especially those designed for state operation—typically face the most severe rate of change in temperature and pressure during startup and shutdown in comparison to the normal operation and therefore remain vulnerable to fire and explosion hazards during such times. Thus these issues should be taken into consideration while calculating the indices just mentioned.

The value brought by a process to the society is another criterion for evaluating social sustainability of a process. Consider a process that can create social benefit in a less populated region by creating job opportunities. This can encourage people to move away from the overpopulated areas and help to develop less populated regions, while the quality of life in both areas would improve [37]. Various categories have been suggested in the open literature for estimating the value brought by a process to society. The Sustainability Reporting Guidelines from the Global Reporting Initiative suggested four categories: Labor Practices and Decent Work, Human Rights, Society, and Product

Responsibility [38]. Each category has several criteria. A number of these criteria can neither be evaluated at the process designs stage nor be quantified. Thus more research needs to be done in this area to come up with measures that can be helpful in decision making during the design stage.

EVALUATION OF DESIGN ALTERNATIVES BY CONSIDERING VARIOUS SUSTAINABILITY MEASURES

Unlikely of the economic sustainability measure, there is no unique environmental or social sustainability measure that is unambiguous, is not affected by user perception, and is quantifiable without dispute. Thus it is not obvious which design to select if one has a superior environmental sustainability but inferior social sustainability. One would think that it would depend on the relative differences, but since the measures are not based on the same scale, it is not obvious how to weigh them. Further, there is no unique way to combine these measures. Applying arbitrary weights such as in the work of Othman et al. [1] can lead to biased estimates. Sikdar has defined three groups of metrics corresponding to the three aspects of sustainability—economic, environmental (ecological), and sociological metrics [39]. These are 1-D metrics since they account for only one aspect of the system [39]. An example of 2-D metrics is eco-efficiency, socio-ecological, and socio-economic metrics. According to the author [39] the true sustainability metric is the intersection of all three aspects. According to the authors, metrics such as nonrenewable energy use, materials use, and pollutant dispersion are 3-D indicators. In the hierarchical approach proposed by the author [39], the cost of manufacturing is first minimized. The author has mentioned that it will improve all 3-D indicators. However, this is not necessarily true because when the cost of manufacturing is minimized, all individual costs contributing to manufacturing do not necessarily reduce, only the overall cost does. Thus energy use can be reduced, but materials use can increase or vice versa. Since it is not guaranteed that all 3-D metrics will decrease (or minimize for an optimal design), there is no obvious way how to weigh the environmental and social impact of increase of one (or more) 3-D metric to the decrease of another (one or more) 3-D metric. Overall, considerable research still needs to be performed in search of universally accepted environmental and social sustainability measures and an unambiguous and unbiased way to combine these measures to get an overall score.

EXAMPLE: BIOETHANOL PROCESS
Corn Dry Grind Versus Corn Wet Milling

As an example of how sustainability considerations can be incorporated along with the economic and profitability analysis, two bioethanol processes, as shown in Figs. 6.1 and 6.2, are evaluated. Corn dry grind and wet milling processes are two most widely

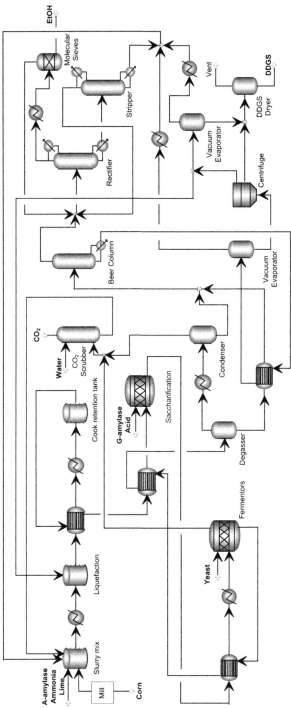

Figure 6.1 Process flow diagram of corn dry grind ethanol production process. *DDGS*, Distillers dried grains with solubles.

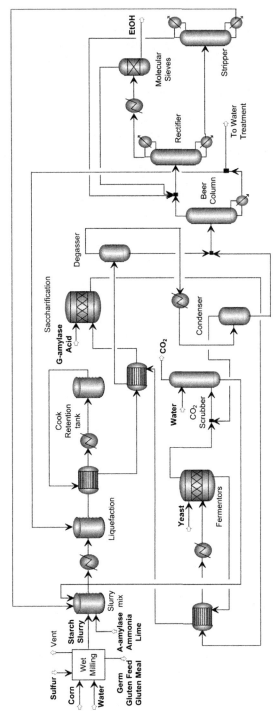

Figure 6.2 Process flow diagram of corn wet milling ethanol production process.

applied technologies for producing ethanol. Current technologies can produce about 2.5 gal (wet mill process) and 2.7 gal (dry grind process) of ethanol from one bushel (1 bushel = 56 lb) of corn [40,41]. In the dry grind process (Fig. 6.1), corn is first milled and liquefied and then sent to a saccharification and fermentation unit to convert starch into ethanol, while oil, fiber, and protein are processed with starch until fermentation, and then separated by distillation and centrifuge to produce coproduct distillers dried grains with solubles. In the wet milling process (Fig. 6.2), corn is first separated into germ, corn gluten meal, corn gluten feed, and starch slurry. Then the starch slurry is sent to the facility similar to the dry grind process for ethanol production. Compared with the dry grind process, the wet milling process is more capital and energy intensive, but produces a number of high-value coproducts [40]. Sixty percent of ethanol used to be produced from wet mills before 1999 [41]. However, farmer's organizations currently favor the dry grind process, because of the simplicity of the process, and the possibility of small-scale operation with less capital investment and a lesser number of personnel. As a result, the majority of US ethanol is produced from the dry grind approach nowadays [40]. Because of the complexity of the biochemical process, the following assumptions have been made to simplify the sustainability assessment. Solid compounds are not considered when analyzing the environmental effect of the process, because they are assumed to have no or negligible negative environmental impact as the hazardous components are locked in a solid mixture [1]. It is also assumed that water has been treated in the water treatment plant before discharging and has no or negligible negative environmental impact [42].

Process Modeling

In this example the steady-state simulation of two bioethanol processes is conducted by modifying the model developed by McAloon et al. using Aspen Plus and Microsoft Excel, where the non-random two liquid thermodynamic model is used [41,43,44]. The plant capacity is considered to be 100 million gal per year (378.5 km^3 per year) for both cases. The simulation results shown in Tables 6.2–6.4 provide information on the overall input and output material, component balance, and utility consumption for both cases.

In the dry grind process, corn is first ground in a hammer mill and sent to a slurry tank mixed with recycled process water, alpha–amylase, ammonia (or urea), and lime. The corn slurry then undergoes liquefaction, continuous saccharification with backset (recycled thin stillage from the centrifuge) and glucoamylase, and fermentations with yeast [41]. The main reactions considered in the process are listed in Table 6.5. 99% conversion of starch is assumed for saccharification reaction, while 100% conversion of glucose is assumed for the fermentation process. The main stages of the dry grind process can be found in Table 6.6 with operating conditions reported in the open literature [41,44]. The saccharification and fermentation reactors are simplified to continuous

Table 6.2 Material Balance of the Two Corn-to-Ethanol Processes

Materials (ton/h)	Dry Grind Input	Dry Grind Output	Wet Milling Input	Wet Milling Output
Corn (15% water)	113.05		122.08	
Water[a]	23.95	38.81	167.06	183.27
Ethanol		35.85		35.85
DDGS (9% water)		37.31		
Dry germ (3% water)				7.81
Gluten feed (10% water)				24.59
Gluten meal (10% water)				7.06
Ethanol yield (gal/bushel)		2.74		2.53

DDGS, Distillers dried grains with solubles.
[a]Including vaporized water from dryers.

Table 6.3 Major Utility Consumption of the Two Corn-to-Ethanol Processes

Utilities	Dry Grind	Wet Milling
150 psi steam (kg/h)	110,723	119,413
Electricity (kW)	8168	15,020
Natural gas (GJ/h)	148.39	239.93

Table 6.4 Component Balance of the Two Corn-to-Ethanol Processes[a]

Component (ton/h)	Dry Grind Input	Dry Grind Output	Wet Milling Input	Wet Milling Output
Water	40.91	38.81	185.37	183.27
Ethanol	0	35.85	0.00	35.85
CO_2	0	34.25	0.00	34.25
NaOH	0.59	0	0.59	0.00
$Ca(OH)_2$	0.14	0	0.14	0.00
NH_3	0.22	0	0.22	0.00
SO_2[b]	0	0	0.00	0.03
S	0	0	0.05	0.00
H_2SO_4	0.22	0	0.22	0.00
Alpha–amylase	0.08	0	0.08	0.00
Glucoamylase	0.11	0	0.11	0.00
Yeast	0.02	0	0.02	0.00
Starch	67.27	0.66	72.64	5.37
C5Poly	5.00	4.98	5.40	5.40
C6Poly	2.88	2.87	3.11	3.11
Proteins	5.57	5.56	6.02	6.02
Oil	3.84	3.83	4.15	4.15
NFDS	7.69	11.27	8.30	8.30
ProtSol	3.84	4.78	4.15	4.15

NFDS, Nonfermentable dissolved solids.
[a]The flow rate of any component is calculated by summing the flow rate of that component in all input and output streams, including raw material, process water, enzymes, products, and nonproducts streams; $M_c = \sum_h M_h x_{c,h}$, where M is the mass flow rate of stream h and x is the mass fraction of component c in stream h.
[b]SO_2 emission of wet milling process is estimated using the work of Watson et al. [45].

Table 6.5 Reactions in the Corn-to-Ethanol Processes

Process	Reactions
Saccharification	Starch + water → glucose
Fermentation	Glucose → 1.9 ethanol + 1.9 CO_2 + 0.06 NFDS
	NFDS → ProtSol

NFDS, Nonfermentable dissolved solids; *ProtSol*, soluble protein.

Table 6.6 Main Stages of the Corn Dry Grind Ethanol Production Process

Process	Purpose
Milling	Reduce the particle size of corn kernels
Liquefaction	Primary step in starch hydrolysis
Saccharification	Convert starch to fermentable sugars
Fermentation	Convert sugars to ethanol and CO_2
Distillation	Concentrate ethanol
Dehydration	Concentrate ethanol
Centrifugation	Separate solids from liquid from beer column bottoms
Evaporation	Concentrate dissolved solids in stillage
Drying	Achieve desired moisture in DDGS

DDGS, Distillers dried grains with solubles.

operations and simulated by the Rstoic model in Aspen Plus. The RadFrac model and SSplit model are used for the distillation and centrifugation process, by specifying the desired component recovery.

In addition to the models developed for the dry grind process, a Sep (component separator) block is specified in the wet milling process to represent the physical separation section based on the material balance provided in the open literature [46,47]. The stages involved in the wet milling process are listed in Table 6.7. Downstream operation is similar to the dry grind process. However, the solid separation system, including centrifugation, evaporation, drying, etc., is not considered, because oil, fiber, and protein are separated before the starch fermentation process [48].

Sustainability Indicators

With the steady-state model developed in Aspen Plus, a techno-economic analysis is conducted using Aspen Process Economic Analyzer (APEA) for calculating the value of economic indicators such as NPV and DCFROR using the approach shown in Fig. 6.3. The operating and manufacturing costs are calculated mainly based on the material and energy balances available from the process model. For the standard equipment items such as heat exchangers, distillation columns, and pumps, APEA is used to size and evaluate the capital cost, while for other equipment items, such as milling equipment, saccharification, and fermentation reactors, vendor costs are obtained from the open literature

Table 6.7 Main Stages of the Corn Wet Milling Ethanol Production Process

Process	Purpose
Grain handling	Remove foreign matter and prepare cleaned corn for steeping
Steeping[a]	Clean corn is soaked in dilute SO_2 solution to remove soluble solids
Germ separation	Separate and wash germ using hydrocyclones and a series of screens
Fiber separation	Recover fiber by grinding to produce gluten feed
Gluten separation	Recover gluten in centrifuges to produce gluten meal
Starch washing	Wash starch in a series of small hydrocyclones to produce starch slurry with 60% moisture and less than 1% of impurities
Starch to ethanol	Include liquefaction, saccharification, fermentation, distillation and dehydration stages as describe in Table 6.6

[a]SO_2 is produced from burning elemental sulfur.

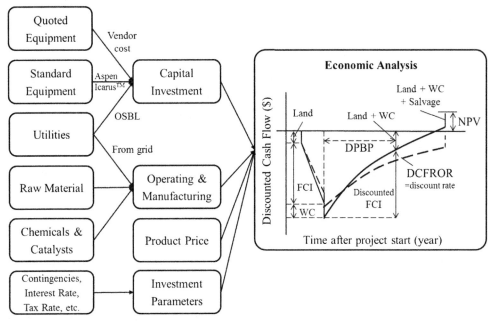

Figure 6.3 Evaluation of economic sustainability. *OSBL*, Outside the battery limits; *WC*, working; *NPV*, net present value; *FCI*, fixed capital investment; *DPBP*, discounted payback period; *DCFROR*, discounted cash flow rate of return.

[41,44,46]. If utilities are generated inside the facility, such as cooling water, the capital cost for outside the battery limits facilities are evaluated by APEA. Using the specified indirect cost, discount rate, tax, and interest, the discounted cash flow and the economic indicators are generated by APEA with the method described in Section "Economic Sustainability Analysis." Parameters for economic evaluations and prices of the raw materials and products are shown in Table 6.8. Utility costs for electricity, steam, and

Table 6.8 Parameters for Economic Evaluations and Prices of the Raw
Materials and Products

Investment Parameter	Value	Material	Cost
Contingency percent	18%	Corn ($/bushel)	3.78
Number of years for analysis	10	Ethanol ($/gal)	2.4
Tax rate	40%	DDGS ($/ton)	190
Project capital escalation (%/year)	1	Dry germ (dry, $/ton)	443
Products escalation (%/year)	1	Gluten feed (dry, $/ton)	170
Raw material escalation (%/year)	1	Gluten meal (dry, $/ton)	740

DDGS, Distillers dried grains with solubles.

Table 6.9 Economic Performance (2014 Pricing Basis)

	Dry Grind	Wet Milling
NPV ($ million)	148.21	130.43
DCFROR (%)	12.00	11.96
TPC ($ million)	95.30	143.42

DCFROR, Discounted cash flow rate of return; NPV, net present value; TPC, total project cost.

fuels are obtained from Turton et al. [7]. Eq. [6.7] shows the scaling calculations, where
0.6 is selected for the equipment scaling exponent [41]. The prices of raw material and
product are specified based on the information available in the open literature [49,50].
Table 6.9 shows the results of economic analysis. Because the configuration of the dry
grind process is relatively simpler than the wet milling process, the capital investment of
the process is significantly reduced. Hence, even though more valuable by-products are
produced in the wet milling process, dry grind technology is economically superior by
comparing the NPV of both the processes.

$$\text{New Cost} = \text{Original Cost} \left(\frac{\text{New Size}}{\text{Original Size}} \right)^{\text{exp}} \qquad [6.7]$$

In this example, the WAR algorithm is applied for environmental performance
assessment based on the flow rate calculation of potential environment impact (PEI)
[1,22]; total rate of PEI output ($I_{\text{out}}^{(t)}$), relative PEI output ($\widehat{I}_{\text{out}}^{(t)}$), total rate PEI generation
($I_{\text{gen}}^{(t)}$), relative PEI generation ($\widehat{I}_{\text{gen}}^{(t)}$) are selected as environmental indicators for sustainability assessment, defined in Eqs. [6.8]–[6.11]. In Eqs. [6.8]–[6.11] $I^{(\text{cp})}$ and $I^{(\text{ep})}$ are
the mass input and output rates of PEI of a chemical process and an energy conversion
process defined in Eqs. [6.12] and [6.13]. The values for the component-specific PEI are
obtained from the MSDS datasheet and classification factor published in the open
literature [1,51]. The component-specific PEI of the chemical process is shown in
Table 6.10, while the component-specific PEI and emission factor of the energy
conversion process are shown in Table 6.11.

Table 6.10 Component-Specific Potential Environment Impact (Chemical Process)[a]

	HTPI	HTPE	ATP	TTP				
Component	LD_{50}	TWA-TLV	LC_{50}	LD_{50}	GWP	POCP	AP	ODP
Ethanol	7060	1000	14,000	7060	0	0.268	0	0
CO_2	0	0	0	0	1	0	0	0
NaOH	4090	2	0	0	0	0	0	0
$Ca(OH)_2$	7340	5	0	7340	0	0	0	0
NH_3	350	25	1390	350	0	0	1.88	0
SO_2	0	2	2520	0	0	0	1	0
S	175	0	0	175	0	0	0	0
H_2SO_4	2140	3	49	2140	0	0	0	0

HTPI, Human toxicity potential by ingestion; *HTPE*, human toxicity potential by inhalation/dermal exposure; *ATP*, aquatic toxicity potential; *TTP*, terrestrial toxicity potential; *TWA*, time-weighted average; *TLV*, threshold limit value; *GWP*, global warming potential; *POCP*, photochemical oxidation potential; *AP*, acidification potential; *ODP*, ozone depletion potential; *CGIH*, Conference of Governmental Industrial Hygienists.
[a]LD_{50} (oral-rate, mg/kg), TWA-TLV (CGIH, ppm), LC_{50} (ppm).

Table 6.11 Emission Factor (EF) and Component-Specific Potential Environment Impact (Energy Conversion Process)[a]

	HTPI	HTPE	ATP	TTP					Electricity[b]	Natural Gas
Component	LD_{50}	TWA-TLV	LC_{50}	LD_{50}	GWP	POCP	AP	ODP	kg/kWh	kg/GJ
SO_x	1.2	0	0	1.2	0	0	1	0	0.00272	0.000253
NO_x	0.78	0	0	0.78	0	0	1.77	0	0.001814	0.001194
CO_2	0	0	0	0	1	0	0	0	0.3719	50.5327
HCl	0	0	0	0	0	0	0.88	0	0.00009	
CH_4	0	0	0	0	35	0.007	0	0	0.4763	
Mercury	0	0.025	500	0	0	0	0	0	4.94E-09	0.000969

HTPI, Human toxicity potential by ingestion; *HTPE*, human toxicity potential by inhalation/dermal exposure; *ATP*, aquatic toxicity potential; *TTP*, terrestrial toxicity potential; *TWA*, time-weighted average; *TLV*, threshold limit value; *GWP*, global warming potential; *POCP*, photochemical oxidation potential; *AP*, acidification potential; *ODP*, ozone depletion potential; *CGIH*, Conference of Governmental Industrial Hygienists.
[a]LD_{50} (oral-rate, mg/kg), TWA-TLV (CGIH, ppm), LC_{50} (ppm).
[b]Electricity from coal-fired power plant.

$$I_{out}^{(t)} = I_{out}^{(cp)} + I_{out}^{(ep)} \qquad [6.8]$$

$$\widehat{I}_{out}^{(t)} = \frac{I_{out}^{(cp)} + I_{out}^{(ep)}}{\sum P_p} \qquad [6.9]$$

$$I_{gen}^{(t)} = I_{out}^{(cp)} + I_{out}^{(ep)} - I_{in}^{(cp)} - I_{in}^{(ep)} \qquad [6.10]$$

$$\widehat{I}_{gen}^{(t)} = \frac{I_{out}^{(cp)} + I_{out}^{(ep)} - I_{in}^{(cp)} - I_{in}^{(ep)}}{\sum P_p} \qquad [6.11]$$

$$I^{(cp)} = \sum_c^{\text{Comp}} \sum_h^{\text{Steams}} M_h x_{c,h} \sum_i^{\text{EnvCat}} \psi_{c,i}^s \qquad [6.12]$$

$$I^{(ep)} = \sum_m E_m^{\text{Direct}} \sum_g^{\text{Comp}} \text{EF}_g^{m,n} \sum_i^{\text{EnvCat}} \psi_{g,i}^s + \sum_n \alpha E_n^{\text{Indirect}} \sum_g^{\text{Comp}} \text{EF}_g^{m,n} \sum_i^{\text{EnvCat}} \psi_{g,i}^s \quad [6.13]$$

where P_p is the mass flow rate of product p; M_h is the mass flow rate of stream h; $x_{c,h}$ is the mass fraction of component c in stream h; $\psi_{c,i}^s$ is the normalized value of the specific PEI of component c and impact category i, defined as $\frac{\text{Score}_{c,i}}{\text{avg}(\text{Score}_c)_i}$; E is the utility consumption because of m direct energy (electricity and natural gas) and n indirect energy (steams) usage; EF is the emission factor for gas pollutants g (in kg/kW for coal-fired power plants and kg/GJ for natural gas combustor); and α is the ratio of electrical energy to steam energy for plant utilities produced through burning the same amount of coal, assumed to be 3 [52]. The results obtained from the WAR algorithm are shown in Table 6.12. Environmental performance of the dry grind process is found to be superior.

For the social criteria, two methods are applied in this study. A more detailed safety-related discussion of the bioethanol process can be found in the open literature [45]. In the first method, safety during operation, operability of the plant, safe startup and shutdown, and location-specific demands are selected as four soft indicators for sustainability assessment. The evaluation and justification of each indicator are provided in Table 6.13, where standard definition and scaling system can be found in Othman et al. [1]. In this approach, the process with the higher score is more desirable. Again the dry grind process is found to be superior.

In the second method, the inherent safety index proposed by Heikkilä [36] is applied for safety analysis, which requires less information and covers lots of safety aspects [53]. Table 6.14 shows the results obtained from the inherent safety index method for both the technologies. In the inherent safety index method, the chemical inherent safety index and process inherent safety index are assigned to analyze the hazards caused by chemicals,

Table 6.12 Environmental Assessment Results of the Two Processes

	Dry Grind	Wet Milling
PEI output rate (10^3 PEI/h)	564	595
PEI generation rate (10^3 PEI/h)	552	578
Relative PEI output (PEI/kg)	7.71	7.88
Relative PEI generation (PEI/kg)	7.55	7.65

PEI, Potential environment impact.

Table 6.13 Values of Social Indicator for Sustainability Assessment

Social Indicators	Score Dry Grind	Score Wet Milling	Justification
Safety during operation	5	4	Ethanol is highly flammable. Corn dust is explosive [54]. Mechanical equipment presents serious entanglement and amputation hazards. Wet milling process involves more dust and mechanical equipment. Wet milling process also has SO_2 emission problem, which the dry grind process does not have
Operability of the plant	5	3	Wet milling process involves more recycling streams [46]
Safe startup and shutdown	5	4	The operating conditions of the ethanol production and purification section of two approaches are similar. However, sulfur burner, used for generating SO_2 for steeping [46] in the wet milling process, may cause more startup and shutdown safety issues
Location-specific demands	10	10	Criteria met. The wet milling process is more manpower intensive, but the effect on this indicator is negligible.
Total	25	21	

equipment, and inventories in the plant. The scoring of each indicator in the inherent safety index method is based on the parameters of the individual components and is well addressed in the open literature [36]. Table 6.15 gives an example of a scoring system for the flammability indicator. In this approach, the process with the lower score is more desirable.

It is noted that the inherent safety index only considers the worst case scenario for scoring regardless of the quantity of the chemicals and equipment. Hence the complexity of the process has negligible impact on the final score. Li et al. [53] have proposed an enhanced inherent safety index method. In this approach, scores are assigned to each component and equipment in the system, using the same system as shown in Tables 6.14 and 6.15. Then the total score of the chemical inherent safety index is calculated by multiplying the severity of the chemicals with the flow rates, while the total scores of the process inherent safety index are given for individual equipment and multiplied by the number of equipment, as shown in Eqs. [6.14] and [6.15], [53]. Table 6.16 lists the scores of selected chemicals and equipment in the two processes. The final results are shown in Table 6.17 using the modified inherent safety index.

Table 6.14 Results From Inherent Safety Index Method [36]

	Symbol	Score	Determination	Dry Grind	Wet Milling	Remarks
Chemical Inherent Safety Index, I_{CI}						
Heat of main reaction	I_{RM}	0–4	ΔH_r	—	—	
Heat of side reaction, max	I_{RS}	0–4	ΔH_r	—	—	
Chemical interaction	I_{INT}	0–4	Unwanted side reaction	—	—	
Flammability	I_{FL}	0–4	Flash point	3	3	Ethanol
Explosiveness	I_{EX}	0–4	UEL-LEL	1	1	Ethanol
Toxic exposure	I_{TOX}	0–4	TLV	4	4	SO_2, NH_3, etc.
Corrosiveness	I_{COR}	0–2	Construction material	1	1	Stainless steel
Process Inherent Safety Index, I_{PI}						
Inventory	I_I	0–5	Equipment size	4	4	
Process temperature, max	I_T	0–4		1	4	Distillation column bottom, 116°C; sulfur burner, 900°C
Process pressure, max	I_P	0–4		0	0	Pressure, 3.5 bar
Equipment safety (ISBL)	I_{EQI}	0–4	Equipment type	3	4	Hammer mills; sulfur burner
Equipment safety (OSBL)	I_{EQO}	0–3		2	2	Cooling tower, ethanol storage tank
Safe process structure	I_{ST}	0–5		0	0	
Total inherent safety index	I_{TI}			19	23	

UEL, upper explosive limit; LEL, lower explosive limit; *TLV*, threshold limit value; *ISBL*, inside the battery limits; *OSBL*, outside the battery limits.

Table 6.15 Determination of the Flammability Subindex I_{FL} [36]

Flammability	Score of I_{FL}
Nonflammable	0
Combustible (flash point > 55°C)	1
Flammable (flash point ≤ 55°C)	2
Easily flammable (flash point < 21°C)	3
Very flammable (flash point <0°C and boiling point ≤ 35°C)	4

Table 6.16 Scores of the Chemical and Process Inherent Indexes of Selected Chemicals and Equipment

Component	I_{FL}	I_{EX}	I_{TOX}	I_{COR}	Equipment	I_I	I_T	I_P	I_{EQ}
Ethanol	3	1	2	0	Mill	3	0	0	3
NH_3	0	0	3	0	Fermenter	4	0	0	2
SO_2	0	0	4	1	Beer column	4	1	0	1
CO_2	0	0	0	0	Sulfur burner	0	4	0	4

Table 6.17 Results of the Modified Inherent Safety Index Method

	Dry Grind	Wet Milling
Chemical inherent safety index, I_{CI}	219.8	219.9
Process inherent safety index, I_{PI}	151	213
Total inherent safety index, I_{TI}	370.8	432.9

$$I_{CI} = \sum_{C}^{Comp} \sum_{i}^{ChemI} M_C I_{C,i} \qquad [6.14]$$

$$I_{PI} = \sum_{P}^{Equp} \sum_{i}^{ProcI} N_P I_{P,i} \qquad [6.15]$$

In Eqs. [6.14] and [6.15] M_C is the flow rate of chemical C; $I_{C,i}$ is the score of the chemical inherent safety subindex of chemical C; N_P is the number of equipment P; and $I_{P,i}$ is the score of the process inherent safety subindex of equipment P. Again the dry grind process is found to be superior. In summary, it is observed that the dry grind process is superior to the wet milling process for all three sustainability measures.

This example shows one approach to evaluation of the economic, environmental, and social sustainability analysis. It is observed that mainly because of the additional operations, such as steeping and separation units in the wet milling process in comparison to the dry grind process, the dry grind process is superior to the wet milling process for all three sustainability measures. Even through more valuable by-product can be produced in the wet milling process, the additional units increase capital investment, emissions, as well as health and safety issues. In this example, all three measures are superior for one technology. However, this may not be the case for many potential technologies for a given product. For those technologies, different measures need to be weighed for selecting the technology that has the superior performance in terms of overall sustainability.

CONCLUDING REMARKS

Traditionally, the economic impact is the main consideration during the process design or technology selection. Future process design should include rigorous sustainability analysis for selecting the final technology or design. Considerable progress has been made in the last several decades in quantifying a number of metrics that can be used for the three main aspects of sustainability. This chapter focused on several methods or approaches that can be applied to evaluate various economic, environmental, and social sustainability measures during process design. For evaluation of economic sustainability the preferred approach is to select the alternative with the highest possible NPV. Another suggested measure for evaluating economic sustainability is the economic value of environmental quality. The leading approaches for evaluating environmental sustainability at the design stage are LCA, exergy analysis, emergy analysis, and evaluation of SPI. Health risk assessment and safety analysis are two important criteria for social sustainability analysis during the design stage. In addition, the value brought by a process to the society is another criterion to be considered for evaluating social sustainability of a process. While economic performance of multiple technologies or designs can be performed unambiguously, this is not necessarily true for the environmental and social sustainability measures. For the environmental and social impact, current approaches strongly rely on user perception, experience, domain knowledge, and completeness of information. For the future, considerable research will be required not only to obtain such measures, but also to combine them such that technology or design can be selected that maximizes the overall sustainability. Such approaches should be developed not only for existing processes, but also for novel technologies or design for which historical data or experience may not be available. Perhaps dividing the process industries into various categories such as petroleum and petrochemicals, biofuels, etc. might be helpful in performing in-depth analysis by proposing measures that uniquely characterize these categories.

NOMENCLATURE

AP Acidification potential
ATP Aquatic toxicity potential
APEA Aspen Process Economic Analyzer
CEI Chemical exposure index
CGIH Conference of Governmental Industrial Hygienists
CPI Consumer price index
CS Compensating surplus
DCFROR Discounted cash flow rate of return
DDGS Distillers dried grains with solubles

DPBP Discounted payback period

ELR Environmental loading ratio

EPA Environmental Protection Agency

ES Equivalent surplus

ESI Emergy sustainability index

EYR Emergy yield ratio

FCI Fixed capital investment

FEI Fire and explosion index

GWP Global warming potential

HTPI Human toxicity potential by ingestion

HTPE Human toxicity potential by inhalation/dermal exposure

IRR Internal rate of return

LCA Life cycle analysis

LEL Lower explosive limit

NFDS Nonfermentable dissolved solids

NPV Net present value

ODP Ozone depletion potential

OSBL Outside the battery limits

PCOP Photochemical oxidation potential

PEI Potential environment impact

SPI Sustainable process index

TLV Threshold limit value

TTP Terrestrial toxicity potential

TWA Time-weighted average

UEL Upper explosive limit

WAR Waste reduction

WC Working capital

REFERENCES

[1] Othman MR, Repke JU, Wozny G, Huang Y. A modular approach to sustainability assessment and decision support in chemical process design. Ind Eng Chem Rec 2010;49:7870–81.

[2] Biodiesel and alternative fuel; claims for 2014; excise tax. Available from: http://www.irs.org [accessed 05.11.15].

[3] Using biofuel tax credits to achieve energy and environmental policy goals. The Congress of the United States, Congressional Budget Office; July 2010. Available from: http://www.cbo.gov [accessed 05.11.15].

[4] Panayotou T. Economic instruments and environmental management and sustainable development. Prepared for the UNEP's consultative expert group meeting on the use and application of economic policy instruments for environmental management and sustainable development, Nairobi; 1995.

[5] Helm D. Climate-change policy: why has so little been achieved. Oxf Rev Econ Policy 2008;24(2):211–38.

[6] American Institute of Chemical Engineers. Dow's fire and explosion index hazard classification guide. 7th ed. New York, NY: AIChE; 1994.

[7] Turton R, Bailie RC, Whiting WB, Shaeiwitz JA, Bhattacharyya D. Analysis, synthesis, and design of chemical process. 4th ed. Upper Saddle River, NJ: Prentice Hall; 2012.

[8] Chang N. Systems analysis for sustainable engineering: theory and applications. New York, NY: McGraw Hill; 2011.

[9] Pintarič ZN, Kravanja Z. Selection of the economic objective function for the optimization of process flow sheets. Ind Eng Chem Res 2006;45:4222–32.

[10] Adams RM, Horst Jr RL. Future directions in air quality research: economic issues. Eviron Int 2003;29:289–302.

[11] Bergstrom JC. Concepts and measures of the evonomic value of environmental quality: a review. J Environ Manage 1990;31:215—28.

[12] Loehman ET, Berg SV, Arroyo AA, Hedinger RA, Schwartz JM, Shaw ME, et al. Distributional analysis of regional benefits and costs of air control. J Environ Eco Manage 1979;6:222—43.

[13] Randall A, Grunewald O, Johnson S, Ausness R, Pagoulates A. Reclaiming coal surface mines in central Appalachia: a case study of benefits & costs. Land Econ 1978;54:472—89.

[14] Thayer MA. Contingent valuation techniques for assessing environmental impacts: further evidence. J Environ Eco Manage 1981;38:27—44.

[15] Azapagic A. Life cycle assessment and its application to process selection, design, and optimization. Chem Eng J 1999;73:1—21.

[16] Finnveden G, Hauschild MZ, Ekvall T, Guinée J, Heijungs R, Hellweg S, et al. Recent developments in life cycle assessment. J Environ Manage 2009;91:1—21.

[17] Čuček L, Klemeš J, Kravanja Z. Overview of environmental footprints. In: Klemeš J, editor. Assessing and measuring environmental impact and sustainability. Oxford: Elsevier; 2015. p. 131—78 [chapter 5].

[18] Cherubini F, Bird ND, Cowie A, Jungmeier G, Schlamadinger B, Woess-Gallasch S. Energy and greenhouse gas-based LCA of biofuel and bioenergy systems: key issues, ranges and recommendation. Resour Conserv Recycl 2009;53(8):434—47.

[19] Rosen MA, Dincer I, Kanoglu M. Role of exergy in increasing efficiency and sustainability and reducing environmental impact. Energy Policy 2008;36:128—37.

[20] Arbault D, Rugani B, Tiruta-Barna L, Benetto E. A semantic study of the emergy sustainability index in the hybrid lifecycle-emergy framework. Ecol Indic 2014;43:252—61.

[21] Krotscheck C, Narodoslawsky M. The sustainable process index—a new dimension in ecological evaluation. Ecol Eng 1996;6(4):241—58.

[22] Young DM, Cabezas H. Designing sustainable processes with simulation: the waste reduction (WAR) algorithm. Compt Chem Eng 1999;23:1477—91.

[23] Young DM, Scharp R, Cabezas H. The waste reduction (WAR) algorithm: environmental impacts, energy consumption, and engineering economics. Waste Manag 2000;20:605—15.

[24] Debenedetti PG. The thermodynamic fundamentals of exergy. Chem Eng Educ 1984;18:116.

[25] Fan LT, Zhang T, Liu J, Schlup JR, Seib PA, Friedler F, et al. Assessment of sustainability-potential: hierarchical approach. Ind Eng Chem Res 2007;46:4506—16.

[26] Cornelissen RL. Thermodynamics and sustainable development [Ph. D. thesis]. The Netherlands: University of Twente; 1997.

[27] Ayres RU, Ayres LW, Martinas K. Exergy, waste accounting, and life-cycle analysis. Energy 1998;23:355—63.

[28] Szargut J, Morris DR, Steward FR. Exergy analysis of thermal, chemical, and, metallurgical processes. New York: Hemisphere; 1988.

[29] Odum HT. Environmental accounting: emergy and environmental decision making. New York: Wiley; 1996.

[30] Tilley DR, Swank WT. EMERGY-based environmental systems assessment of a multi-purpose temperature mixed-forest watershed of the southern Appalachain Mountains, USA. J Environ Manage 2003;69:213—27.

[31] Narodoslawsky M. Sustainable process index. In: Klemeš J, editor. Assessing and measuring environmental impact and sustainability. Oxford: Elsevier; 2015. p. 73—84 [chapter 3].

[32] IChemE. The Sustainability Metrics. The Institution of Chemical Engineers, Rugby. Available from: https://www.icheme.org; 2002 [accessed 03.15.15].

[33] Martins AA, Mata TM, Costa CAV, Sikdar SK. Framework for sustainability metrics. Ind Eng Chem Res 2007;46:2962—73.

[34] American Institute of Chemical Engineers. Dow's chemical exposure index guide. 1st ed. New York, NY: AIChE; 1998.

[35] Herder PM, Weijnen MPC. Quality criteria for process design in the design process-industrial case studies and an expert panel. Comput Chem Eng 1998;22:S513—20.

[36] Heikkilä AM. Inherent safety in process plant design, an index-based approach [Dissertation]. Helsinki: Helsinki University of Technology; 1999.

[37] Mota B, Gomes MI, Carvalho A, Barbosa-Povoa AP. Towards supply chain sustainability: economic, environmental and social design and planning. J Clean Prod 2015;105:14—7.

[38] GRI. G4 sustainability reporting guidelines. Amsterdam: Global Reporting Initiative; 2013.

[39] Sikdar SK. Sustainable development and sustainability metrics. AIChE J 2003;49:1928—2003.

[40] Bothast RJ, Schlicher MA. Biotechnological processes for conversion of corn into ethanol. Appl Microbiol Biotechnol 2005;67:19—25.

[41] McAloon A, Taylor F, Yee W. Determining the cost of producing ethanol from corn starch and lignocellulosic feedstocks. Technical report, NREL/TP-580—28893. Golden, CO: National Renewable Energy Laboratory; October 2000.

[42] GE. Water and process solutions for the corn milling industry. GEA18908A. General Electric Company; 2012. Available from: http://www.ge.com/water.

[43] Aspen Tech. Aspen plus model of bioethanol from corn model. Burlington, MA: Aspen Technology Inc.; 2008. Available from: http://www.aspentech.com.

[44] Kwiatkowski JR, McAloon AJ, Taylor F, Johnston DB. Modeling the process and costs of fuel ethanol production by the corn dry-grind process. Ind Crops Prod 2006;23:288—96.

[45] Watson AP, Smith JG, Elmore JL. A survey of potential health and safety hazards of commercial-scale ethanol production facilities. DOE Contract No. W-7405-eng-26. Tennessee: Oak Ridge National Laboratory; 1982.

[46] Ramirez EC, Johnston DB, McAloon AJ, Yee W, Shigh V. Engineering process and cost model for a conventional corn wet milling facility. Ind Crops Prod 2008;27:91—7.

[47] Ramirez EC, Johnston DB, McAloon AJ, Shigh V. Enzymatic corn wet milling: engineering process and cost model. Biotechnol Biofuels 2009;2:2.

[48] Johnston DB. New enzymatic advances in the dry grind (grain) ethanol. In: M. Al-Dahhan (Washington University Bioenergy), K. Hicks (ERRC/ARS/USDA), Abbas CA (Archer Daniels Midland Company), Eds. Bioenergy-I: from concept to commercial processes, ECI symposium series, vol. P08; 2007. Available from: http://dc.engconfintl.org/bioenergy_i/4.

[49] USDA. Agricultural prices. 2014. Available from: http://www.nass.usda.gov [accessed 03.03.15].

[50] US Grains Council. 2014. Available from: http://www.grains.org [accessed 03.03.15].

[51] Heijungs R, Huppes G, Lankreijer RM, Udo de Hayes HA, Wegenersleeswijk A. Environmental life cycle assessment of products guide. Leiden: Centre of Environmental Science; 1992.

[52] Othman MR. Sustainability assessment and decision making in chemical process design [Dissertation]. Technical University Berlin; 2011.

[53] Li X, Zanwar A, Jayswal A, Lou HH, Huang Y. Incorporating exergy analysis and inherent safety analysis for sustainability assessment of biofuels. Ind Eng Chem Res 2011;50:2981—93.

[54] Nara LA. The need for process safety in the biofuels industry. In: Proceeding AIChE 2011 Spring meeting & 7th global Congress on process safety, Chicago, Illinois, March 13—16, 2011.

CHAPTER SEVEN

Managing Conflicts Among Decision Makers in Multiobjective Design and Operations*

V.M. Zavala

Argonne National Laboratory, Argonne, IL, United States

INTRODUCTION

Almost any decision-making activity involves multiple decision makers (stakeholders). For instance, in the design of a process system, stakeholders must trade off myriad economic, environmental, and safety metrics (objectives) [1]. The stakeholders likely will disagree on which metrics should be used and on how they should be prioritized. If disagreements are not systematically managed, they can leave a subset of stakeholders strongly dissatisfied, a situation that can ultimately delay consensus reaching and lead to arbitrary decisions.

The most popular approach for dealing with multiple objectives in a decision-making process is to compute a Pareto front and then use *expert* knowledge to make a final decision by choosing a suitable point along the front. This approach has two important disadvantages: (1) it is ambiguous in that it assumes that a single decision maker (expert) makes the final decision (which is often not the case), and (2) the complexity of forming the Pareto front is exponential in the number of objectives, and choosing a particular point along the front is cumbersome in multiple dimensions. Another approach commonly used in multiobjective decision making is to give equal priority to all objectives. This approach is equivalent to picking a specific point along the Pareto front and therefore is also ambiguous. Moreover, the approach is unreliable because, depending on the shape of the Pareto front, a slight modification of the weights can yield drastically different solutions [2]. In other words, this approach does not capture the shape of the Pareto front and thus might neglect solutions that yield high returns for one objective with few sacrifices for others. Another popular approach in multiobjective decision making is to prioritize objectives, as proposed in [3]. This approach, however, also assumes that a single decision maker is involved in creating the priority hierarchy, and consequently it is ambiguous.

---

*Argonne National Laboratory, operated by UChicago Argonne, LLC, under Contract No. DE-AC02-06CH11357 with the US Department of Energy.

Ambiguity can be mitigated by considering the opinion of multiple stakeholders when choosing a suitable Pareto solution (compromise solution). An interesting multistakeholder approach was presented in [4]. Here the authors assume that stakeholders are polled to provide *priority rules* to be followed. From these rules, a unique set of weights that satisfy such rules is computed. A disadvantage of this approach is that it can yield situations in which no unique feasible weights can be obtained that satisfy all the stakeholders' rules. In addition, this approach does not provide insights into the level of dissatisfaction of the stakeholders with a given compromise decision.

In this work we present an optimization framework that systematically quantifies and mitigates dissatisfactions among stakeholders. The idea consists of factoring the opinion of the multiple stakeholders in the form of weights (instead of rules). Consequently, compared with the approach presented in [4], the proposed framework provides more flexibility. The framework is an extension of the robust optimization approach proposed in [5] in which a compromise decision is obtained by minimizing the maximum dissatisfaction among stakeholders. A key advantage of the robust approach is that it provides a measured stakeholder disagreement. In addition, it does not require the computation of a Pareto front and can thus be used to address problems with many objectives and stakeholders. We generalize this approach by considering average and conditional-value-at-risk (CVaR) metrics. This generalization enables us to shape the distributions of the stakeholder dissatisfactions and capture the statistics of the stakeholder population more effectively. We argue that this feature is advantageous in certain applications. In addition, generalizing the robust approach using CVaR and average metrics enables us to provide utopia-tracking interpretations of the different metrics in a common setting. The proposed approach provides a systematic procedure to inform decision makers about the influence of their opinions on the final decision and can help decision makers *reassess* their priorities and thus resolve and quantify the *cost of conflict*. Examples are presented to illustrate the concepts.

APPROACH

Consider a set of objectives functions $\mathcal{O} := \{1...O\}$ and the corresponding objective function vector $\mathbf{f}(x)^T = [f_1(x), f_2(x)..., f_O(x)]^T$. Consider also a set of stakeholders $\mathcal{S} := \{1..S\}$ and the case in which each stakeholder $s \in \mathcal{S}$ prioritizes the objectives according to a weight vector $w_s \in \Re^O$. We define the elements of the weight vectors as $w_{s,i}$, $i \in \mathcal{O}$, and we assume that the weight vectors satisfy $\sum_{i \in \mathcal{O}} w_{s,i} = 1$, $i \in \mathcal{O}$.

A key observation that we make is that, if the stakeholder population is finite, we can interpret the weight vectors as samples from a probability distribution with finite support. In other words, the weight vectors can be interpreted as samples from the population of stakeholders. Each stakeholder $s \in \mathcal{S}$ seeks to solve its individual weighted optimization problem

$$\min_{x} \mathbf{w}_s^T \mathbf{f}(x) = \sum_{i \in \mathcal{O}} w_{s,i} f_i(x) \qquad [7.1a]$$

$$\text{s.t. } g(x) \leq 0. \qquad [7.1b]$$

Here the constraint vector $g(x)$ includes operational constraints and/or system models. The solution of problem [7.1] will yield an optimal solution x_s^* and a weighted cost for stakeholder s that we denote as $\mathbf{w}_s^T \mathbf{f}_s^* := \mathbf{w}_s^T \mathbf{f}(x_s^*)$. This weighted cost is ideal, or utopian, in the sense that it assumes that stakeholder s does not have to compromise with the rest of the stakeholders. When compromise is needed, as is often the case, we define the dissatisfaction of stakeholder s at an arbitrary compromise decision x as $d_s(x) := \mathbf{w}_s^T (\mathbf{f}(x) - \mathbf{f}_s^*)$. Note that from optimality of x_s^* and associated weighted cost $\mathbf{w}_s^T \mathbf{f}_s^*$, we have that $d_s(x) \geq 0$ for all x and for all $s \in \mathcal{S}$.

Consider now that two arbitrary decisions $\bar{x}, x$ yield $d_s(\bar{x}) < d_s(x)$. Thus stakeholder s will be more satisfied under decision $\bar{x}$ than under decision x. Because of disagreement, however, another stakeholder s' might be less satisfied under decision $\bar{x}$ than under decision x (i.e., $d_{s'}(\bar{x}) > d_{s'}(x)$). We thus have that, given a compromise decision x, we can measure the disagreement among stakeholders by using a measure of the dissatisfactions $d_s(x)$, $s \in \mathcal{S}$. Note that the ideal case with no disagreement at decision x occurs only when $d_s(x) = 0$ for all $s \in \mathcal{S}$. In the presence of disagreements among stakeholders, however, this situation cannot occur.

Our objective is thus to find a compromise decision x that minimizes a measure of the dissatisfactions $d_s(x)$, $s \in \mathcal{S}$. We can think of this problem as one of shaping the distribution of the dissatisfactions. For convenience, we define the vector of dissatisfactions $\mathbf{d}(x)^T := [d_1(x), d_2(x), ..., d_S(x)]^T$.

The most straightforward alternative for managing disagreements consists of minimizing the average dissatisfaction among the stakeholders. This is done by solving the problem

$$\min_{x} \frac{1}{|\mathcal{S}|} \sum_{s \in \mathcal{S}} \mathbf{w}_s^T (\mathbf{f}(x) - \mathbf{f}_s^*) \qquad [7.2a]$$

$$\text{s.t. } g(x) \leq 0. \qquad [7.2b]$$

Because $d_s(x) \geq 0$ for all $s \in \mathcal{S}$ and x, we have that problem [7.2] is also equivalent to

$$\min_x \frac{1}{|\mathcal{S}|} \|\mathbf{d}(x)\|_1 = \frac{1}{|\mathcal{S}|} \sum_{s \in \mathcal{S}} d_s(x) \qquad [7.3a]$$

$$\text{s.t. } g(x) \leq 0. \qquad [7.3b]$$

In other words, the solution of problem [7.2] can be interpreted as a compromise solution relative to a utopia point given by the collection of the ideal stakeholder weighted costs $\mathbf{w}_s^T \mathbf{f}_s^*$. This definition of utopia point is not to be confused with the traditional definition used in multiobjective optimization [6].

Another way to address disagreement consists of minimizing the worst (largest) dissatisfaction among the stakeholders. This is done by solving the robust optimization problem

$$\min_x \max_{s \in \mathcal{S}} \left\{ \mathbf{w}_s^T \left(\mathbf{f}(x) - \mathbf{f}_s^* \right) \right\} \qquad [7.4a]$$

$$\text{s.t. } g(x) \leq 0. \qquad [7.4b]$$

This formulation was proposed in [5]. It is well known that the min−max problem [7.4] can be reformulated as

$$\min_x \quad \eta \qquad [7.5a]$$

$$\text{s.t. } \mathbf{w}_s^T \left(\mathbf{f}(x) - \mathbf{f}_s^* \right) \leq \eta, s \in \mathcal{S} \qquad [7.5b]$$

$$g(x) \leq 0. \qquad [7.5c]$$

The optimal value of η is the worst dissatisfaction. Because $d_s(x) \geq 0$, a solution x of problem [7.4] also solves the problem

$$\min_x \frac{1}{|\mathcal{S}|} \|\mathbf{d}(x)\|_\infty = \frac{1}{|\mathcal{S}|} \max_{s \in \mathcal{S}} \left\{ d_s(x) \right\} \qquad [7.6a]$$

$$\text{s.t. } g(x) \leq 0. \qquad [7.6b]$$

Because we can assume that the stakeholders' polls are obtained from a finite distribution, we can measure the disagreement by using a risk metric such as CVaR. To this end we solve the following problem:

$$\min_x \text{CVaR}_\alpha \left[\mathbf{w}_s^T \left(\mathbf{f}(x) - \mathbf{f}_s^* \right) \right] \qquad [7.7a]$$

$$\text{s.t. } g(x) \leq 0. \qquad [7.7b]$$

Here $\alpha \in [0, 1]$ is the probability level. This problem can be reformulated as [7]

$$\min_{x,v,\phi_s} \frac{1}{|\mathcal{S}|} \sum_{s \in \mathcal{S}} \left(\frac{1}{1-\alpha} \phi_s + v \right) \qquad [7.8a]$$

$$\text{s.t. } \mathbf{w}_s^T \left(\mathbf{f}(x) - \mathbf{f}_s^* \right) - v \leq \phi_s, s \in \mathcal{S} \qquad [7.8b]$$

$$\phi_s \geq 0, \ s \in \mathcal{S} \qquad [7.8c]$$

$$g(x) \leq 0. \qquad [7.8d]$$

This approach penalizes the large dissatisfactions in the $(1 - \alpha)$ tail of the distribution. One can show that the CVaR solution converges to the robust solution as $\alpha \to 1$ and to the average solution as $\alpha \to 0$ [8]. Consequently, the CVaR solution covers the spectrum of solutions between the average and robust solutions.

ILLUSTRATIVE EXAMPLES

In this section we present examples to demonstrate the applicability of the presented concepts.

Generation Expansion

Consider a decision-making setting in which a community (stakeholders) needs to decide among three technologies (denoted as I, II, and III) for power generation. In doing so, the community must satisfy a given demand while trading off three objectives: minimize electricity cost (denoted as C), minimize carbon emissions (denoted as E), and minimize land use (denoted as L). Table 7.1 lists the coefficients for cost, emissions, and land use for the three technologies.

The coefficients are adimensional and are used only to represent relative magnitudes of different technologies. Technology I has high emissions, low cost, and high land use (relative to the others). Technology II has low emissions, high cost, and medium land use. Technology III has medium emissions, medium cost, and low

Table 7.1 Emissions, Cost, and Land Use for Each Technology

Technology	E	C	L
I	100	10	100
II	50	50	50
III	75	50	25

land use. The weighted multiobjective optimization problem can be formulated as follows:

$$\min w_C C + w_E E + w_L L \tag{7.9a}$$

$$\text{s.t. } C = \gamma_I C_I + \gamma_{II} C_{II} + \gamma_{III} C_{III} \tag{7.9b}$$

$$E = \gamma_I E_I + \gamma_{II} E_{II} + \gamma_{III} E_{III} \tag{7.9c}$$

$$L = \gamma_I L_I + \gamma_{II} L_{II} + \gamma_{III} L_{III} \tag{7.9d}$$

$$D = \gamma_I P_I + \gamma_{II} P_{II} + \gamma_{III} P_{III} \tag{7.9e}$$

$$\gamma_I, \gamma_{II}, \gamma_{III} \in \{0, 1\}. \tag{7.9f}$$

Here, γ_I, γ_{II}, and γ_{III} denote the decisions to install technology I, II, or III, respectively. Symbol D denotes the electricity demand, and P_I, P_{II}, and P_{III} denote the power supplied by each technology. For simplicity we assume that $P_I = P_{II} = P_{III} = 10$, and we set $D = 10$. Note that the demand constraint [7.9e] implies that only one technology must be installed. All the objectives (C, E, L) are normalized by their best and worst possible values (these can be obtained from Table 7.1) so as to lie in the range [0,1].

In Table 7.2 we present the average and worst-case solutions under four different polls from 100 stakeholders. We assume that the polls are designed in such a way that the stakeholders express four different opinions: (1) their only priority is emissions, (2) their only priority is cost, (3) their only priority is land use, and (4) all three objectives are equally important. In a first poll we have {50%, 50%, 0%, 0%}, in a second poll we have

Table 7.2 Compromise Solutions Under Different Polls

Poll	Strategy	Compromise Solution
{50%,50%,0%,0%}	Average	γ_I; γ_{II}
	Robust	γ_I; γ_{II}; γ_{III}
{49%,51%,0%,0%}	Average	γ_I
	Robust	γ_I; γ_{II}; γ_{III}
{25%,25%,25%,25%}	Average	γ_{II}
	Robust	γ_I; γ_{II}; γ_{III}
{0%,0%,0%,100%}	Average	γ_{II}
	Robust	γ_{II}

{49%, 51%, 0%, 0%}, in a third poll we have {25%, 25%, 25%, 25%}, and in a fourth poll we have {0%, 0%, 0%, 100%}. The first poll indicates that 50% of stakeholders give full priority to minimize emissions and 50% give full priority to minimize cost. In the second poll the number of stakeholders giving full priority to minimize cost dominates by 1% the number of stakeholders giving full priority to minimize emissions. In the third poll 25% of the stakeholders give full priority to emissions, 25% give full priority to cost, 25% give full priority to land use, and 25% give equal priority to all objectives. The fourth poll corresponds to the special case in which all stakeholders give equal priority to minimize all objectives. In other words, in the fourth poll we have perfect agreement among stakeholders.

From the first three polls we can see that the robust strategy achieves the same worst-case dissatisfaction for all technologies. In other words, the three technologies are optimal in the sense that they minimize the worst-case dissatisfaction. With perfect agreement (fourth poll), on the other hand, technology II is optimal, and the worst-case dissatisfaction is zero. From the first poll we can see that the average strategy predicts that technologies I and II are equally optimal. This result is expected because the number of stakeholders giving priority to emissions and cost is the same so the solutions are indistinguishable. From the second poll we see that technology II is optimal because the number of stakeholders giving priority to cost is larger (by one vote) than the number giving priority to emissions. From the third poll we have the less obvious result that technology II is optimal.

By comparing the results for the robust and average strategies we can obtain some interesting insights. First, we note that in the fourth poll with perfect agreement the robust and the average compromise solutions are equal, as expected. From the rest of the polls, however, the robust solution can be insensitive to the statistics of the polls. While this feature might seem desirable at first sight, it is not likely to be accepted by stakeholders because it implies that their opinions do not influence the solution (even if there is a majority of stakeholders). In fact, in this example, even a poll with a distribution of {1%, 99%, 0%, 0%} will give the same compromise solution under the robust approach. The average approach, on the other hand, does account for the statistics of the stakeholder polls; but it cannot guarantee minimization of the worst dissatisfaction, as the robust approach does.

Energy–Comfort Management in Buildings

One of the objectives of an energy management system is to minimize energy subject to thermal comfort constraints of a population of occupants (stakeholders) [2,9]. Thermal comfort is difficult to enforce because perceptions vary significantly from individual to individual as a result of variations in factors such as metabolic rate (e.g., activity, gender, race, age), building location (e.g., next to air damper, next to window), and clothing level. To address this disagreement we can poll the occupants about their temperature preferences. Consider the following stakeholder problem in which we seek to minimize energy demand while satisfying the stakeholder j temperature constraint:

$$\min E(T) \tag{7.10a}$$

$$\text{s.t. } T \geq T_j, (\lambda_j). \tag{7.10b}$$

Here $E(\cdot)$ is the building energy that is a function of occupant's j temperature requirement T_j. Clearly, $-\lambda_j = \frac{\partial E}{\partial T_j}$ and $-\lambda_j > 0$ if the objective and the comfort constraint are in conflict. Consequently, λ_j can be interpreted as a comfort price. We can thus formulate the weighted multiobjective problem

$$\min w_{j,1} E(T) + w_{j,2} T, \tag{7.11}$$

with $w_{j,1} := \frac{1}{1+\lambda_j}$ and $w_{j,2} := \frac{\lambda_j}{1+\lambda_j}$. For simplicity we assume that energy is a quadratic function of the difference between the building temperature and the ambient temperature,

$$E(T) := (T - T_{\text{amb}})^2 \tag{7.12}$$

and we set the ambient temperature to 35°C.

We consider a poll of temperature preferences for $S = 1000$ occupants in a building. The average preference is 22°C, the minimum temperature preference is 15°C, and the maximum preference is 29°C. All temperatures are below ambient temperature, and we thus simulate a situation in which energy is used for cooling. Thus if comfort is not a concern, the energy required will be zero, and the building will be set at ambient temperature. This also implies that, as T_j is increased, the comfort price λ_j will decrease and will be zero at T_{amb}. We thus have that the weight $w_{j,2}$ will tend to zero and $w_{j,1}$, reflecting the fact that a larger T_j implies a lower priority on comfort. Consequently, polling temperature preferences can be interpreted as polling priorities of energy and comfort among occupants.

In Fig. 7.1 we present the corresponding comfort weights $w_T(w_{j,2})$ for the occupants. In Fig. 7.2 we present the Pareto curve of temperature against energy demand. Each point along the front is obtained by solving problem [7.10] for each stakeholder s. Note that these points represent the ideal (nonachievable) situation in which each stakeholder can reach the desired temperature preference without having to compromise with the rest of the stakeholders. In the figure we also present the solution of different compromise decisions. The vertical line indicates the solution in which the stakeholders compromise naively by averaging their temperature

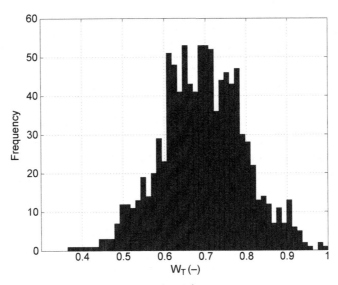

Figure 7.1 Occupants' weights on temperature.

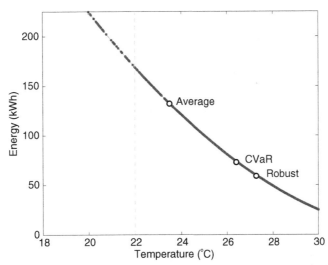

Figure 7.2 Pareto front and compromise decisions. Vertical line is naive approach. CVaR, conditional-value-at-risk.

preferences (22°C). This solution corresponds to solving the energy minimization problem

$$\min E(T) \qquad [7.13a]$$

$$\text{s.t. } T \geq \frac{1}{|\mathcal{S}|} \sum_{j \in \mathcal{S}} T_j. \qquad [7.13b]$$

Note that this naive approach does not capture energy in the stakeholders' opinions. The black dot next to the naive solution represents the solution obtained by compromising based on the minimization of the average dissatisfaction given by [7.2]. The solutions obtained by averaging preferences and minimizing the average dissatisfaction do not coincide. The reason is that averaging temperatures is not equivalent to averaging dissatisfactions: dissatisfactions factor in energy and not only temperature. The black dots at the end of the Pareto front represent the decision obtained by minimizing CVaR for $\alpha = 95\%$ given by [7.7] and the decision obtained by minimizing the worst dissatisfaction among the stakeholders given by [7.4]. Note that the compromise decisions move toward the maximum temperature preference (warmer building) as robustness is increased. Consequently, the CVaR and robust approaches yield much lower energy demands than do the naive and average approaches. From the naive approach perspective (without factoring in energy in the opinions) this result is counterintuitive because one would expect that at a higher temperature more people would be dissatisfied. From a dissatisfaction perspective, however, a colder building yields much larger dissatisfactions.

These results illustrate how a more systematic management of stakeholder opinions can yield more efficient (and nonintuitive) solutions. In Fig. 7.3 we present histograms

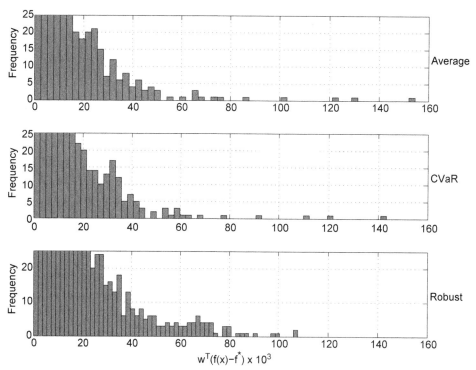

Figure 7.3 Dissatisfaction of stakeholders under different compromise decisions. CVaR, conditional-value-at-risk.

for the dissatisfactions of all the stakeholders. In the top graph we present the dissatisfactions when the stakeholders compromise by minimizing the average dissatisfaction. Note the pronounced tail of large dissatisfactions. In the middle graph we present the dissatisfactions when the stakeholders compromise by minimizing CVaR, and in the lower graph we present the dissatisfactions when the stakeholders compromise by minimizing the worst dissatisfaction. Note that the tail of the distribution of dissatisfactions is reduced by CVaR and that the robust approach reduces the tail further by penalizing the largest dissatisfaction. We can thus see that in this application the robust approach provides important benefits compared with the average approach.

Arguably, the need to choose among average, CVaR, and worst-case metrics in the proposed approach introduces some ambiguity in the decision-making process in that the stakeholders must also agree that such a metric is the appropriate one. The level of ambiguity, however, is significantly reduced compared with standard multiobjective approaches that assume a single decision maker. Moreover, the proposed approach has the additional advantage that it can manage many opinions and objective functions in a systematic manner.

CONCLUSIONS

We have presented a framework to manage conflicts among multiple decision makers. The framework enables the computation of compromise solutions in the presence of many objectives without having to form a Pareto front. The framework also provides a systematic procedure to manage conflicts by using quantifiable metrics of disagreement among stakeholders.

We highlight that the framework proposed can manage objective functions in either deterministic or stochastic settings. For instance, one can trade off mean profit and profit variance. Our framework, however, does not account for situations in which stakeholders can change their preferences based on possible scenarios, as discussed in [6]. We will extend our framework to consider this possibility in future work. It is also necessary to consider formulations under which the stakeholders provide their preferences not only in terms of weights but also in terms of goals. This will give rise to interesting goal-oriented multiobjective formulations. We are also interested in understanding under what conditions the CVaR compromise solution gives a Pareto solution. This is motivated from the fact that the average and robust metrics have utopia-tracking interpretations (under different norms). We will look for a similar definition in the CVaR case.

ACKNOWLEDGMENTS

This material was based upon work supported by the US Department of Energy, under contract no. DE AC02-06CH11357. The author thanks Sanjay Mehrotra for suggesting the building energy management example.

REFERENCES

[1] Ruiz-Mercado GJ, Smith RL, Gonzalez MA. Sustainability indicators for chemical processes, I: Taxonomy. Ind Eng Chem Res 2012;51(5):2309—28.

[2] Zavala VM. Real-time resolution of conflicting objectives in building energy management: an utopia-tracking approach. In: Proceedings of the 5th National Conference of IBPSA-USA; 2012. p. 1—6.

[3] Mendoza GA, Prabhu R. Multiple criteria decision-making approaches to assessing forest sustainability using criteria and indicators: a case study. For Ecol Manag 2000;131(1):107—26.

[4] Smith RL, Ruiz-Mercado GJ. A method for decision making using sustainability indicators. Clean Technol Environ Policy 2014;16(4):749—55.

[5] Hu J, Mehrotra S. Robust and stochastically weighted multiobjective optimization models and reformulations. Oper Res 2012;60(4):936—53.

[6] Zavala VM, Flores-Tlacuahuac A. Stability of multiobjective predictive control: a utopia-tracking approach. Automatica 2012;48(10):2627—32.

[7] Tyrrell Rockafellar R, Uryasev S. Optimization of conditional value-at-risk. J Risk 2000;2:21—42.

[8] Tyrrell Rockafellar R, Uryasev S. Conditional value-at-risk for general loss distributions. J Bank Finance 2002;26(7):1443—71.

[9] Morales-Valdez P, Flores-Tlacuahuac A, Zavala VM. Analyzing the effects of comfort relaxation on energy demand flexibility of buildings: a multiobjective optimization approach. Energy Build 2014;85(0):416—26.

CHAPTER EIGHT

Sustainable System Dynamics: A Complex Network Analysis

U. Diwekar

Vishwamitra Research Institute, Crystal Lake, IL, United States

INTRODUCTION

The concept of sustainability applies to integrated systems comprising humans and nature. The structures and operation of the human component (in terms of society, economy, government, etc.) must be such that these reinforce or promote the persistence of the structures and operation of the natural component (in terms of ecosystem trophic linkages, biodiversity, biogeochemical cycles, etc.), and vice versa [1]. In order to study sustainability, we have to look at an integrated system where sustainability is linked to ecology, society, economy, technology, and time. In summary, as Cabezas [2] defines sustainability it is a path through multiple dimensions as shown in Fig. 8.1. A system cannot be sustainable with a major subsystem (economy, ecology, law, etc.) operating without regard to the rest of the system any more than a person can live with a malfunctioning major organ.

The idea of sustainability is to make sure that the system trajectory is following a sustainable path for a long time. In order to study the sustainability of our planet, some rudimentary ecosystem model with links to ecology, society, economy, technology, and time is necessary. Sustainability of the system then can be tested using various projected future scenarios like human population explosion and increase of per capita consumption. We describe three such models with increasing complexity in this work. These models consist of compartments of different living species coupled with economic compartments like industrial sector (IS) and energy sector. All three models show the system to be sustainable if we remain at current conditions. However, the price of economic developments like per capita consumption increase is the extinction of various compartments suggesting loss of sustainability. In order to make this system sustainable for a relatively long time, we need to derive technological and socio-economic time-dependent policies. Optimal control theory with its forecasting mechanism can help in deriving these time-dependent policies. However, for optimal control theory to be effective it is important to see whether the system is controllable and how many decision nodes or variables (policy variables) are necessary for controlling such a system. Further,

Sustainability in the Design, Synthesis and Analysis of Chemical Engineering Processes

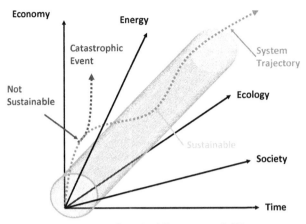

Figure 8.1 Sustainability as a path [3].

it is important to define a single objective function that describes the sustainability of such complex system. Cabezas and coworkers [1,3] have proposed use of information theory for this purpose. This is because Fisher information (FI) collapses the many state variables (defining compartments or nodes) into one index that can measure the system order or organizational change. Therefore we use FI and optimal control theory to derive technological and socio-economic policies for these three models.

The chapter is arranged in this order. The first section presents the three models representing the functioning of our ecosystem (in an increasing order of complexity). We present a state-of-the-art approach in finding controllability of these three systems based on a complex systems approach in the second section. The controllability defined in terms of number of nodes or decision variables is then used to select important decision variables for the three systems respectively in the third section. The derivation of technological and socio-economic policies based on optimal control theory is presented in the same section. A summary of the chapter is presented at the end.

SUSTAINABLE SYSTEM DYNAMIC MODELS

Mathematical models featuring the critical components of a real ecosystem can aid in the formal study of sustainability. For instance, ecologists have used mathematical models to determine the stability of a community based on the size and connectivity of the food web in the face of fluctuations of the system (i.e., invasion of new species, population dynamics, etc.) [4]. Lotka—Volterra models used to describe the predator—prey interactions [5] represent one classical theory used to describe the dynamics of these biological systems. These equations represent interactions between the predator and prey by first order, nonlinear, differential equations. In addition, there have been ecological models that have incorporated varying degrees of socio-economic features. Acutt and

Mason [6] and Heal [7] proposed valuing the various environmental services in an ecosystem in their decision-making models. Some of the other works that included largely simplified economic features in their ecological models are the ones developed by [8–11]. Models accounting for processes such as resource extraction, waste assimilation, recycling, and pollution in an integrated ecological economic framework have also been proposed in the literature [12]. A comprehensive review of some of these models can be obtained from [13]. The model proposed by Whitmore and coworkers [13] integrated an economy under imperfect competition with a widely used 12-cell ecological model [1]. This 12-cell ecological model developed by Cabezas et al. [1] mimics a general ecosystem with a very rudimentary social system that regulates the flows of mass according to its own criteria and is the first model that we present later. The intermediate model includes an IS and rudimentary economic system [3]. The third more comprehensive model is an integrated model including ecology, economics, social system, technology, and time [14]. The food web is modeled by Lotka–Volterra-type expressions whereas the economy is represented by a price-setting model wherein firms and human households (HH) attempt to maximize their well-being.

The basic model in a conceptual form is given next.

Food web model equation:

$$\frac{dy_i}{dt} = f(y_k, P_j) \qquad [8.1]$$

where y_i represent the population of each compartment and P_j are the parameters of the model. These equations are presented in supplementary information for the intermediate model. In the simple model there is no industrial activity and in the third integrated model there is a macroeconomic model included in the system as well as a compartment for energy producers (EP) [14].

Macroconomic model equation:

$$Ec_{l_t} = e(y_k, c_{li}, ds_{lk},) \qquad [8.2]$$

where Ec_{l_t} are the economic outputs of the model, c_{li} are the economic parameters, and ds_{lk} is the demand and supply variables. Apart from these equations, there are algebraic equations related to factors such as seasonal variations. This model can be found in [13] and [14]. The MATLAB code for the integrated model is also presented in the supplementary information.

Model 1: The Simple Model

Fig. 8.2 depicts the first ecosystem model described in [1]. As it can be seen, there are a total of 12 compartments that comprise: four plants (P1, P2, P3, and P4), two of which can represent cultivated crops such as corn and soybean and the others any species of plants; three groups of herbivore animals (H1, H2, and H3) from mammals, birds, and

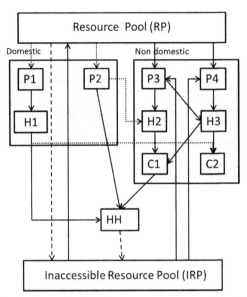

Figure 8.2 Simple model. This figure depicts the compartment model, which is comprised of: plants (P$_i$), which are the primary producers, herbivores (H$_i$), carnivores (C$_i$), human households (HH), and a resource pool and inaccessible resource pool (RP, IRP).

reptiles one of which is domesticated; two groups of carnivore animals (C1 and C2); an HH; a resource pool (RP); and inaccessible resource pool (IRP). The RP represents all biological resources needed by plants (e.g., water, nitrogen, carbon dioxide, etc.). Primary producers or plants (P1…P4) make available mass and energy from the accessible RP to the rest of the food web. The IRP corresponds to biologically unavailable resources that result from human activity (e.g., polluted water, which is biologically unusable unless it gets purified by slow biological process). The arrows represent the mass flows from one compartment (i.e., origination) to another compartment (i.e., termination), and all living compartments have an implied mass flow back to the RP that represents death. In this figure, there are three types of arrows: the solid arrows describe transfer of mass caused by biological or geological drivers (e.g., species of P1 are consumed by species of H1), the dashed arrows represent the transfer of mass from the RP to the inaccessible pool that occurs as a by-product of human activities necessary to support agriculture and product distribution (e.g., water pollution, paving, etc.), and the dotted arrows are the human-influenced mass flows. The dashed arrow connecting RP to IRP represents mass, which becomes biologically unavailable as a by-product of activities associated with P1, P2, and H1. The dashed arrow connecting HH to IRP represents mass made biologically unavailable by human consumption. The primary producers feed on RP and use solar energy to make this mass available to the rest of the system. A small amount of mass from IRP is recycled back into the system by P3 and P4, which symbolizes degradation by the actions of microorganisms. All nine biological compartments recycle mass back to RP through death. The details of the mathematical model can be found in Cabezas et al. [1].

Model 2: The Intermediate Integrated Model

The intermediate model is an extended version of the first model since it includes the IS [3]. This new sector represents—at a very elementary level—generic human industrial activity that offers a benefit to the human population. As it is shown in Fig. 8.3, this model also contains the same 12 compartments shown in Fig. 8.2 that track the flows of mass. However, IS is not considered as a compartment since no mass is resident. The IS simply takes mass from compartments P1, RP, and H3 and combines it to form a product. This plays an important role in the economic model of this system. Therefore the intermediate model consists of the ecosystem dynamic model and a very simple economic model based on a rudimentary IS. The IS is meant to represent at a very elementary level a generic human industrial activity that offers a benefit to the human population. The industrial process simply takes mass from three compartments (P1, RP, and H3) in different proportions and combines it to form a product.

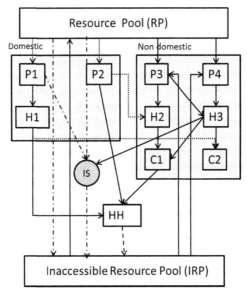

Figure 8.3 Intermediate model. This figure depicts the compartment model, which is comprised of: plants (P_i), which are the primary producers, herbivores (H_i), carnivores (C_i), human households (HH), a resource pool and inaccessible resource pool (RP, IRP), and industrial system (IS) with rudimentary economics.

Consumption of the product reduces the mortality rate of the human compartment (HH) according to mHH/(1 + αIPF). The larger the IPF (IPF is the rate of flow of mass through the industrial process [no mass is resident] flow) the lower the human mortality rate. Here, α is a constant meant to reflect the effectiveness or efficiency of the industrial process in reducing the human mortality rate. For details, please refer to [3]. In this model, the solid arrows represent transfers of mass solely because of biological or geological drivers, with no interference from humans. The dashed arrows represent a transfer of mass from the RP to the IRP that occurs as a by-product of human activities as before, including consumption (HH → IRP) and production (RP → IRP and IS → IRP). The mass flows represented by the dashed arrows are under some level of human control.

Model 3: The Integrated Ecological and Economic Model

The third model is a further enhanced version of the two previous models minus one plant species. In this case the model incorporates some crucial and representative elements of the real world social, economic, technological, and biological system. This model is based on the model presented by Whitmore and coworkers [13]; however, despite the unique features of Whitmore's model, it has a very limiting assumption as it is presumed that an infinitive amount of energy was available without any cost to the IS and various components of the ecological system. This assumption does not reflect the real world since factors related to energy not only have a cost but geopolitical ramification. They also cause enormous stress on certain components of the ecosystem. Therefore this new version of the model, although still a simplified version of any real system, considers one or more aspects related to the production and utilization of energy. The production of energy considers various types of energy sources (ES). As shown in Fig. 8.4, this modified model contains three plants (P1, P2, and P3), three herbivores (H1, H2, and H3), two carnivores (C1 and C2), an HH, an IS, ES, EP, the RP, and IRP resulting in 14 compartments. The compartments HH, IS, EP, P1, and H1 describe the economic perspective of this model. The price-setting macroeconomic model determines flows of economic goods and labor that govern the dynamic (decisions) of these five compartments. This price-setting model is based on four firms (IS, EP, P1, and H1) and the fact that HH attempts to maximize humans' well-being. For instance, the energy source represents a finite nonrenewable resource. The EP is a firm that uses labor to transform mass (ES) into a usable form of energy. This energy is produced from fossil fuels and bioenergy. This energy is supplied to HH and IS. EP is also capable of producing energy using P1, and this would represent the production of energy using biomass. For example, sugar cane is used to produce bioethanol or soybean oil is used to produce biodiesel. The IS produces products valuable to HH using P1 and RP. The use of the IS products does not increase the mass of HH, but it instead passes through and is discarded to increase the mass of IRP. Similarly the use of mass by the EP to produce energy results in waste that

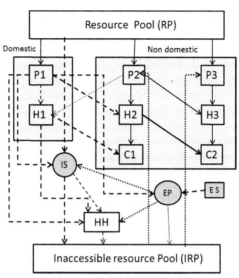

Figure 8.4 The integrated economic—ecological model. This figure depicts the compartment model, which is comprised of: plants (P$_i$), which are the primary producers, herbivores (H$_i$), carnivores (C$_i$), human households (HH), and a resource pool and inaccessible resource pool (RP, IRP). The *arrows* represent the mass flows from one compartment (origination) to another compartment (termination), and all living compartments have an implied flow back to the resource pool that represents death. IS is the industrial sector, whereas EP and ES are the energy producers and the energy source compartment, respectively [14].

increases the mass of IRP, and a corresponding decrease in the mass of ES. For details, please refer to Kotecha et al. [14].

As shown in Fig. 8.4 the dashed lines indicate mass flows that occur under anthropogenic influence. The dotted lines indicate the flow of energy from EP to HH and IS. The square dotted lines between P2, P3, and IRP indicate slow transfers of mass as a result of microbial activity. The formulation was presented in [13,14]. For this model it is assumed that 30% of the total energy demand by the integrated system is being provided by the biomass. The 30% figure is on the higher side and represents a future scenario. If a sufficient amount of biomass is not available, the maximum available biomass is used for the production of energy and the remaining energy is produced from the nonrenewable energy source. Other important parameters involved in the formulation of this model are the demands for goods and labor: the wages rate is set by the IS, and demands of various products are set by HH. Note that there is a cost to IS and EP for the generation of biologically unavailable mass that is discharged into IRP.

Table 8.1 shows a summary of the state variables (i.e., yi) involved in the mathematical model of the three socioecological—economic systems presented before. Each of these variables represents a different node of the network. For instance, $yP1$ represents the state variable for plant 1 or node P1 while $yH1$ is the node for compartment H1. As it can be

Table 8.1 Summary of State Variables for Each Model

Description	Model 1: Simple Model	Model 2: Intermediate Model	Model 3: Integrated Economic–Ecological Model
Primary producers	$\gamma P1, \gamma P2, \gamma P3, \gamma P4$	$\gamma P1, \gamma P2, \gamma P3, \gamma P4$	$\gamma P1, \gamma P2, \gamma P3$
Herbivorous animals	$\gamma H1, \gamma H2, \gamma H3$	$\gamma H1, \gamma H2, \gamma H3$	$\gamma H1, \gamma H2, \gamma H3$
Carnivorous animals	$\gamma C1, \gamma C2$	$\gamma C1, \gamma C2$	$\gamma C1, \gamma C2$
Human households	γHH	γHH	γHH
Industries	—	—	$\gamma IS, \gamma EP$
Natural resources	γRP	γRP	γRP
Inaccessible resources	γIRP	γIRP	γIRP
Deficits	—	—	$\gamma P1H1d, \gamma P1ISd, \gamma P1HHd$ $\gamma H1HH, \gamma ISHHd$
Human population	—	—	γNHH

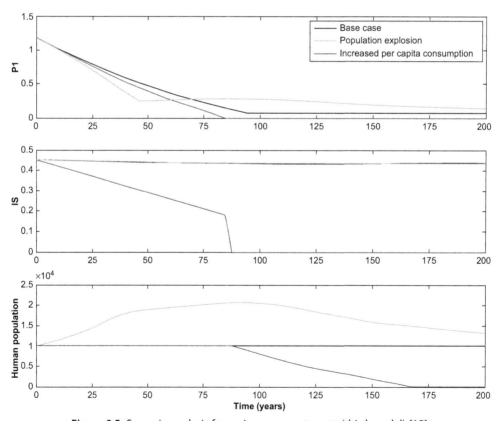

Figure 8.5 Scenario analysis for various compartments (third model) [15].

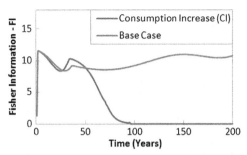

Figure 8.6 Transient Fisher information for the two scenarios: base case with *blue* (gray in print versions) *color* representing the top graph; consumption increase scenario *red* (light gray in print versions) *color* representing the bottom graph.

observed, model 3 has eight more nodes as a result of the integration of the price-setting economic model: $\gamma P1H1d$, $\gamma P1ISd$, $\gamma P1HHd$, $\gamma H1HH$, and $\gamma ISHHd$. These represent the state variables involved with the mass flows. Model-based scenario analyses are an integral part of systems theory and help in understanding the dynamics of a system under various simulated scenarios without disturbing the actual system. Scenarios have been considered as plausible, challenging, and relevant situations about the evolution of the future, which can be described in qualitative as well as quantitative terms [15]. These models have been used to study particularly two scenarios that we will be facing in the near future, namely, the population growth and economic growth resulting in an increase in per capita consumption. Fig. 8.5 presents this scenario analysis as compared to the current state, which is the base case for model 3. It can be seen that particularly for increased consumption scenario, the compartments start to vanish and the system becomes unsustainable. Cabezas et al. [1] proposed that information theory can be used to construct a basic theory of sustainability for ecological applications [16]. According to Frieden [17], FI is a measure of the state of order or organization of a system. In a dynamically sustainable regime, the FI remains constant over time or oscillating about a constant value. If the FI is increasing over time, the system is migrating toward a state of increasing self-organization and is in a potentially sustainable state. But if FI is decreasing, the system's loss of organization indicates that it is unsustainable or transitioning to a different regime away from sustainability [16]. This can be seen from Fig. 8.6. Fig. 8.6 presents FI for the base case (stable regime) and the consumption increase scenario presented in Fig. 8.5.

CONTROLLABILITY ANALYSIS

According to control theory, controllability is the ability to guide a system's behavior from any initial state to any desired final state through the appropriate manipulation of a suitable choice of inputs [18]. We are interested in finding

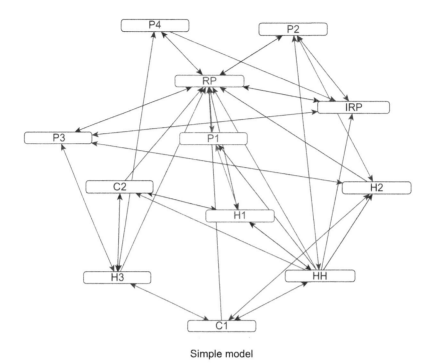

Simple model

Figure 8.7 Network representation of model 1 showing directions of the nodes: simple model [19].

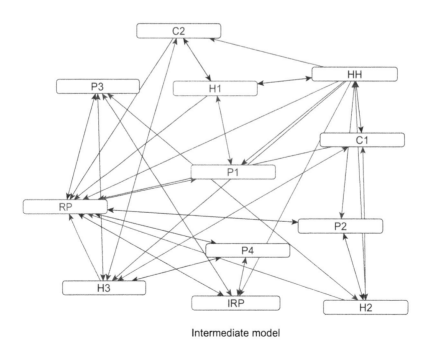

Intermediate model

Figure 8.8 Network representation of model 2: intermediate model [19].

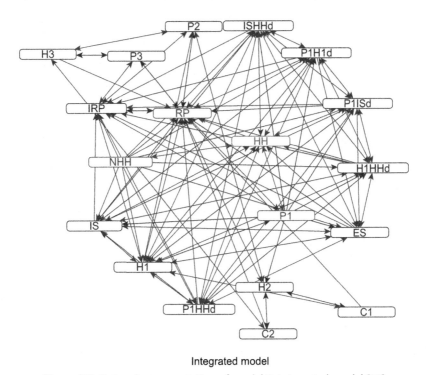

Integrated model

Figure 8.9 Network representation of model 3: integrated model [19].

technological and socio-economic policies that can make the system sustainable. Control theory provides a path toward sustainability by manipulating these parameters. However, sustainable system models are complex involving a number of state variables representing different compartments linked with differential algebraic equations. In order to apply control theory it is important to see if the system is controllable, and if it is controllable then how many variables or nodes are required for controlling this system. Controllability analysis of the system provides answers to these questions. However, in order to study controllability of a complex system we need to study system architecture and dynamic rules that capture the time-dependent interactions. Benavides et al. [19] presented controllability analysis of the three models presented earlier based on the approach presented by Liu et al. [18]. This analysis is presented briefly in this section.

When complex systems like the systems represented in Figs. 8.2–8.4 are represented as networks, the nodes are the different elements of systems and the links are the set connections that represent the interaction among these elements. Figs. 8.7–8.9 present the network representations of the simple, intermediate, and integrated model, respectively, derived using the software the Network Workbench (NWB) [20].

Some interesting features of networks that can be used for network analysis are, for instance, degree of a node and degree distributions. The degree of a node (k) is the number of edges that are connected to that node. On the other hand, the degree

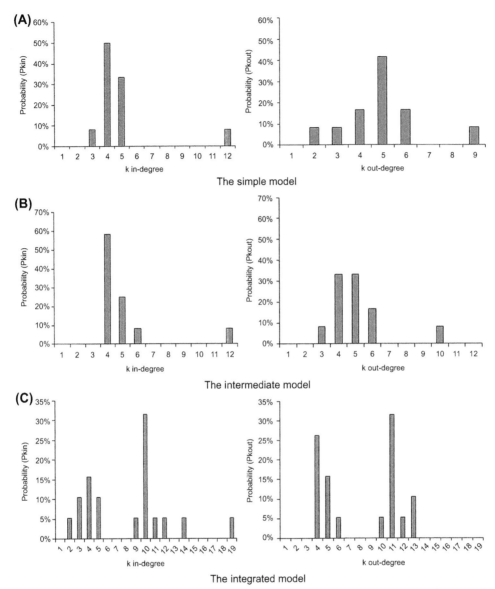

Figure 8.10 Degree distributions for the two networks: (A) simple model; (B) intermediate model; (C) integrated model [19].

distribution is a statistical property that expresses the probability $P(k)$ of nodes in the network having degree k. This property depends on the type of network, that is, whether the network is directed or undirected. In a directed network, links go in only one direction while in an undirected network links go in both directions [21]. From this network representation, it can be found that all three networks are directed networks.

Analytical tools [18] developed to study the controllability of an arbitrary complex network were developed for directed networks using the degree distribution for calculation of controllability. Fig. 8.10 shows the in–degree and out-degree distributions for the three models calculated using the network representations and dynamic equations for the network using the NTB software. From these degree distributions, it can be seen that model 3 degree distributions have more variability and hence a higher degree of heterogeneity.

From the in- and out-degree distributions the controllability defined in terms of minimum number of nodes (N_D) needed to control the system can be calculated using the cavity method. The cavity method is a powerful method employed to compute the properties of the ground state in condensed matter physics and optimization problems [22]. The cavity method is important for counting matching in a graph or network. Therefore, based on these ideas, Liu et al. [18] developed a set of self-consistent equations whose inputs are the probabilities from degree distributions and whose outcome is the average number n_D. In other words, the minimum number of driver nodes (N_D) is determined by the number of incoming and outgoing links of each node, which are described through the in-degree and out-degree distributions $P(k_{in})$ and $P(k_{out})$, respectively. The set of equations needed to calculate n_D derived by Liu et al. [18] are given in Eq. [8.3].

$$N_D = N * n_D \qquad [8.3]$$

where N is the number of nodes (i.e., state variables) involved in the network and n_D is the minimum density of unmatched nodes or equivalently the minimum density of driver nodes computed from Eq. [8.4]:

$$n_D = \frac{1}{2}\left[\left(G(\widehat{w2}) + G(1 - \widehat{w1}) - 1 \right) + \left(\widehat{G}(w2) + \widehat{G}(1 - w1) - 1 \right) \right.$$
$$\left. + \frac{\langle k \rangle}{2} \left(\widehat{w1}(1 - w2) + w1(1 - \widehat{w2}) \right) \right] \qquad [8.4]$$

where each expression of G and W are functions that are defined by:

$$G(\widehat{w2}) = \sum_{k_{out}=0}^{\infty} P(k_{out})(\widehat{w2})^{k_{out}} \qquad [8.5]$$

$$G(1 - \widehat{w1}) = \sum_{k_{out}=0}^{\infty} P(k_{out})(1 - \widehat{w1})^{k_{out}} \qquad [8.6]$$

$$\widehat{G}(w2) = \sum_{k_{\text{in}}=0}^{\infty} P(k_{\text{in}})(w2)^{k_{\text{in}}} \qquad [8.7]$$

$$\widehat{G}(1-w1) = \sum_{k_{\text{in}}=0}^{\infty} P(k_{\text{in}})(1-w1)^{k_{\text{in}}} \qquad [8.8]$$

$$\widehat{w1}(1-w2) = \sum_{k_{\text{in}}=0}^{\infty} \frac{(k_{\text{in}}+1)P(k_{\text{in}}+1)}{\langle k \rangle}(1-w2)^{k_{\text{in}}} \qquad [8.9]$$

$$w1(1-\widehat{w2}) = \sum_{k_{\text{out}}=0}^{\infty} \frac{(k_{\text{out}}+1)P(k_{\text{out}}+1)}{\langle k \rangle}(1-\widehat{w2})^{k_{\text{out}}} \qquad [8.10]$$

However, the values of $w1$, $w2$, $\widehat{w1}$, and $\widehat{w2}$ are the weights (probabilities) of cavity fields computed by:

$$w1 = \sum_{k_{\text{out}}=0}^{\infty} \frac{(k_{\text{out}}+1)P(k_{\text{out}}+1)}{\langle k \rangle}(\widehat{w2})^{k_{\text{out}}} \qquad [8.11]$$

$$w2 = 1 - \sum_{k_{\text{out}}=0}^{\infty} \frac{(k_{\text{out}}+1)P(k_{\text{out}}+1)}{\langle k \rangle}(1-\widehat{w1})^{k_{\text{out}}} \qquad [8.12]$$

$$\widehat{w1} = \sum_{k_{\text{in}}=0}^{\infty} \frac{(k_{\text{in}}+1)P(k_{\text{in}}+1)}{\langle k \rangle}(w2)^{k_{\text{in}}} \qquad [8.13]$$

$$\widehat{w2} = 1 - \sum_{k_{\text{in}}=0}^{\infty} \frac{(k_{\text{in}}+1)P(k_{\text{in}}+1)}{\langle k \rangle}(1-w1)^{k_{\text{in}}} \qquad [8.14]$$

$$0 \le n_{\text{D}} \le 1 \qquad [8.15]$$

where the expressions of $P(k_{\text{in}})$ and $P(k_{\text{out}})$ represent the in-degree and out-degree probability, respectively, taken from Fig. 8.10, depending on which model we were studying. Since these equations involve equality (Eqs. [8.4] to [8.14]) and inequalities (Eq. [8.15]), this needs to be solved using nonlinear programming (NLP) optimization [23]. In NLP the square of error between $\left(n_{\text{D,assumed}} - n_{\text{D,calculated using equation 8.4}}\right)$ is minimized to find the value of n_{D}. Remember that although N is discrete, n_{D} is a continuous variable. The results of network and controllability analysis are presented in

Table 8.2 Results of Network and Controllability Analysis

Model	Number of Nodes (N)	Number of Links (L)	Minimum Density of Driver Nodes (n_D)	Number of Nodes to Be Controlled (N_D)
Simple model	12	59	0.12	2
Intermediate model	12	61	0.12	2
Integrated economic– ecological model	19	155	0.59	12

Table 8.2. From the table it can be seen that the first two models need two nodes or variables to control the system, while the third more comprehensive model needed 12 nodes. This information is used to derive optimal control profiles for technological and socio-economic parameters of the three models. This is presented in the next section.

OPTIMAL CONTROL FOR DERIVING TECHNO-SOCIO-ECONOMIC POLICIES

Optimal control problems in optimization involve vector decision variables like the technological and socio-economic profiles to be determined for sustainability. It involves integral objective function and the underlying model is a differential algebraic system. Shastri and Diwekar [24] presented a mathematical definition of the sustainability hypothesis proposed by Cabezas and Fath [16] based on FI. They assumed a system with n species, and calculated the time average FI_t using Eq. [8.16].

$$FI_t = \frac{1}{T_c} \int_0^{T_c} \frac{a(t)^2}{v(t)^4} dt \qquad [8.16]$$

$$v(t) = \sqrt{\sum_{i=1}^{n} \left(\frac{dy_i}{dt}\right)^2} \quad (i = 1, 2 \ldots n) \qquad [8.17]$$

$$a(t) = \frac{1}{v(t)} \left[\sum_{i=1}^{n} \frac{dy_i}{dt} \frac{d^2 y_i}{dt^2} \right] \quad (i = 1, 2 \ldots n) \qquad [8.18]$$

In Eq. [8.16], T_c is the time cycle, $v(t)$ and $a(t)$ are the velocity and acceleration terms, defined by Eqs. [8.17] and [8.18], respectively. Here the terms $\frac{dy_i}{dt}$ and $\frac{d^2 y_i}{dt^2}$ represent changes found in the n state variables given in Eq. [8.1].

As seen in Eq. [8.19] the objective function minimizes the deviation of the system's FI from the FI of a sustainable base case [15,25]. The minimization function was chosen in order to follow the sustainable regime hypothesis, which states, "the system in a stable dynamic regime has an FI which is constant with time" [24]. Furthermore, the maximization of FI may result in a regime change; although the system may be stable, the alternative system's characteristics may not be favorable to human existence. For this reason, we chose to minimize FI variance for the objective function.

$$J = \text{Min}_{DV} \sum_{i=1}^{10} (\text{FI}(i) - \text{FI}_c(i))^2 \qquad [8.19]$$

Subject to: Model Eqs. [8.1], [8.2], and [8.16–8.18].

Here, J is the objective function, $\text{FI}_c(i)$ is the current FI, and $\text{FI}(i)$ is the ideal FI for a sustainable system. The differential algebraic system model for this optimal control problem involves Eqs. [8.1], [8.2], and [8.16–8.18], and the decision variables (DV) are technological and socio-economic parameters of the model.

The optimal control problem represents one of the most difficult optimization problems as it involves determination of optimal variables, which are vectors. There are three methods to solve these problems, namely, calculus of variation, which results in second-order differential equations, maximum principle, which adds adjoint variables and adjoint equations, and dynamic programming, which involves partial differential equations. For details of these methods, please refer to [23]. If we can discretize the whole system or use the model as a black box, then we can use NLP techniques. However, this results in discontinuous profiles. Since we need to manipulate the techno-socio-economic policy, we can consider the intermediate and integrated model for this purpose as it includes economics in the sustainability models. As stated earlier, when we study the increase in per capita consumption, the system becomes unsustainable. Here we present the derivation of techno-socio-economic policies using optimal control applied to the two models.

Model 2: The Intermediate Model

We use the differential algebraic system presented in Eqs. [8.1], [8.2], and [8.16–8.18] for evaluation of the objective function stated in Eq. [8.19]. As stated earlier, we have to identify two nodes that can make this system sustainable. Shastri et al. [25] evaluated various parameters of the model for this purpose. Maximum principle was used to obtain the optimal control profiles. They also evaluated single-variable optimal control profiles and found that (as indicated by the controllability analysis in last section) single-variable profiles could not make the system sustainable. The two parameter profiles, which they found made the system sustainable, were human mortality rate, α, and the coefficient of mass transfer from RP to P1, g_{RPP1}. We are assuming here that we can affect the mortality rate by some technological changes in the system. The mass transfer coefficients from RP to various plants govern the consumption patterns of natural

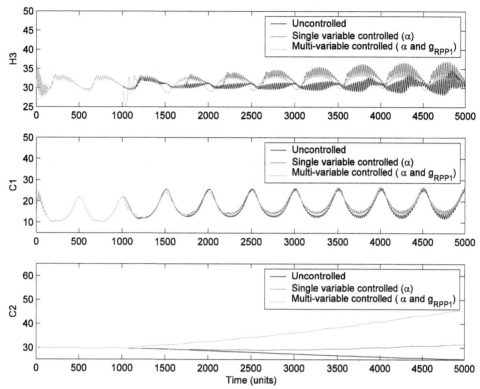

Figure 8.11 Single-variable and multivariable control for sustainability using the intermediate model. *Bottommost black line* represents uncontrolled variation, *blue* (grey in print versions) *lines* (*top line* in top two figures and *second from bottom line* for the bottom figure) represent the single-variable control, and *red* (light grey in print versions) *lines* (*second from bottom* for top two figures and *topmost* for the bottom figure) represent multivariable control [25].

resources. Since P1 and P2 are domesticated (agriculture), manipulation of g_{RPP1} reflects changes in the intensity of agricultural activities. Fig. 8.11 shows the profiles of the three compartments for the increase in consumption scenario. It can be seen that the system is unsustainable without intervention but single-variable and two-variable optimal control profiles make the system sustainable for a long run. The two optimal profiles are shown in Fig. 8.12. The changes in variables range from 0% to 13%, which is not a drastic change.

Model 3: The Integrated Ecological and Economic Model

As this is a more realistic model, sustainability studies for this model are very important. The controllability analysis for this model shows that we have to control a significant amount of nodes, to be precise 12, to make this system sustainable. For more details of

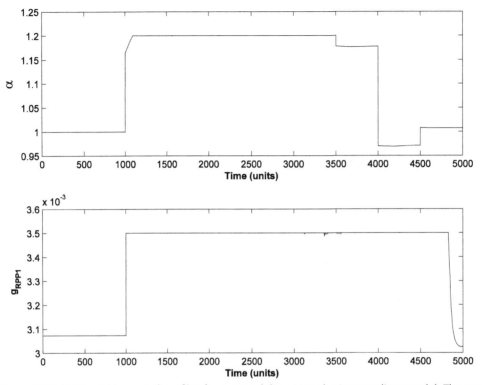

Figure 8.12 Multivariable control profiles for sustainability using the intermediate model. The two parameter profiles, which were found to make the system sustainable, were human mortality rate, α, and coefficient of mass transfer from RP to P1, g_{RPP1} [25].

this study, please refer to [26]. Table 8.3 shows the control variables chosen for our study. However, we found that some of the parameters are not sensitive to the objective function and hence we carried out the study with one to six variables. Since this model was very complex we could not use maximum principle to solve this problem. Therefore we discretized the profile in intervals of 20 years each and used black box representation of the model. This means we will change the policy every 20 years. The NLP optimization technique was then used to obtain optimal control profiles. It was found that with six variables we could make the system sustainable for 200 years, as can be seen from Fig. 8.13 compartmental profiles. However, when we looked at the optimal control profile results (shown in Fig. 8.14), it could be seen that drastic changes needed to be made to these variables to make the system sustainable for the first 200 years. This is consistent with the controllability analysis, which shows that a minimum 12 parameters are needed to control the system without drastic changes in parameter values.

Table 8.3 Candidate Policy Options for Sensitivity Analysis

Category	Name	Parameter	Mass Flow	Meaning	Selected?
Regulatory variables	Discharge fee	d_{fee}	IS → IRP	Discharge fee imposed by government to the industry sector for waste disposal	Yes
	Energy production fee	d_{EE}^{fee}	EP → IRP	Discharge fee imposed by government to the energy producers	No
Economic variables	Plant consumption	k_{P1}	P1 → HH	Parameter relating the price of P1 to the demand of P1 by human households	Yes
	Animal consumption	l_{H1}	H1 → HH	Parameter relating the price of H1 to the demand of H1 by human households	Yes
	Agricultural fee	F_{P1H1}	P1 → H1	Parameter relating the price of P1 and the demand of P1 by H1	No
	Plant material fee	S_{P1IS}	P1 → IS	Parameter relating the price of P1 to the demand of IS	No
	Industrial energy consumption	S_{EEIS}	EP → IS	Parameter relating the price of energy to the demand of IS	No
	Household energy consumption	O_{EE}	EP → HH	Parameter relating the price of energy to the demand of energy by human household	No
	Household biofuel consumption	K_{EE}	P1 → EP	Parameter relating the price of P1 to the demand of energy by human household	No
Technologic variables	Theta	θ_{P1}	P1 → IS	Amount of P1 required to produce a unit of industrial product IS	Yes
	Khat	$\hat{k}$	P2 → H1	Constant value specified by the government. Represents the grazing of P2 by H1	Yes
	Biomass to biofuel conversion rate	λ_{biom}	P1 → EP	Amount of biomass (P1) needed to produce a unit of biofuel	Yes

HH, Human household; *RP*, resource pool; *IRP*, inaccessible resource pool; *IS*, industrial sector; *EP*, energy producer.

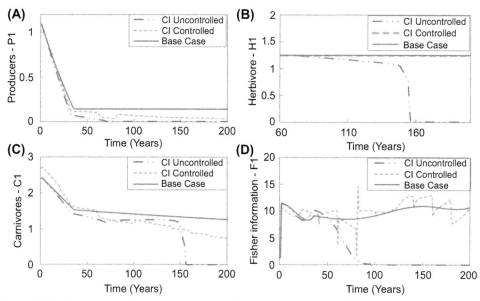

Figure 8.13 The consumption increase scenario for the integrated model with three scenarios: base case, consumption increase scenario with no control, and consumption increase scenario with control [26].

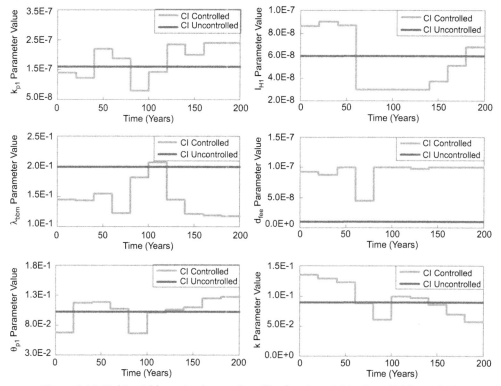

Figure 8.14 Multivariable optimal control profiles for six variables from Table 8.3 [26].

SUMMARY

The concept of sustainability is applied to relations between humans, ecosystem, economics, technology, and social systems. In order to study sustainability, integrated models, which involve these systems, are necessary. Scenario analysis using these models provides future forecasts for our planet. It has been shown that economic growth can cause loss of species in the ecosystem making the system unsustainable. In order to make the system sustainable for a long time, we need to derive time-dependent techno-socio-economic policies. Optimal control theory with its forecasting ability can help in this regard. However, in order to obtain these policies, controllability of the system needs to be studied. In this chapter the controllability analysis is presented for the three sustainability models based on complex network analysis. It has been found that in order to make these systems sustainable, multivariable control is necessary. Among the three systems, the third system presented in this chapter is closer to reality and requires a large number of nodes to make the system sustainable. If one cannot identify enough nodes to control the system, then drastic changes to the techno-socio-economic parameters are necessary.

REFERENCES

[1] Cabezas H, Pawlowski C, Mayer A, Hoaglad T. Sustainability: ecological, social, economic, technological, and systems perspectives. Clean Techn Environ Policy 2003;5:167.

[2] Cabezas H. In: Cabezas H, Diwekar U, editors. Sustainability indicators and metrics in sustainabiliy: a multi-disciplinary perspective. Bentham e-Books; 2003. p. 197.

[3] Cabezas H, Pawlowski C, Mayer A, Hoaglad T. Simulated experiments with complex sustainable systems: ecology and technology. Resour Conserv Recycl 2005;44:279—91.

[4] Proulx SR, Promislow DEL, Phillips P. Network thinking in ecology and evolution. Trends Ecol Evol 2005;20:345—53.

[5] Hoppensteadt F. Predator-prey model. Scholarpedia 2003;10:1563.

[6] Acutt M, Mason P. Environmental valuation, economic policy and sustainability: recent advances and environmental economics. Cheltenham: Edward Elgar; 1998.

[7] Heal G. Valuing the future: economic theory and sustainability. New York: Columbia University Press; 1998.

[8] Ludwig D, Carpenter S, Brock W. Optimal phosphorous loading for a potentially eutrophic lake. Ecol Appl 2003;13:1135.

[9] Brock W, Xepapadeas A. Optimal ecosystem management when species compete for limiting resources. J Environ Econ Manag 2002;33:189.

[10] Costanza R, Voinov A, Boumans R, Maxwell T, Villa F, Wainger L, et al. Integrated ecological economic modeling of the Patuxent River watershed. Md Ecol Monogr 2002;72:203.

[11] Carpenter S, Brock W, Hanson P. Ecological and social dynamics in simple models of ecosystem management. Conserv Ecol 1999;3.

[12] van den Bergh J. A multisectoral growth model with materials flows and economic-ecological interactions. In: Ecological economics and sustainable development. Cheltenham, UK: Edward Elgar; 1996. p. 147.

[13] Whitmore H, Pawlowski C, Cabezas H. Integration of an economy under imperfect competition with a twelve-cell ecological model. Technical report EPA/600/R-06/046. 2006.

[14] Kotecha P, Diwekar U, Cabezas H. Model based approach to study the impact of biofuels on sustainability of an integrated system. Clean Technol Environ Policy 2013;15:21.

[15] Shastri Y, Diwekar U, Cabezas H, Williamson J. Is sustaibanility achievable? Exploring the limits of sustainability with model systems. Environ Sci Technol 2008;42:6710.

[16] Cabezas H, Fath BD. Towards a theory of sustainable systems. Fluid Phase Equilibria 2002;194−197:3.

[17] Frieden BR. Physics from Fisher information: a unification. Cambridge, UK: Cambridge University Press; 1998.

[18] Liu YY, Slotine JJ, Barabasi A. Controllability of complex networks. Nature 2011;473:167.

[19] Benavides P, Diwekar U, Cabezas H. Controllability of complex networks for sustainable system dynamics. J Complex Netw 2015:1.

[20] NWBTeam. Network workbench tool. Indiana University, Northeastern University and University of Michigan; 2006. Retrieved from: http://nwb.slis.indiana.edu.

[21] Newman M. The structure and function of complex network. SIAM Rev 2003;45:167.

[22] Mezard M, Parisi G. The cavity method at zero temperature. J Stat Phys 2003;111:1.

[23] Diwekar U. Introduction to applied optimization. 2nd ed. Springer; 2008.

[24] Shastri Y, Diwekar U. Sustainable ecosystem management using optimal control theory: Part 1 (deterministic systems). J Theor Biol 2006;24:506.

[25] Shastri Y, Diwekar U, Cabezas H. Optimal control theory for sustainable enviromental management. Environ Sci Technol 2008;42:5322.

[26] Doshi R, Diwekar U, Benavides P, Yenkie K, Cabezas H. Maximizing sustainability of ecosystem model through socio-economic policies derived from multivariable optimal control theory. Clean Technol Environ Policy 2015;17:1573.

CHAPTER NINE

Process Synthesis by the P-Graph Framework Involving Sustainability

B. Bertok, I. Heckl
University of Pannonia, Veszprem, Hungary

INTRODUCTION

Process synthesis is the creative act of constructing the optimal structure or flowsheet of a process system. It is reported to be capable of reducing energy consumption by 50% and net present cost by 35% [1]. In chemical engineering applications it involves the determination of the functional units performing the operations within the system as well as their configurations and capacities [2,3]. Each functional unit denoted by a block in the flowsheet represents a subsystem. It comprises a set of processing equipment or unit operations. In the simplest case a single piece of equipment can form a functional unit or block in the flowsheet of the system. Process synthesis can select the best technology among the alternatives and determine the optimal flowsheet, i.e., interconnection of the functional units, as well as their volumes. Process synthesis is often utilized as a tool for evaluating the competitiveness of a new technology [4,5], estimating its payback period, or analyzing its sensitivity for uncertain parameters like raw material prices or availability [6].

A simplified schema of the activities typically preceding process synthesis in chemical process design is summarized in Fig. 9.1. First, information is to be gathered on demands for products and availability of raw materials in the market of interest. Then, best practices and emerging technologies are to be identified, which can potentially perform certain steps on the way leading to the desired products from the available raw materials. Based on this information, market potentials are to be estimated including which products could be sold and which raw materials can be utilized at what expected price and annual amount. At the second step, simulation software aids the determination of the sizes and design parameters of equipment units involved in the technologies identified as having the potential to produce the required products or intermediate products yielding the final products at the estimated annual amount. At this point, potential alternative technologies are simulated one by one, each of which can play the role of a functional unit in the process to be set up. To ensure that the alternative technologies can sufficiently substitute each other performing the same

Sustainability in the Design, Synthesis and Analysis of Chemical Engineering Processes

203

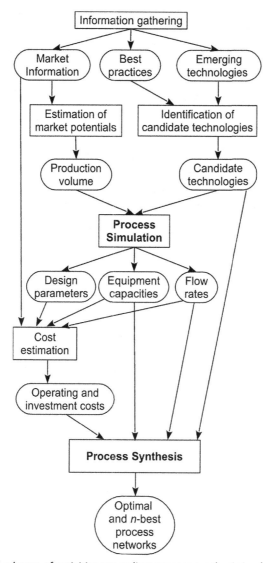

Figure 9.1 Simplified schema of activities preceding process synthesis in chemical process design.

function, the acceptable ranges for their quality parameters, for example, composition of inlet and outlet streams, are specified prior to the simulation or harmonized during the simulation. As a result, in addition to the qualities of their inputs and outputs, for each functional unit in the flowsheet to be constructed a set of alternative technologies is specified by the sizes and design parameters of the equipment units involved. Third, according to the capacities and design parameters, investment and operating costs for the equipment units and then for the alternative technologies are estimated typically by another software. Usually, energy demand greatly influences the operating costs and is a function of the material flow rates. Consequently, material flow rates calculated by simulation are also utilized during cost estimation. Prior to process synthesis, preferably the costs related to the alternative technologies are given not only for a single volume of realization but for multiple volumes or as a function of the volume. The cost function is usually given by a linear function of the mass load with fixed charge. Fitting a line to the points representing the costs of the technologies at different discrete volumes, parameters of a fixed charged linear cost function can easily be calculated by linear regression [6]. Finally, as the fourth step, process synthesis can algorithmically determine the optimal network of technologies as well as their optimal volumes by software. This step can never be done by process simulation software since simulation cannot answer the question of determining the cost optimal design between two or more qualities of materials. Moreover, without the help of a systematic process synthesis method, trial-and-error experiments have a very low chance of yielding in a near-optimal overall network of operations.

Two major classes of systematic methods for process synthesis are routed in heuristics and mathematical programming. Usually, heuristics are straightforward to implement, but they are effective only at the local level. Globally, optimal solutions are often missed by the heuristic methods [7]. Evolutionary methods belong to heuristic methods but they aim to generate a feasible solution or feasible solutions early and try to improve them with a well-defined set of instructions [8]. Conventional algorithmic methods based on mathematical programming can find the optimal solution for a mathematical programming model; however, they assume the existence of the mathematical model, and also that the mathematical programming model properly represents each of the design alternatives for the engineering system to be synthesized [9]. Mathematical programming methods are difficult to solve computationally if the problem is large and/or detailed. It is also difficult to formulate some features (e.g., sustainability issues) into a mathematical model, thus multiobjective optimization offers a set of solutions [10]. In practice the problem definition for process synthesis specifies the available raw materials, candidate operating units, and desired products as well as their major parameters including prices, costs, limitations on their volumes, and flow rates. In order to safely utilize the power of mathematical programming a model generation procedure is required, which is mathematically rigorous, preferably axiomatic, and algorithmically implementable efficiently on computers. This has been achieved by resorting to graph theory, which can be regarded as a branch of combinatorics, thus giving rise to graph theoretic, algorithmic methods.

The graph theoretic method presented herein introduces a unique class of graphs, by which the structures of process networks can be unambiguously represented formally as well as graphically, and their general combinatorial properties can be stated formally in the form of axioms. Each of these axioms is inherent in feasible processes and thus gives rise to three algorithms [11−14] applicable to a wide range of process synthesis problems. First, algorithm MSG (maximal structure generation) reduces the initial structure to the maximal number of the building blocks, which can directly or indirectly contribute to the production of at least one of the desired products. Mostly it helps checking whether the initial set of operating units is entered correctly as the input for process synthesis, or a large number of operating units cannot contribute to the production of any product, regardless of whether it was included in the problem definition as a candidate to be incorporated in the process under development. Algorithm SSG (solution structure generation) systematically enumerates each combinatorially feasible alternative process structure or flowsheet. It often highlights if some of the functional units considered to be alternatives in the flowsheet are accidentally interconnected in a way that they rely on each other, and thus cannot be incorporated one without the other. Finally, algorithm ABB (accelerated branch and bound) provides the optimal process structure. Algorithm ABB has a major advantage compared to general-purpose optimizers. It provides not only the globally optimal, but also the n-best suboptimal, structures or flowsheets. The number n is given by the user prior to executing algorithm ABB. A structure is defined to be suboptimal if it does not involve a better substructure.

The method, originally developed by Friedler and his collaborators [11−15], has been repeatedly shown to be extraordinarily efficient in process network synthesis (PNS) since its inception [16−19]. The method has been increasingly gaining utility in other areas. Some of these areas include representation of process structures in developing decision support systems for process operations [20,21,25]; the identification of catalytic or metabolic pathways [22,24,26]; and environmentally friendly system design [23,27]. Specifically, the efficacy of the proposed holistic approach based on the P-graph (process graph) is demonstrated in this chapter in the retrofitting of the downstream-processing system for biochemical production of butanol, ethanol, and acetone by incorporating one of the unconventional separating units, namely, adsorption, into the upper segment of the system.

ILLUSTRATIVE EXAMPLE

For illustration purposes, during this chapter the best flowsheets are determined for producing butanol, ethanol, and acetone for grains by fermentation and separation of the fermentation broth. The process has been gaining substantial interest. Related contributions have been summarized by Liu et al. and Fan et al. [3,4].

The initial structure represented in Fig. 9.2 has two functional parts. One removes the water content and the other separates the products, i.e., butanol, ethanol, and acetone. Candidate technologies for the first part involve conventional operating units including distillation and azeotropic distillation. Later in the elaboration of the illustrative

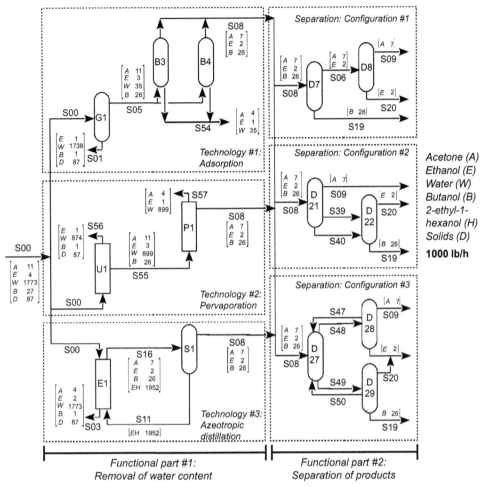

Figure 9.2 Initial flowsheet involving alternative technologies and design configurations for both functional parts: removal of water content and production of butanol, ethanol, and acetone from fermentation broth of grains.

example, competitiveness of emerging technologies involving adsorption and pervaporation are evaluated. For the second part, alternative configurations of distillation columns are taken into account. The inlet stream to each of the technologies in the first part is the fermentation broth, while the products of any configurations in the second part are almost pure acetone, ethanol, and butanol. The interface between the two functional parts is a stream transporting a mixture of approximately 7000 lb/h acetone, 2000 lb/h ethanol, and 26,000 lb/h butanol. Since the inlet and outlet streams are synchronized among the alternative technologies each of them can sufficiently substitute the other, i.e., any of them is equally applicable in the final process.

Unlike conventional approaches, the proposed procedure totally resynthesizes the entire process by incorporating the operating units with enhanced performances. As such it can take into account all possible outcomes, including the inevitable restructuring of the flowsheet's network structure. Design parameters for each of the technologies have been identified by the simulation software Aspen Plus, then the cost parameters have been estimated by Aspen Process Evaluator Icarus.

BASICS OF THE P-GRAPH FRAMEWORK

Functional units in a process network performing various operations, such as mixing, reacting, and separating, are termed operating units. These operating units, corresponding to the blocks in the flowsheet of the process, transform the physical and/or chemical states of materials being processed or transported. Such transformation is carried out by one or more pieces of processing equipment, for example, mixers, reactors and separators, in the operating units. Overall the chemical process converts raw materials into the desired products; meanwhile, some by-products generated can be usefully recovered or regarded as waste to be treated.

In PNS a material is uniquely defined by its components and their concentrations, i.e., by its composition, which is, for simplicity, denoted by a symbol and is considered as the identifier of the material. Two classes of materials or material streams are associated with any operating unit. The first is the input materials and the second is the output materials. Operating units are defined when their input and output materials are specified. Moreover, in general, the output materials from one operating unit can be the input materials to all other operating units.

At the outset of the flowsheeting process, we know what final products are to be manufactured from what raw materials. Then the first step of PNS is to identify all plausible operating units and concomitant intermediate materials, which can participate in implementing the transformation of the raw materials to the final products.

Structural Representation: P-Graph

A P-graph [11–14] is a directed bipartite graph comprising two types of vertices or nodes. As depicted in Fig. 9.3, one type with circles as their symbols is of the M-type and represents materials, and the other type with horizontal bars as their symbols is of the O-type representing operating units. An arc, with an arrow indicating the direction of flow of a material stream, is either from a vertex signifying a material to that signifying an operating unit or vice versa. An operating unit in a P-graph can represent a single equipment unit with related flows or a group of interconnected equipment units; operating unit G3 in Fig. 9.3 represents a single gas stripping unit, while the group of adsorption units B3 and B4 are represented by a single operating unit B3_B4 in Fig. 9.3.

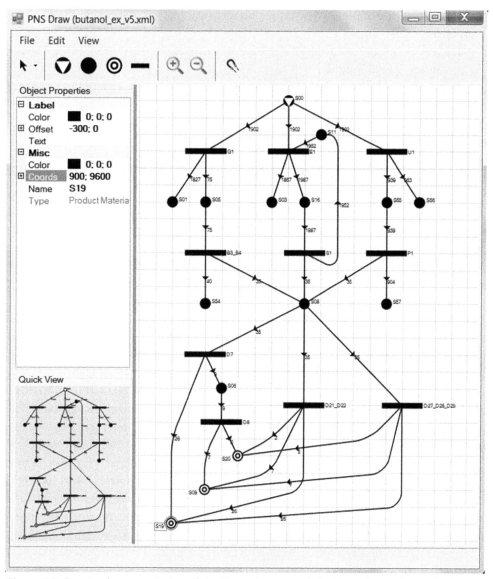

Figure 9.3 Structural representation of the illustrative example by P-graph in software PNS Draw.

Structurally Feasible Process Networks

A P-graph is defined to be structurally feasible or to be a solution structure of synthesis problem if it satisfies the following axioms [11—14]:

(S1) Every final product is represented in the graph.

(S2) A vertex of the M-type has no input if and only if it represents a raw material.

(S3) Every vertex of the O-type represents an operating unit defined in the synthesis problem.

(S4) Every vertex of the O-type has at least one path leading to a vertex of the M-type representing a final product.

(S5) If a vertex of the M-type belongs to the graph, it must be an input to or output from at least one vertex of the O-type in the graph.

Axiom (S1) implies that each product is produced by at least one of the operating units of the system; axiom (S2) implies that a material is not produced by any operating unit of the system if and only if this material is a raw material; axiom (S3) implies that only the plausible operating units of the problem are taken into account in the synthesis; axiom (S4) implies that any operating unit of the system has a series of connections eventually leading to the operating unit generating at least one of the products; and axiom (S5) implies that each material appearing in the system is an input to or an output from at least one operating unit of the system.

These axioms express necessary structural properties of process flowsheets to be feasible and serve two purposes. First, they help to analyze the initial structure whether it contains operating units or material flows, which can never contribute to the production of a product. Later on, axioms are utilized to take into account each potential alternative structure with minimal computational effort.

Algorithms MSG, SSG, and ABB

A polynomial algorithm based on the five axioms, algorithm MSG [11,12] yields a mathematically rigorous but the simplest superstructure, i.e., the maximal structure. The maximal structure of synthesis problem comprises all the combinatorially feasible structures capable of yielding the specified products from the specified raw materials. Certainly the optimal network or structure is among these feasible structures. These flowsheets range from the simplest to the most complex, or complete, which is represented by the maximal structure itself. Obviously, the optimal structure in terms of a specific objective function, often the cost, is contained in the maximal structure; nevertheless, the simplest is not necessarily the optimal.

Algorithm SSG [13] renders it possible to generate all the solution structures, i.e., it gives rise to the computational procedure for generating the solution structures. In other words, algorithm SSG unveils every feasible flowsheet of the process of interest. Algorithm SSG generates all the solution structures representing the combinatorially feasible flowsheets from the maximal structure. Moreover, these flowsheets are deemed feasible if they can be optimized in terms of the profit or any other appropriate objective function. Consequently, they can be ranked according to the magnitude of the objective function.

When the number of solution structures generated by algorithm SSG is exceedingly large, however, their optimization can be indeed time-consuming. In practice, only a finite number of the optimal and near optimal solution structures should be of interest to the designer, and thus they should be ranked, thereby eliminating the need to generate all other solution structures. It is now possible to do so with the advent of algorithm ABB [14]. As its name implies (accelerated branch and bound), this algorithm resorts to the notion of branch and bound even though it differs substantially from the conventional branch-and-bound methods.

SOFTWARE: PNS DRAW AND PNS STUDIO

The software implementations of algorithms MSG, SSG, and ABB are freely available on the website www.p-graph.com with additional software for computer-aided process synthesis. They require at least MS Windows XP, 4 GB RAM, and 1 GHz or faster ×86- or ×64-bit processor. PNS Draw is a software to define initial structures for process synthesis graphically. First, the components of the structure have to be placed to the white field by the help of a mouse, touchpad, or touch screen using drag-and-drop. Then, the components can be interconnected according to the direction of the represented material flows. PNS Draw automatically ensures that an operating unit can only be connected to a material and vice versa, i.e., it does not allow to connect operating units to operating units or materials to materials. Finally, parameters and properties of the operating units (e.g., name, fixed cost, and proportional cost) and materials (e.g., name, type, and required flow) can be given. The flow rates between materials and operating units can be set on the edges. Note that all these parameters are optional; one can create the structure if the process is of interest only to analyze its structural properties. Structures created in PNS Draw can be exported either to.pns format or to.png and .svg image formats. The former can be fed to PNS Studio for further modeling or calculations while the latter can be used for illustration purposes.

Software PNS Studio implements algorithms MSG, SSG, and ABB, and therefore it is primarily a solver and also a model analyzer for process synthesis problems. Furthermore, it is also capable of constructing process synthesis models from the beginning. A "tree-view" provides a clear overview of the actual problem under consideration and makes it possible to edit the properties of multiple materials and operating units in parallel (Fig. 9.4). Measurement unit management is aided with automated conversions. As a solver, PNS Studio can generate the maximal structure, the combinatorial feasible structures, and the globally optimal and suboptimal solutions of the problem. In the latter case the objective can be either cost minimization or profit maximization. If required annual production rates are defined for each of the products, then the overall cost is minimized. If no required flow rate is given for the products but prices, then profit is maximized. If the optimal cost becomes negative, then the process is profitable. PNS Studio provides a double pane view of solutions to compare alternatives, as depicted in Fig. 9.5.

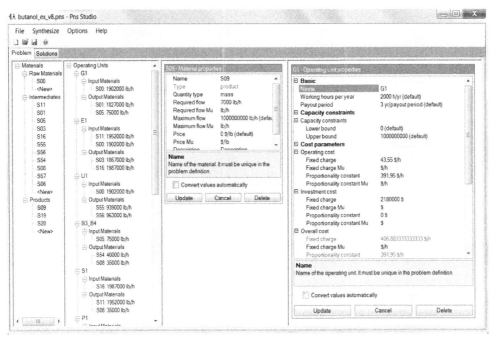

Figure 9.4 Parameters of a process synthesis problem in software PNS Studio.

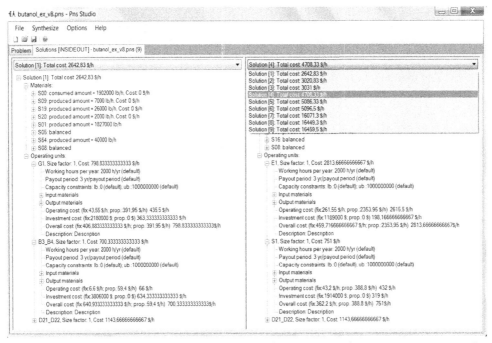

Figure 9.5 Double pane view in software PNS Studio for comparing alternative process networks.

For those interested in the overview or modification of the mathematical programming model itself, it can be exported from software PNS Studio in ZIMPL format. Moreover, brief or more detailed reports on both the problem and its alternative solutions can be exported to Microsoft Excel.

Model Development by P-Graphs

At the outset, the set of candidate operating units needs to be given by their input and output materials, as well as their capacities, i.e., the flow rates of their inlet and outlet streams. The upper bound on the availability of the raw materials and the lower bound on the required amount of desired products can be defined as well. For each intermediate material or by-product stream, the gross production must be nonnegative, i.e., at least the amount consumed must be produced. If the upper bound on the gross production of an intermediate is defined to be zero, then producing it at a higher rate than consuming it is not allowed, i.e., waste cannot be generated.

The P-graph representation expresses the structural or combinatorial properties of a PNS problem and resultant structures unambiguously. For instance, if an operating unit has multiple inlet streams, each of the streams needs to be provided for the operation of the unit, which is a logical AND constraint. If a material can be produced by two or more operating units, then any combination of them can be sufficient for the production of the material, which is a logical OR condition. Thus a mixer is never represented by an O-type node, which requires each of its inlet streams, but by M-type node requiring at least one of its multiple inlet streams, but not necessarily all of them to be present, only if the material identified is further utilized as a final product or input to an operating unit.

The sizes of processing equipment in the operating units can be estimated directly from the information available in former studies or through simulation by software, for example, Aspen Plus. The costs of chemicals, processing equipment, and reactors are estimated based on the most currently available market information in conjunction with the method of cost estimation covered in chapter 6. Subsequently, the total cost is obtained as the sum of the operating and capital costs by algorithm ABB. For each operating unit, PNS studio considers a linear function with a fixed charge expressing both the expanses arising regardless of the volume of the activity and the increase in cost proportional to the growth of mass load. If the costs are not defined as a single value but cost function parameters and the operating and investment costs are distinguished, then not only the best technology can be selected as most appropriate for a single set of parameters, but further examinations resulting in the optimal flowsheets for different payout periods and different estimated average loads of process can be determined. The payout period and the annual working hours can be set or changed at the menu Options/Default Measurement Units in PNS Studio.

Structural Analysis

One of the major advantages of the P-graph framework and PNS Studio compared to conventional mathematical programming and related computer aid is that by structural analysis of the initial structure, the algorithm MSG or its software implementation can highlight inconsistencies in the initial structure defining the synthesis problem. Three different kinds of such modeling mistakes are illustrated herein.

In Fig. 9.6 the execution of algorithm MSG in software PNS Studio excludes operation unit D8 from consideration. The modeling mistake identified is that only the major product of the process material S19 is labeled as product, while the other two, S09 and S20, are not. As a consequence, operating unit D8 is eliminated in the light of axiom (S4).

The second modeling mistake illustrated is the definition of the starting raw material as intermediate material, thus eliminated according to axiom (S2) because of the absence of any operating unit producing it. Subsequently, all the initial structure is excluded from the consideration indicated as "ERROR: There is no maximal structure." by algorithm MSG in software PNS Studio in Fig. 9.7.

The most common mistake taken during entering a flowsheet to software is missing an interconnection among the flowsheet elements. As an example the flow from

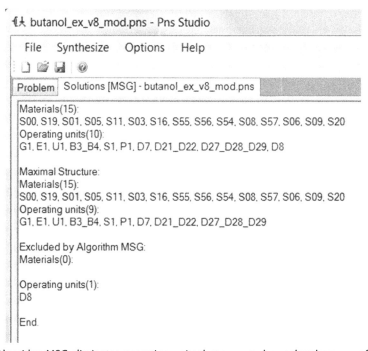

Figure 9.6 Algorithm MSG eliminates operating units that seem to be useless because of incomplete definition of the set of final products.

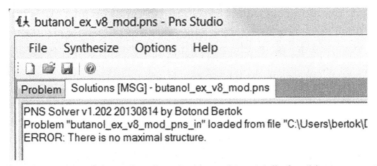

Figure 9.7 Algorithm MSG indicates that there is no combinatorially feasible structure in the initial structure.

operating unit S1 to material S11 is missing. Thus E1 cannot be started because of the absence of one of its inlet streams S11 eliminated in the light of axiom (S2). Consequently, operating S1 is eliminated as well, since its input material S16 cannot be provided by operating unit E1 (Fig. 9.8).

The elimination of one or more materials or operating units from the sets defined in the synthesis problem by algorithm MSG in practice implies that the input is incomplete

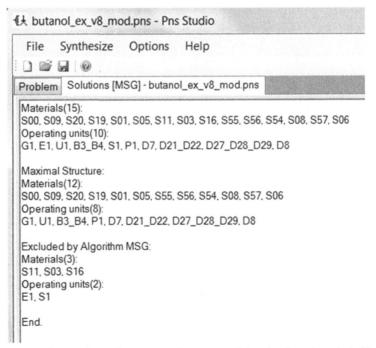

Figure 9.8 A group of operating units representing a potential technology is excluded by algorithm MSG because of a missing interconnection in the initial structure. S1 does not produce S11.

or inconsistent. Executing algorithm MSG in PNS Studio renders it possible to verify whether the input is entered correctly. Note that typically no algorithmic support is available for general-purpose mathematical modeling tools to ascertain if part of the model will never be required in the solution. As a consequence, without preliminary analysis of the model constructed for process synthesis or optimization, one cannot be sure that those process elements never appearing in the optimal solution are excluded on the basis of their costs or input mistakes. Most of such preliminary analysis is available by algorithm MSG in software PNS Studio in the light of the previously introduced axioms of structurally feasible process flowsheets.

Generate Structurally Feasible Flowsheets by Algorithm SSG

An important forthcoming step of validating the input data, i.e., the problem definition in PNS Studio, is to check whether the major overall flowsheet alternatives assumed by the designer are included in the optimization. Algorithm SSG serves this target by enumerating the structurally feasible process flowsheets. First, we have to ensure that the resulting number of structures is not limited to a small number is Options/Default values in software PNS Studio (Fig. 9.9).

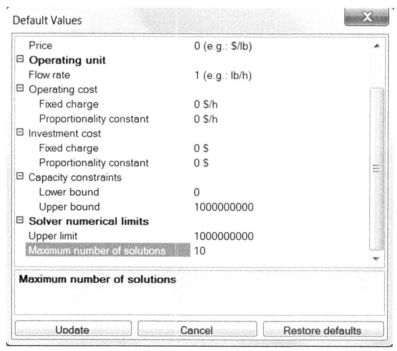

Figure 9.9 Set default values in software PNS Studio including the maximal number of expected solutions.

A usual logical mistake at entering a process synthesis problem is to require joint involvement of alternative operating units by interconnecting them as different inputs to a single forthcoming operating unit instead of connecting them as potential alternative producers of a single intermediate material. As mentioned previously, the best practice to define multiple alternative realizations of the same functional unit is to define the expected inlet and outlet streams of a functional unit by material IDs representing a range of their required qualities, and then connect each potentially alternative realization to these materials. In our illustrative example of the functional unit removing the water content from the fermentation broth, the inlet and outlet streams are given by the approximate compositions and flow rates identified as S00 and S08, respectively. Each of the alternative realizations of this function, namely, Technology #1, #2, and #3, are required to be capable of consuming S00 and produce S08 (Fig. 9.2). These alternative technologies are adsorption, pervaporation, and azeotropic distillation in the illustrative example.

If design alternatives are entered properly, each of their possible combinations appears among the combinatorially feasible structures involving their disjoint or joint appearance. For the illustrative example of interest, Technology #1, #2, and #3 can form seven combinations for the functional part removing the water content: #1; #2; #3; #1 in combination with #2; #1 with #3; #2 with #3; and, finally, the combination involving #1, #2, and #3. Separation configurations as represented in Fig. 9.3 can form 10 combinatorially feasible configurations. They can form seven similar combinations as the technologies: moreover, three other when either configuration #2, #3, or both is extended with operating unit D7 without the presence of D8. Finally, altogether $7 \times 10 = 70$ combinations of the alternative realizations of the two functional units are generated as structurally feasible alternative flowsheets for the overall process (Fig. 9.10). Note that algorithm SSG takes into account axioms (S1) through (S5) of combinatorially feasible process structures only and none of the continuous parameters such as required and maximum flows of materials or limitations on the volumes of the operating units. Taking into account all the continuous parameters, the number of feasible flowsheets will be less. However, axioms (S1) trough (S5) are necessary properties for any process structure to be feasible, and thus each of the feasible process structures including the optimal one is included in the list of structures generated by algorithm SSG.

Economical and Ecological Analysis

In process synthesis, the best network structures are constructed by interconnecting subsets of the operating units defined in the problem. For determining the best structures and the optimal loads of the operating units, algorithm ABB is executed in PNS Studio. The objective is to minimize the overall annual cost of the process. The overall cost is the sum of the costs of the operating units and prices of raw materials. The annual cost of an operating unit is the sum of its operating cost and annualized investment cost. For convenience, default measurement units can be set under menu Options/Default

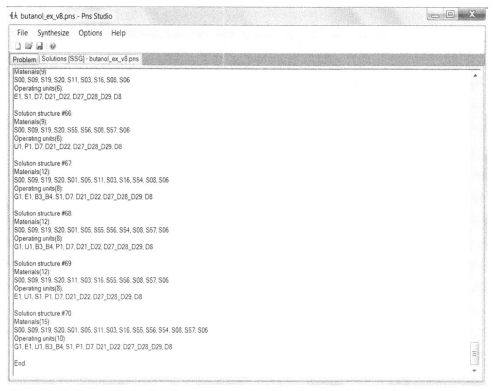

Figure 9.10 Structurally feasible flowsheets generated by algorithm SSG in software PNS Studio.

Measurement Units. To be able to compare annual investment costs and hourly operating costs the annual number of working hours have to set (Fig. 9.11).

Nevertheless, after constructing an appropriate mathematical programming problem, for example, exported from PNS Studio as depicted in Figs. 9.12 and 9.13, the optimal flowsheet can be generated by any general-purpose optimization software, it can provide a single optimal solution only and result in a difficult-to-read list of values of the design variables (Fig. 9.14). In contrast, the P-graph algorithms and also their implementations in PNS Studio provides not only a single optimal solution but the series of n-best structurally different suboptimal flowsheets, where n can be given in the options of the software (Fig. 9.9). For the best structures, the optimal load of the operating units is computed as well besides the set of operating units incorporated. It is assumed that the flow rates of the inlet and outlet streams increase and decrease proportionally to the load. The resulting alternative processes can be compared either in a two pane view in PNS Studio or in the form of a spreadsheet (Figs. 9.5 and 9.12, respectively). For the illustrative example of interest, each of the three water removal technologies is paired with each of the three separation configurations, resulting in nine feasible networks by the

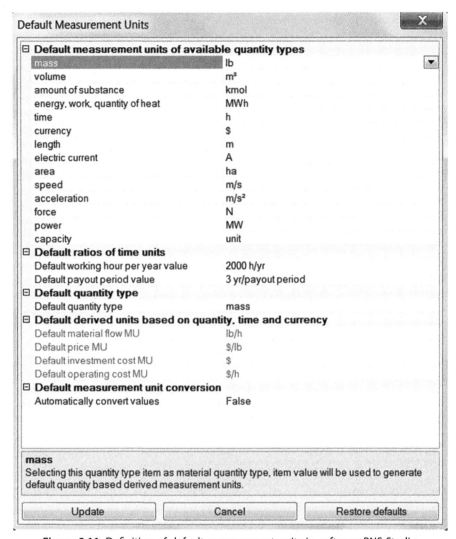

Figure 9.11 Definition of default measurement units in software PNS Studio.

implementation of the algorithm ABB. Algorithm ABB in software PNS Studio provides such structures only, which have no better feasible subnetworks. This is the reason why combinations of the alternative technologies and combinations of alternative separation configurations are not listed. Typically, combinations of alternatives come into play in two cases. First, if a good alternative has such limitations on its capacity, which is not sufficient to satisfy all the demands, then another alternative has to fulfill the remaining needs. Second, if one of the alternatives has no fixed cost, then it can be associated to any other without extra penalty.

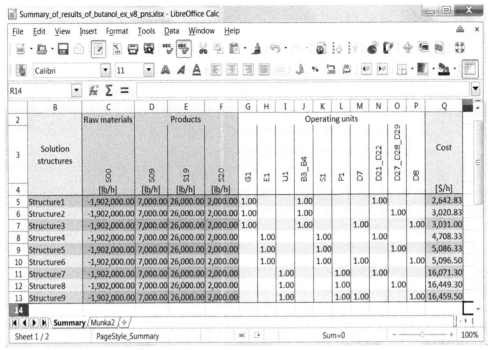

Figure 9.12 Overview of the optimal and alternative suboptimal process networks in a spreadsheet generated by PNS Studio.

```
var exist[operating_units] binary;
var size[<o> in operating_units] real >= 0 <= capacity_upper_bound[o];
minimize cost:
   (sum<o> in operating_units: size[o]*
     (proportional_cost[o] -
       sum<m> in (raw_materials union products):
         price[m]*material_to_operating_unit_flow_rates[m, o]
       )
   ) +
   (sum<o> in operating_units:
      exist[o] * fix_cost[o]);
subto size_ub:
   forall <o> in operating_units do
     size[o] <= exist[o] * capacity_upper_bound[o];
subto size_lb:
   forall <o> in operating_units do
     size[o] >= exist[o] * capacity_lower_bound[o];
subto raw_lb:
   forall <r> in raw_materials do
     sum<o> in operating_units: -1 * material_to_operating_unit_flow_rates[r, o] * size[o] >= flow_rate_lower_bound[r];
subto inter_lb:
   forall <i> in intermediates do
     sum<o> in operating_units: material_to_operating_unit_flow_rates[i, o] * size[o] >= 0;
subto prod_lb:
   forall <p> in products do
     sum<o> in operating_units: material_to_operating_unit_flow_rates[p, o] * size[o] >= flow_rate_lower_bound[p];
subto raw_ub:
   forall <r> in raw_materials do
     sum<o> in operating_units: -1 * material_to_operating_unit_flow_rates[r, o] * size[o] <= flow_rate_upper_bound[r];
subto inter_ub:
   forall <i> in intermediates do
     sum<o> in operating_units: material_to_operating_unit_flow_rates[i, o] * size[o] <= flow_rate_upper_bound[i];
subto prod_ub:
   forall <p> in products do
     sum<o> in operating_units: material_to_operating_unit_flow_rates[p, o] * size[o] <= flow_rate_upper_bound[p];
```

Figure 9.13 Mathematical programming model exported from PNS Studio.

```
Optimal - objective value          2642.8333
       0 exist$G1                            1
       1 exist$E1                            0
       2 exist$U1                            0
       3 exist$B3_B4                         1
       4 exist$S1                            0
       5 exist$P1                            0
       6 exist$D7                            0
       7 exist$D21_D22                       1
       8 exist$D27_D28_D29                   0
       9 exist$D8                            0
      10 size$G1                             1
      11 size$E1                             0
      12 size$U1                             0
      13 size$B3_B4                          1
      14 size$S1                             0
      15 size$P1                             0
      16 size$D7                             0
      17 size$D21_D22                        1
      18 size$D27_D28_D29                    0
      19 size$D8                             0
```

Figure 9.14 Single optimal solution of the mathematical programming model generated by software COIN-OR's CBC.

Sensitivity Analysis of the Best Flowsheets

After generating the optimal configuration of a process network, PNS can easily determine if the risk of potentially decreasing demands resulting in partial load operation entails an alternative design. The operating units are sized to be able to accommodate the maximum flow rates defined in the initial structure, thus their investment cost is fixed. In contrast, their operating cost is considered to be mainly proportional to the mass load, i.e., 90% of their operating cost is assumed to be proportional to their load and 10% as fixed cost. It means that at half load, for example, the operating cost decreases by 45%. Besides these conditions the cost of best networks can easily be resynthesized for smaller required flow rates of products representing decreased average load for the system. The outcome is visible in Fig. 9.15 for the average load of 70%. It shows that even if the optimal structure remained invariant, the order of forthcoming, for example, second and third best, structures changed. This points our attention to the fact that costs and order of structures may greatly rely on the average load of the overall process.

Evaluation of the Competitiveness of Emerging Technologies

As we saw in the previous examinations among technologies #1, #2, and #3 for the removal of water from the fermentation broth, the most promising is absorption, then azeotropic distillation, and finally pervaporation. Process synthesis is also capable of

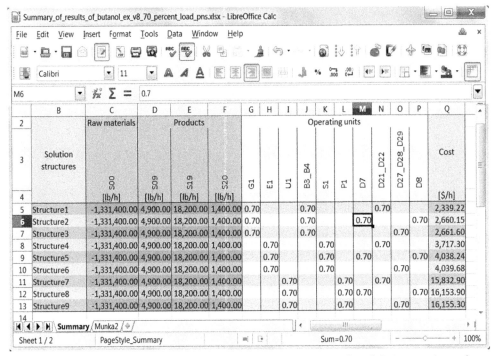

Figure 9.15 List of the nine best solution structures in decreasing order of their annual cost for an average load of 70%.

determining the rate of reduction in the investment cost of pervaporation, which makes it economically viable. After some trials, a 95% reduction was needed in the investment cost of pervaporation to be ranked as the number one technology for water removal (Fig. 9.16).

Sustainability as Alternative Objective for Process Synthesis

The actual implementation of the algorithm ABB in software PNS studio considers a single objective function only. However, besides cost minimization, this single objective can the best sustainability or combination of the previous two as well. The sustainability metric, for example, footprint or emergy, can be taken into account by representing the environment as a resource with price or penalty and upper bound on its utilization. This resource is connected to any of the operating units having an effect on sustainability while the rate of the effect is expressed by a sustainability metric assigned to the interconnecting arc (Fig. 9.17). Note that the environmental impact is typically proportional to the volume of an operation, thus it can be modeled similarly to material flows. If the price or penalty is assigned to only the resource representing the environment, then the best networks will be generated in decreasing order of their environmental impact. In some real-life situations a limited environmental impact has a

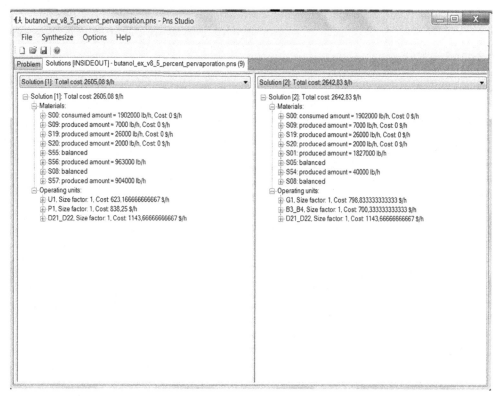

Figure 9.16 The best and second best network after 95% reduction in the investment cost of pervaporation.

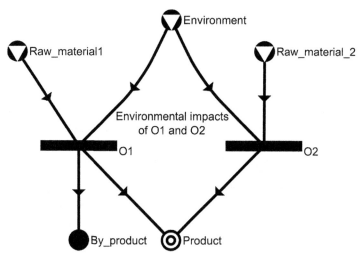

Figure 9.17 Representing the environment as a resource in P-graph and the environmental impacts of the operations as weights of interconnections between the operations and the environment.

real cost, which can be added to other costs of the process resulting in a combined economical—ecological objective function. Such examples are the price of CO_2 quota or environmental fees.

SUMMARY

The P-graph framework, implemented in software PNS Draw, PNS Studio, or the very recently developed P-Graph Studio, provides us with a mathematically rigorous approach for formulating and solving process synthesis problems, as well as analyzing the resultant flowsheets with the aid of the built-in optimizer. Algorithm MSG can verify if a problem is entered without the loss of essential interconnections among the candidate operating units and related material streams. The best structures computed by algorithm ABB render it possible to select the most appropriate flowsheet under various scenarios representing expected or unexpected situations.

The P-graph framework is an effective tool for solving various engineering optimization problems. Originally it was developed for conceptual design of chemical processes, and later become widely applied in formulating network synthesis problems ranging from metabolic reaction pathways to supply chains to predict and improve energy efficiency, as well as sustainability. Progressive growth of the number of applications is expected in the near future.

REFERENCES

[1] Siirola J. Industrial applications of chemical process synthesis. In: Anderson JL, editor. Advances in chemical engineering, vol. 23. San Diego, CA: Academic Press; 1996. p. 1—62.

[2] Westerberg AW. A review of process synthesis. In: Squires RG, Reklaitis GV, editors. ACS symposium series, 124. Washington DC: American Chemical Society; 1980. p. 53—87.

[3] Nishida N, Stephanopoulos G, Westerberg AW. A review of process synthesis. AIChE J 1981;27:321—51.

[4] Liu J, Fan LT, Seib PA, Friedler F, Bertok B. Holistic approach to process retrofitting: application to downstream process for biochemical production of organics. Ind Eng Chem Res 2006;45:4200—7.

[5] Fan LT, Zhang T, Liu J, Seib PA, Friedler F, Bertok B. Price-targeting through repeated flowsheet re-synthesis in developing novel processing equipment: pervaporation. Ind Eng Chem Res 2008;47:1556—61.

[6] Bertok B, Barany M, Friedler F. Generating and analyzing mathematical programming models of conceptual process design by p-graph software. Ind Eng Chem Res 2013;52(1):166—71.

[7] Feng G, Fan LT. On stream splitting in separation system sequencing. Ind Eng Chem Res 1996;35:1951—8.

[8] Douglas JM. A hierarchical decision procedure for process synthesis. AIChE J 1985;31(3):353—62.

[9] Grossmann IE, Santibanez J. Applications of mixed-integer linear programming in process synthesis. Comp Chem Eng 1980;4:205—14.

[10] Azapagic A, Clift R. The application of life cycle assessment to process optimization. Comp Chem Eng 1999;23:1509—26.

[11] Friedler F, Tarjan K, Huang YW, Fan LT. Combinatorial algorithms for process synthesis. Comp Chem Eng 1992;16:S313—20.

[12] Friedler F, Tarjan K, Huang YW, Fan LT. Graph-theoretic approach to process synthesis: polynomial algorithm for maximal structure generation. Comp Chem Eng 1993;17:929—42.

[13] Friedler F, Varga JB, Fan LT. Decision mapping: a tool for consistent and complete decisions in process synthesis. Chem Eng Sci 1995;50:1755—68.

[14] Friedler F, Varga JB, Feher E, Fan LT. Combinatorially accelerated branch-and-bound method for solving the MIP model of process network synthesis. In: Floudas CA, Pardalos PM, editors. State of the art in global optimization: nonconvex optimization and its applications. Computational methods and applications. Norwell, MA: Kluwer Academic Publishers; 1996. p. 609—26.

[15] Brendel MH, Friedler F, Fan LT. Combinatorial foundation for logical formulation in process network synthesis. Comp Chem Eng 2000;24:1859—64.

[16] Friedler F, Varga JB, Fan LT. Algorithmic approach to the integration of total flowsheet synthesis and waste minimization. In: El-Halwagi MM, Petrides DP, editors. AIChE symposium series: pollution prevention via process and product modifications, 90; 1995. p. 86—97.

[17] Partin LR. Combinatorial analysis application for flowsheet synthesis of chemical plants. Maple Tech News 1998;5:15—26.

[18] Keller GE, Bryan PF. Process engineering moving in new directions. Chem Eng Prog 2000;96(1):41—50.

[19] Sargent R. Process systems engineering: a retrospective view with questions for the future. Comp Chem Eng 2005;29:1237—41.

[20] Halim I, Srinivisan R. Systematic waste minimization in chemical processes. 1. Methodology. Ind Eng Chem Res 2002;41:196—207.

[21] Halim I, Srinivisan R. Systematic waste minimization in chemical processes. 2. Intelligent decision support system. Ind Eng Chem Res 2002;41:208—10.

[22] Lee DY, Fan LT, Park S, Lee SY, Shafie S, Bertok B, et al. Complementary identification of multiple flux distributions and multiple metabolic pathways. Metab Eng 2005;7:182—200.

[23] Niemetz N, Kettl KH, Eder M, Narodoslawsky M. RegiOpt conceptual planner—identifying possible energy network solutions for regions. Chem Eng Trans 2012;29:517—22.

[24] Seo H, Lee DY, Park S, Fan LT, Shafie S, Bertok B, et al. Graph-theoretical identification of pathways for biochemical reactions. Biotechnol Lett 2001;23:1551—7.

[25] Rossello R, Valiente G. Graph transformation in molecular biology. Lect Notes Comp Sci 2005;3393:116—33.

[26] Tan RR, Cayamanda CD, Aviso KB. P-graph approach to optimal operational adjustment in poly-generation plants under conditions of process inoperability. Appl Energy 2014;135:402—6.

[27] Xu W, Diwekar UM. Environmentally friendly heterogeneous azeotropic distillation system design: Integration of EBS selection and IPS recycling. Ind Eng Chem Res 2005;44:4061—7.

CHAPTER TEN

Sustainability Assessment and Performance Improvement of Electroplating Process Systems

H. Song, N. Bhadbhade, Y. Huang
Wayne State University, Detroit, MI, United States

INTRODUCTION

Electroplating is widely used for coating metal objects with a thin layer of a different metal in various industries, such as the aerospace, appliances, automotive, electronics, and heavy equipment industries. The surface-finished parts (i.e., workpieces) possess an aesthetic appearance and corrosion resistance capability, and demonstrate various engineering functionalities [1,2]. A typical electroplating process is composed of cleaning, rinsing, and plating operational steps [3]. In production, parts are cleaned, etched, electroplated, and finished in operating units that contain a combination of corrosive, metal, and/or chemical solutions. Various chemicals are used in cleaning units. Electrolytic plating, electroless plating, and electrochemical conversion processes are typically used for parts plating [4].

The electroplating industry is considered one of the most polluting industries, largely because of the emission of hazardous chemicals and toxic waste in different forms. Toxic chemicals, such as cyanide, acid, and alkaline, are widely used in the cleaning and plating processes, while heavy metals, such as zinc, copper, silver, chrome, and nickel, are plated on the work piece surface [3]. More than 100 different toxic chemicals, metals, and other regulated pollutants are generated during operation [5]. The electroplating process consumes a huge amount of fresh water in multiple rinsing steps, which are installed after each cleaning and plating step. Energy is mainly used for heating cleaning baths and ensuring direct deposition of metal ions to the surface of products in plating units.

Development of a deep understanding of process science and engineering aspects is essential to addressing sustainability issues with electroplating. Huang and associates developed a set of mathematical models to characterize the dynamic behavior of different types of cleaning and different configurations of rinsing operations [3]. The models have been used to quantitatively evaluate dirt removal, chemical and water consumption, and

Sustainability in the Design, Synthesis and Analysis of Chemical Engineering Processes

227

waste generation if the operating condition and initial condition of parts are known. Luo and Huang applied artificial intelligence techniques to study chemical-containing drag-out reduction and process wastewater minimization [5]. Luo et al. proposed a sludge model to evaluate sludge generation during parts cleaning and rinsing [6]. Yang et al. introduced a water reuse network design method for maximizing the reuse of rinsing water in rinsing steps [7,8].

In 2000, Lou and Huang introduced the Profitable Pollution Prevention (P3) concept and conducted a fundamental study on the mechanisms for achieving P3 in the electroplating industry [9]. A series of P3 technologies were then developed to minimize chemical, water, and energy consumption as well as hazardous waste emission in electroplating operations. A distinct feature of the P3 is the promise of gaining economic benefits while minimizing adverse environmental impact. Xiao and Huang demonstrated successful industrial applications using a number of P3 technologies [10]. The P3 theory can be further extended to the inclusion of social sustainability aspects. Piluso and Huang introduced a new concept called collaborative profitable pollution prevention (CP3) [11] to address the sustainability concern for large industrial zones, where the social component is restricted to the synergy among the participating organizations. Needless to say, many more social aspects should be considered.

To help the electroplating industry develop and achieve sustainable development goals, effective methods for comprehensive sustainability assessment and accurate evaluation of available development options are needed. As of today a number of sustainability metrics systems have already been created for performing sustainability assessment. For instance, the IChemE and AIChE sustainability metrics are widely adopted in the chemical and allied industries; both contain three sets of metrics for assessing economic, environmental, and social sustainability [12—14]. The assessment utilizes the system information provided by system models or other means (e.g., direct and/or indirect measurements). Many other types of sustainability metrics are also available. The Dow Jones Sustainability Indices assess corporate business sustainability, which creates global indexes tracking the financial performance of leading sustainability-driven companies [15]. BASF has created and implemented eco-efficiency sustainability metrics, which mainly focuses on economic and environmental performances [16—18]. However, the known sustainability metrics systems are not directly designed for the sustainability assessment of electroplating systems. Therefore it is of great importance to generate a specific metrics system particularly for evaluating sustainability performance of the electroplating industry.

This chapter presents a framework for systematically investigating electroplating sustainability problems through introducing a family of sustainability indicators, conducting sustainability assessment, and identifying solutions for sustainability performance enhancement. Case studies are provided to demonstrate the efficacy of the framework.

FUNDAMENTALS FOR PROCESS SUSTAINABILITY

The aim of the study is to enhance the performance of electroplating systems by applying sustainability principles and methods. More specifically it is to simultaneously achieve waste reduction, production improvement, and social satisfaction (Fig. 10.1); electroplating system sustainability (ESS) can be expressed as:

$$\text{ESS} = \text{Waste}\downarrow + \text{Production}\uparrow + \text{Satisfaction}\uparrow \qquad [10.1]$$

Waste reduction is the best approach to mitigate the negative impact on the environment. The reduction of waste emission and hazardous chemical consumption can also significantly alleviate the health burden on human beings. In the meantime, minimization of water and energy consumption can directly contribute to environmental sustainability. The waste reduction in Eq. [10.1] can be elaborated as:

$$\text{Waste}\downarrow = \text{Sludge}\downarrow + \text{Chemicals}\downarrow + \text{Water}\downarrow + \text{Energy}\downarrow \qquad [10.2]$$

According to the P3 principle, it is possible to make profits through reducing waste from electroplating systems. In addition, the assurance of product quality requires lowering the product defect rate, which can boost the revenue in return. The production rise in Eq. [10.1] can be expressed as:

$$\text{Production}\uparrow = \text{Product quality}\uparrow + \text{Production rate}\uparrow + \text{Operating cost}\downarrow$$
$$+ \text{Capital cost}\downarrow + \text{Chemical cost}\downarrow \qquad [10.3]$$

Note that the reduction of hazardous chemical consumption and a safer electroplating process can lead to the improvement of employee satisfaction. External satisfaction takes into consideration both the local community and customers. The reduction of hazardous waste can also satisfy the local community while high product

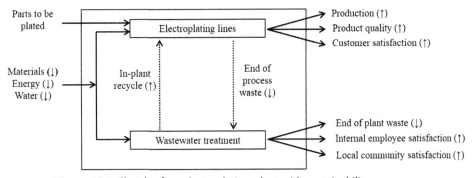

Figure 10.1 Sketch of an electroplating plant with sustainability concerns.

quality can gain satisfaction from customers. Therefore the social aspects can include the following:

$$\text{Social sustainability} \uparrow = \text{Customer satisfaction} \uparrow + \text{Employee satisfaction} \uparrow$$
$$+ \text{Local community satisfaction} \uparrow \qquad [10.4]$$

SUSTAINABILITY METRICS SYSTEM

An effective sustainability metrics system should be capable of providing a deep understanding of sustainability performance of electroplating systems. The desired sustainability metrics system should establish an appropriate measurement that can address the stakeholder's economic interest, severe environmental concerns, as well as social impact simultaneously. Selection of sustainability indicators is challenging. In this study an investigation of the electroplating system from the perspective of the supply chain is used to generate proper sustainability indicators. Fig. 10.2 depicts the position of the electroplating industry in the supply chain of product manufacturing. The electroplating industry mainly plays the role of an intermediate service sector, which receives unfinished parts from suppliers and transforms them into components with expected functionalities and appearance for downstream industries. Electroplating operations consume various types of chemicals supplied by the chemical industry. In production, waste streams generated are pretreated in plants. In this section we introduce a sustainability metrics system to evaluate the sustainability performance of electroplating systems.

Economic Sustainability

The main economic interests of the electroplating industry are the profit that comes from the sales of finished product, the cost that includes the expenses for maintaining normal production, and the investment for technology innovation and application, employee's

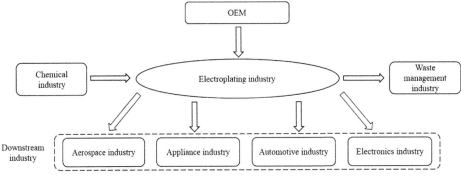

Figure 10.2 Electroplating industry-centered supply chain.

training and education, etc. Note that product quality is always a key factor in plants, and product defect rate during production and product return rate after shipment should also be seriously considered. Thus economic sustainability indicators should cover the assessment of all these factors associated with production. A set of economic sustainability indicators is listed in Table 10.1.

Environmental Sustainability

As source reduction is a top priority of plants, environmental sustainability indicators must evaluate waste generation in different forms emitted from the plants, and assess the impact on human health. Note that a good fraction of chemicals used for cleaning and plating is usually dragged out by parts and enters the rinsing systems or evaporates in the working environment. Heat loss occurs in cleaning and plating operations. Fresh water is used mainly to rinse off the remaining dirt and toxic chemicals on the surface of parts. Possible water recycle may come from water reuse within manufacturing processes or recycling from the wastewater treatment facility. A set of indicators suitable for assessing environmental sustainability of the electroplating systems is provided in Table 10.2.

Social Sustainability

In industries, social sustainability is usually reflected by the working environment within an industrial organization and the surrounding communities, including those along the

Table 10.1 Economic Sustainability Indicators

	Economic Sustainability Indicator	Unit
Profit, value, and tax	Value added	$/year
	Value added per unit value of sales	$/year
	Value added per direct employee	$/year
	Value added per kg product	$/kg
	Net income	$/year
	Net income per direct employee	$
	Return on average capital employed	%/year
Investments	Percentage increase in capital employed	%/year
	Employees with postschool qualification	%
	New appointments/number of direct employees	%/year
	Training expense as percentage of payroll expense	%
	Ratio of indirect jobs/number of direct employees	%
	Investment in education/employee training expense	$/year
	Investment in technologies to improve product quantity and process safety	$
Product quality	Product return rate after shipment	%
	Product defect rate during production	%
	Percentage of finished product delivered on time	%

Table 10.2 Environmental Sustainability Indicators

	Environmental Sustainability Indicator	Unit
Energy	Total net primary energy usage rate	kJ/year
	Percentage total net primary energy sourced from renewable	%
	Total net primary energy usage per kg product	kJ/kg
	Total net primary energy usage per finishing line	kJ
	Total net primary energy usage per unit value added	kJ/$
Material (excluding fuel and water)	Total cleaning chemical usage	kg/kg
	Total cleaning chemical usage per kg product	kg/kg
	Total cleaning chemical usage per unit value added	kg/$
	Total plating chemical usage	kg/year
	Total plating chemical usage per kg product	kg/kg
	Total plating chemical usage per unit value added	kg/$
	Percentage of chemical recycled from wastewater treatment facility	%
Water	Total water consumption	m^3/year
	Net water consumed per unit kg product	m^3/kg
	New water consumed per unit value added	m^3/$
	Fraction of water recycled within the company	%
Emission	Hazardous liquid waste per unit value added	kg/$
	Hazardous liquid waste per kg product	kg/kg
	Percentage of wastewater treated within the company	%
	Total other hazardous waste per unit value added	kg/$
	Total other hazardous waste per kg product	kg/kg
	Human health burden per unit value added	kg/$
	Nonhazardous waste generated	kg/year

supply chain. Different from the assessment of economic and environmental sustainability for which indicators can be relatively readily quantifiable, the social aspects in sustainability assessment are usually difficult to quantify; many such indicators are evaluated subjectively. In practice, process safety and human resources as well as the satisfaction of customers and local communities are usually the main categories of information for social sustainability assessment. Table 10.3 shows a set of social sustainability indicators for the electroplating industry.

Note that the quality of sustainability assessment relies largely on the comprehensiveness of the sustainability indicators used and data availability and accuracy. Thus it is imperatively important that an electroplating organization select a set of sustainability indicators suitable for assessing economic, environmental, and social sustainability where data and information are accessible and the uncertainties associated with them are manageable.

Table 10.3 Social Sustainability Indicators

	Social Sustainability Indicator	Unit
Workplace	Benefits as percentage of payroll expense	%
	Employee turnover (resigned + redundant/number employed)	%
	Promotion rate (number of promotions/number employed)	%
	Working hours lost as percent of total hours worked	%
Safety	Process safety index	
	Number of process safety analyses required annually	/year
	Number of process maintenance required annually	/year
Society	Number of stakeholder meetings per unit value added	/$
	Indirect community benefit per unit value added	$/$
	Number of complaints from local community per unit value added	/$
	Number of complaints from downstream customers	/year
	Percentage of finished product delivered on time	%
	Number of legal actions per unit value added	/$

SUSTAINABILITY ASSESSMENT FRAMEWORK

This work adopts a sustainability assessment approach introduced by Liu and Huang [19,20]. For an electroplating system of interest, a selected sustainability metrics set for the sustainability assessment is denoted as:

$$S = \{E, \quad V, \quad L\} \qquad [10.5]$$

where $E = \{E_i | i = 1, 2, \cdots, F\}$ is a set of economic sustainability indicators; $V = \{V_i | i = 1, 2, \cdots, G\}$ is a set of environmental sustainability indicators; and $L = \{L_i | i = 1, 2, \cdots, H\}$ is a set of social sustainability indicators.

For each categorized sustainability, the selected sustainability indicators can be combined to generate a composite indicator. In this process, all the indicator values should be normalized. The composite sustainability indices are expressed as follows:

$$E = \frac{\sum_{i=1}^{F} a_i E_i}{\sum_{i=1}^{F} a_i} \qquad [10.6]$$

$$V = \frac{\sum_{i=1}^{G} b_i V_i}{\sum_{i=1}^{G} b_i} \qquad [10.7]$$

$$L = \frac{\sum_{i=1}^{H} c_i L_i}{\sum_{i=1}^{H} c_i} \qquad [10.8]$$

where a_i, b_i, and $c_i \in [1, 10]$ are the weighting factors associated with the corresponding indices, reflecting the relative importance of the individual indices in overall assessment of the relevant categorized sustainability. Obviously, the result of sustainability assessment is greatly affected by the selection of weighting factors. Note that neither a rigorous mathematical framework nor universal rules are available for weighting factor determination as the relevant importance of indicators for categorized sustainability assessment is always debatable. The best practice in industries is to let a group of experienced engineers and management personnel jointly determine the weighting factor values. This group should have the best understanding of the organization's development goals and strategies, data availability and quality, and relative importance of each indicator in overall assessment.

The overall sustainability performance of the system, $S(P)$, can be evaluated using the composite indices, $E(P)$, $V(P)$, and $L(P)$, i.e.,

$$S(P) = \frac{\|(\alpha E(P), \quad \beta V(P), \quad \gamma L(P))\|}{\|(\alpha, \quad \beta, \quad \gamma)\|} \qquad [10.9]$$

where α, β, and γ are the weighting factors with the value range of 0 and 1.

Technology Evaluation

There is always a need to consider how to select appropriate design and manufacturing technologies to improve organizational sustainability performance. Identification of such technologies requires not only a deep understanding of the processes, but more importantly the potential contributions of the technologies to sustainability. The selected sustainability indicators used for assessing process systems should also be used to evaluate the candidate technologies. Note that different technologies may share the same or similar functions to some extent. For instance, a cleaning operation optimization technology can keep the chemical concentration in a cleaning unit to the lowest possible level without compromising workpiece cleaning quality, while a drag-out minimization technology can reduce the chemical loss from the cleaning unit without compromising production rate. These two technologies all contribute to chemical consumption reduction; they may also interfere with each other as the former technology might require a longer cleaning time because of a lower chemical concentration for part cleaning. However, a drag-out minimization technology seeks a longer drainage time while lifting a barrel of parts from the cleaning unit, which will affect the production rate. Thus how to evaluate the true amount of chemicals to be saved and the production rate using both technologies at same time could be a challenge. In such cases, the evaluation of the total sustainability performance by the two technologies should not be simply an addition of the sustainability performance values of two individual technologies. The combined benefits of the two technologies should be evaluated through either physical experiments with the cleaning system or using computational experiments using a proven process simulator. It is certainly more helpful if engineers participate in evaluation.

Given that a technology set, T, including m technologies is selected from N technology candidates, the categorized sustainability improvement evaluation on them, i.e., the economic sustainability performance, $E(T)$, the environmental sustainability performance, $V(T)$, and the social sustainability performance, $L(T)$, are used to evaluate the overall sustainability status, $S(T)$, after implementing the selected technology set based on the evaluation of the selected indicators, i.e.,

$$S(T) = \frac{\|(\alpha E(T), \quad \beta V(T), \quad \gamma L(T))\|}{\|(\alpha, \quad \beta, \quad \gamma)\|} \qquad [10.10]$$

Investment Assessment

The adoption and implementation of new technologies could be costly. Thus the affordability for using attractive technologies is always a concern by industrial organizations, which is part of the decision-making process for sustainability improvement [20]. Practically, if the technologies and services are provided by different companies, the total investment on those will be a simple sum of the costs for each individual technology. However, if two or more technologies and services are provided by the same company, then the total cost will be lower than the sum of the costs of individuals. Given the cost for using individual technology, B_i, the total cost for using m technologies, B_t, can be estimated as:

$$B_t = p \sum_{i=1}^{m} B(T_i); \qquad [10.11]$$

where p is the coefficient reflecting a discount for using more than one technology.

Note that the investment efficiency, I_{eff}, for sustainability improvement can be estimated as:

$$I_{\text{eff}} = \frac{\Delta S_t(T)}{B_t} \qquad [10.12]$$

where $\Delta S_t(T)$ is the overall improvement of sustainability performance after implementing all selected technologies, $\Delta S_t(T) = S(T) - S(P)$.

Goal Setting and Need for Sustainability Performance Improvement

An electroplating plant can determine if actions are needed for sustainability performance improvement, based on sustainability assessment result and the plant's development goal, which can be expressed as:

E^{sp} = the economic sustainability goal

V^{sp} = the environmental sustainability goal

L^{sp} = the social sustainability goal

Thus the plant's overall sustainable development goal can be evaluated as:

$$S^{sp} = \frac{\|(\alpha E^{sp}, \quad \beta V^{sp}, \quad \gamma L^{sp})\|}{\|(\alpha, \quad \beta, \quad \gamma)\|},$$

[10.13]

where α, β, and γ take the same values as those used in Eq. [10.9] for consistency.

If the budget limit, B^{lim}, of the plant is known, the technology-based sustainability performance improvement task can be described as:

$$S(T) > S^{SP}$$

[10.14]

$$E(T) \geq E^{SP}$$

[10.15]

$$V(T) \geq V^{SP}$$

[10.16]

$$L(T) \geq L^{SP}$$

[10.17]

$$B_t \leq B^{lim}$$

[10.18]

This means that after implementing a selected technology set, at least one categorized sustainability performance is better than a preset goal, while the performance of other sustainability categories is not better than the preset goal. In this way, the overall sustainability performance, $S(T)$, will be guaranteed better than the defined improvement goal, S^{SP}, while the investment is under the budget limit.

Technology Selection

As stated earlier, sustainability performance improvement of an industrial system is technology based in this work. In this section a technology selection procedure is described to help electroplating plants to select the most appropriate technologies based on the results of sustainability assessment as well as a known budget limit.

(a) Set a sustainability improvement goal by the plant, i.e., E^{sp}, V^{sp}, L^{sp}, and thus S^{sp} and the values of α, β, and γ in Eq. [10.9] should be determined.

(b) Evaluate the current sustainability status of an electroplating system. In this step, sustainability indicators should be selected, and Eqs. [10.5]−[10.9] should be used, and the values of $E(P)$, $V(P)$, and $L(P)$, and thus $S(P)$ can be obtained.

(c) If the current sustainability performance of the system is satisfactory, i.e., the values of $E(P)$, $V(P)$, and $L(P)$, and thus $S(P)$ are at least not smaller than the values of E^{sp}, V^{sp}, L^{sp}, and thus S^{sp}, respectively, then performance improvement action is not needed; otherwise proceed to the following steps.

(d) Generate all of the improvement options based on the availability of technologies. For instance, 2^{N-1} technology sets can be obtained if there are N technologies for selection.

(e) Evaluate each set of technologies about its capacity for the improvement of economic, environmental, social, and overall sustainability. This can be accomplished using Eqs. [10.6]–[10.8] and [10.10].

(f) Calculate the total cost for using each set of technologies using Eq. [10.11], as well as the investment efficiency I_{eff} using Eq. [10.12].

(g) Eliminate the technology sets using the criteria shown in Eqs. [10.14]–[10.18], as they are either too expensive or unable to help the company to achieve the preset sustainability goal.

(h) Order the remaining technology sets based on the company needs, for example, in the order of investment efficiency. It is also possible that the company has some other preference, for example, the maximum improvement of economic sustainability performance. Note that all the technology sets on the list after step (g) are satisfactory. The company has a freedom to choose any set if they have any additional consideration.

CASE STUDY

An electroplating company with a number of zinc plating lines is selected to study the applicability of the introduced sustainability metrics system and performance improvement method. A representative zinc plating line is selected, which has a production capacity of six barrels of parts per hour, 110 kg/barrel, and the plant operates 300 days/year. Fig. 10.3 shows a flowsheet of the plating process. The purchase price of unfinished parts and the sale price of plated products are $4/kg and $4.8/kg, respectively. Electricity is the only energy source for the line and the annual energy consumption is 4.02×10^6 kWh/year. Fresh water consumption is at 1.33×10^5 m³/year. The alkaline solvent used for part cleaning is consumed at the rate of 0.0062 kg/kg-part; the plating chemical (zinc chloride) is consumed at the rate of 0.025 kg/kg-part. The total hazardous waste emission is 0.04 kg/kg-part. The parts return rate is 8%, based on the company's record. The company receives about 20 complaints per year from the local community and end-use companies. The process safety is rated on a scale of 0–100 with 0 being no safety and 100 the safest. Based on the feedback from a group of process and environmental experts, the current process safety is rated at 65. The process safety analysis is conducted once a month. It is assumed that 30 employees are hired for production of three shifts per day. The average annual salary of employees is approximately $45,000.

A small set of selected sustainability metrics is listed in Table 10.4 and is used to evaluate process sustainability performance. The assessment result is shown in the column "Current" of Table 10.5. Note that the normalized values of the assessment

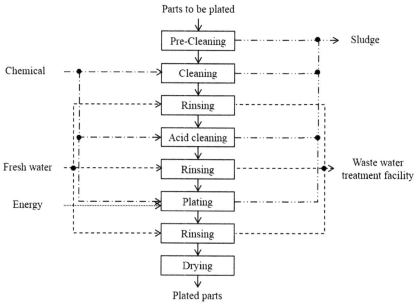

Figure 10.3 Typical electroplating process.

Table 10.4 Selected Sustainability Metrics

Metrics		Indicators	Value
Economic sustainability	E1	Value added	$
	E2	Value added per direct employee	$
	E3	Net income	$
	E4	Capital investment on new technology	$
	E5	Product defect rate	%
Environmental sustainability	V1	Total net energy usage per unit value added	kWh/$
	V2	Total net energy usage per kg product	kWh/kg
	V3	Hazardous cleaning chemical usage per kg product	kg/kg
	V4	Hazardous plating chemical usage per kg product	kg/kg
	V5	Net water consumed per kg product	m^3/kg
	V6	Hazardous liquid waste per unit value added	kg/$
Social sustainability	S1	Number of complaints	/year
	S2	Number of process safety analyses	/year
	S3	Process safety index	

shown in the column "Normalized" of the same table are obtained based on the company's historical data, which are classified as "Worst" and "Best" in the same table. In the company a project team is formed to determine the weighting factors used for sustainability assessment. The agreed weighting factor values for the five economic

Table 10.5 Result of System Sustainability Assessment

Metrics		Current	Worst	Best	Normalized
Economic sustainability	E1	3.8×10^6	1.0×10^6	5.0×10^6	0.70
	E2	1.27×10^5	7.0×10^4	3.0×10^5	0.25
	E3	8.89×10^5	0	2.0×10^6	0.44
	E4	0	0	3.50×10^6	0.00
	E5	8%	15%	2%	0.54
Environmental sustainability	V1	1.06	1.6	0.5	0.49
	V2	0.85	1.6	0.4	0.63
	V3	0.0062	0.0085	0.0004	0.28
	V4	0.025	0.05	0.008	0.60
	V5	0.028	0.1	0.0025	0.74
	V6	0.06	0.1	0.005	0.63
Social sustainability	S1	20	100	0	0.80
	S2	12	0	52	0.23
	S3	65	0	100	0.65

indicators, six environmental indicators, and three social indicators are (1, 2, 1, 1, 3), (1, 1, 2, 2, 2, 5), and (1, 1, 1), respectively. The categorized performance of economic, environmental, and social sustainability is 0.41, 0.58, and 0.56, respectively, while the overall sustainability performance is 0.52, assuming that the company considers the equal importance of economic, environmental, and social sustainability.

Technology Candidate Selection

Huang and associates developed various P3 technologies for process performance improvement [9,10]. Four of them are considered as the candidates for sustainability performance enhancement in this work, which are described in the following.

Technology 1: The Cleaning and Rinse Operation Optimization Technology

In any plating line, the cleaning operation (e.g., presoaking, soaking, electrocleaning, and acid cleaning) is always followed by one or two rinsing steps. Chemical and water consumption are largely dependent on the settings of chemical concentrations in cleaning units, the chemical feeding policy, the rinsing water flow rate, as well as the cleaning and rinse times. In the industry, the parts are treated in the cleaning and rinsing units without taking into consideration the dynamic chemical concentration in the units because of chemical reactions between the cleaning chemicals and the dirt on parts. In practice, the concentration of cleaning chemicals in the unit is adjusted periodically rather than continuously. Thus the cleaning time, if set as a constant, often leads to overcleaned parts immediately after the chemicals are added, which also gives rise to a higher chemical loss from the cleaning units to the rinsing units, and in turn the rinse water consumption becomes higher. Those parts to be cleaned later become undercleaned as the chemical concentration in the cleaning units becomes lower, which may cause off-specification parts before plating. Based on a two-layered hierarchical dynamic optimization technique, the optimal settings for chemical concentration and rinsing water flow rate are identified for unit-based consumption minimization in the lower layer of this technology. In the upper layer the processing time distributions for all the cleaning and rinse operations are adjusted so as to explore the global opportunities of minimizing the overall operating cost and waste generation. The developed technology is capable of generating a dynamically adjustable cleaning and rinsing operation, based on the evaluation of job order change, waste generation in different process units, chemical and energy consumption, etc. This technology can contribute significantly to the minimization of the quantity and toxicity of wastewater while maintaining the production rate [3].

Fig. 10.4 depicts the change of dirt residue on the parts before and after implementing this technology. An electroplating process with conventional operating approach is shown in Fig. 10.4(A). Because of the consumption of cleaning chemical over time, the parts entering the cleaning unit at the beginning would have an overcleaning issue while the ones cleaned in the end would not receive sufficient cleaning if constant treatment time is applied. Both scenarios may lead to product quality issues. With the application of this technology, parts are equally cleaned while the reduction of chemical and water usage as well as a rise of production rate are achieved simultaneously (Fig. 10.4(B)).

The adoption of this technology will lead to a substantial reduction in the usage of cleaning chemicals and fresh water, which also results in a significant reduction of hazardous waste emissions. The production rate will have a small rise while energy

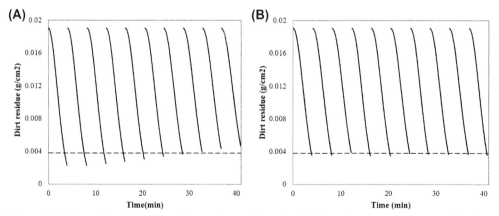

Figure 10.4 Dynamics of the dirt residue on the surface of parts through a cleaning process: (A) using a conventional cleaning technique and (B) using an optimized cleaning technique.

consumption slightly decreases. Note that a process safety check and analysis is needed to avoid any accidents, especially when frequent feeding of chemicals into cleaning units is manually accomplished.

Technology 2: The Optimal Water Use and Reuse Network Design Technology

In a plating line, fresh water is fed to different rinse units for rinsing off the dirt and solution residues on the surface of parts. It is possible that the water used in specific rinsing units can be either partially or entirely reused by other rinsing units without compromising product quality. By this technology an optimal water allocation network can be designed for a plating line of any production capacity, and the optimal operation strategy for the network can also be developed based on rinsing water flow dynamics [7,21]. Fig. 10.5(B) describes a modified water use and reuse network based on this technology. By comparing to the traditional electroplating process in Fig. 10.5(A), this technology maximizes the use and reuse of water, which leads to substantial sustainable development.

The major advantage of Technology 2 is the reduction of water consumption and thus the load of wastewater in the wastewater treatment facility in the company. A possible risk of the use of this technology may occur when the initial dirt on the surface of parts when engineering the plating company varies significantly. This requires a conservative setting of the rinse water flow rates in different rinsing units; otherwise, some parts initially very dirty on their surface may not be rinsed enough by used rinse water. Thus more frequent checking of the rinse quality of parts will be required.

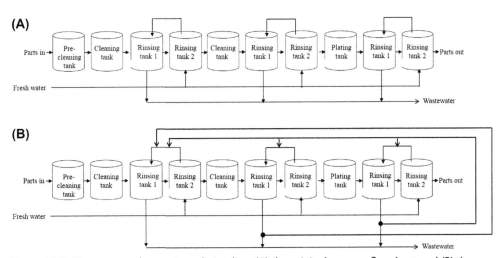

Figure 10.5 Water use and reuse in a plating line: (A) the original process flowsheet and (B) the new process flowsheet with an embedded optimal water use and reuse network design technology.

Technology 3: The Near-Zero Chemical and Metal Discharge Technology

In the plating line, chemical solvents and plating solutions are consumed excessively. This technology is developed for designing an effective direct recovery system based on a reverse drag-out concept that can minimize drag-out-related chemical/metal loss [22]. Fig. 10.6(A) depicts a traditional plating process, in which the plated parts are processed in a series of rinsing units with flow rinsing water to wash off the remaining plating solution. A modified process using this technology is depicted in Fig. 10.6(B). Static rinsing allows solution recovery where fresh water is periodically fed into rinse unit R_N first, and then the solution-containing rinse water in R_N flows to R_{N-1} until R_1 periodically. Finally, the solution containing rinse water in R_1 is periodically pumped back to plating unit E to maximize the use of plating solution. The process modification can also be applied to the cleaning process to maximize the use of cleaning chemicals and avoid unnecessary drag-out.

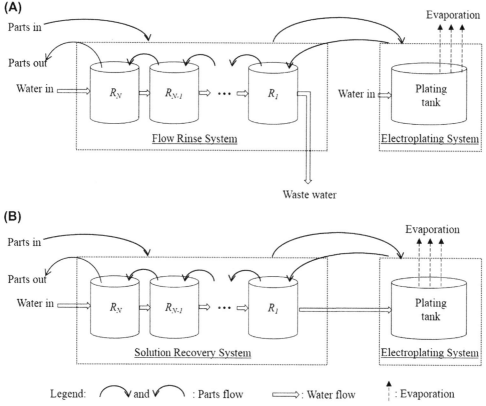

Figure 10.6 Design schemes for electroplating and rinsing: (A) the original electroplating process with a flow rinse system and (B) the modified electroplating process with a solution recover system using a static rinse system.

Using this technology, water consumption can be reduced dramatically because of the static rinsing. The usage of cleaning and plating chemicals can also be reduced accordingly. However, the drag-out minimization strategy if a longer drainage time is adopted could cause a decrease of production rate and an increase of energy consumption because of additional processing time. In the meanwhile, more frequent process checks are also needed to ensure process safety.

Technology 4: The Environmentally Conscious Dynamic Hoist Scheduling Technology [23]

Source reduction can also be achieved through dynamic hoist scheduling in production. With the unit-based minimization of chemical and water consumption while meeting the requirement of product quality, the amount of waste to be generated in each unit can then be calculated. An optimal hoist schedule with maximum production rate and minimum waste generation can be identified, consequently using various techniques, such as graph-assisted search algorithm [24].

With the application of this technology, a dynamically adjustable production schedule can be obtained based on the evaluation of job order change, waste generation in different process units, and chemical and energy consumption. In the meanwhile, parts are cleaned and plated with the reduction of chemical and water usage simultaneously. The dynamic control of the plating process using the technology results in a significant reduction of nonvalue-added time and increase of production rate thereafter. The usage of cleaning and plating chemicals as well as energy consumption can also be minimized. However, it requires more frequent product quality analysis.

Sustainability Assessment of Technologies

The four selected technologies can improve system performance in different ways, which can be demonstrated through sustainability assessment using appropriate sustainability indicators. For the four technologies, there are a total of 15 different technology sets, including four sets with one technology each, six sets with two different technologies each, four sets with three different technologies each, and one set of all four technologies.

Sustainability assessment of the four individual technologies and sustainability improvement potential are conducted first. Tables 10.6–10.9 show the evaluation of the sustainability performance after application of each technology in the process. Note that the application of technologies may result in a decrease of production rate or an increase of energy consumption and process safety index, and thus the improvement of certain categories of sustainability can be negative. The investments for the use of the four technologies are 9.45×10^4, 6.55×10^4, 7.75×10^4, and 1.26×10^5, for Technologies 1, 2, 3, and 4, respectively. As one of the key indicators in economic sustainability, the investment on technology is also included in the evaluation (E4). The

Table 10.6 Sustainability Assessment of Technology 1

Metrics		Technology 1	Normalized Result	Improvement
Economic sustainability	E1	3.86×10^6	0.72	0.02
	E2	1.29×10^5	0.26	0.01
	E3	9.60×10^5	0.48	0.04
	E4	9.45×10^4	0.24	0.24
	E5	4%	0.85	0.31
Environmental sustainability	V1	1.05	0.50	0.00
	V2	0.84	0.63	0.00
	V3	0.003	0.68	0.40
	V4	0.02	0.71	0.12
	V5	0.0264	0.75	0.02
	V6	0.035	0.68	0.05
Social sustainability	S1	20	0.80	0.00
	S2	24	0.46	0.23
	S3	50	0.50	−0.15

Table 10.7 Sustainability Assessment of Technology 2

Metrics		Technology 2	Normalized Result	Improvement
Economic sustainability	E1	3.75×10^6	0.69	−0.01
	E2	1.25×10^5	0.24	−0.01
	E3	8.71×10^5	0.44	−0.01
	E4	6.55×10^4	0.16	0.16
	E5	7%	0.62	0.08
Environmental sustainability	V1	1.10	0.45	−0.04
	V2	0.88	0.60	−0.03
	V3	0.005	0.43	0.15
	V4	0.025	0.60	0.00
	V5	0.0211	0.81	0.07
	V6	0.042	0.61	−0.02
Social sustainability	S1	40	0.60	−0.20
	S2	24	0.46	0.23
	S3	45	0.45	−0.20

sustainability assessment is shown in Table 10.10. The efficiency of capital investment is listed in the last column of Table 10.10.

Technology Recommendation

It is assumed that this plating company sets its economic, environmental, and social sustainability goals, E^{sp}, V^{sp}, and L^{sp}, to 0.51, 0.72, and 0.61, respectively. The overall sustainable development goal is thus 0.62, according to Eq. [10.13]. The overall limit of investment is defined as $\$2.2 \times 10^5$ at the same time.

Table 10.8 Sustainability Assessment of Technology 3

Metrics		Technology 3	Normalized Result	Improvement
Economic sustainability	E1	3.70×10^6	0.68	−0.02
	E2	1.23×10^5	0.23	−0.01
	E3	8.70×10^5	0.43	−0.01
	E4	7.75×10^4	0.19	0.19
	E5	6%	0.69	0.15
Environmental sustainability	V1	1.11	0.44	−0.05
	V2	0.89	0.59	−0.04
	V3	0.004	0.56	0.27
	V4	0.022	0.67	0.07
	V5	0.0254	0.77	0.03
	V6	0.01	0.95	0.32
Social sustainability	S1	15	0.85	0.05
	S2	24	0.46	0.23
	S3	57	0.57	−0.08

Table 10.9 Sustainability Assessment of Technology 4

Metrics		Technology 4	Normalized Result	Improvement
Economic sustainability	E1	4.07×10^6	0.77	0.07
	E2	1.36×10^5	0.29	0.04
	E3	1.08×10^6	0.54	0.09
	E4	1.26×10^5	0.31	0.31
	E5	3%	0.92	0.38
Environmental sustainability	V1	1.01	0.54	0.04
	V2	0.81	0.66	0.03
	V3	0.002	0.80	0.52
	V4	0.017	0.79	0.19
	V5	0.026	0.76	0.02
	V6	0.035	0.68	0.05
Social sustainability	S1	20	0.80	0.00
	S2	24	0.46	0.23
	S3	60	0.60	−0.05

According to step (e) of the technology identification procedure, technology set Nos. 12, 13, 14, and 15 are eliminated at the first place because of excess capital investment compared to the budget limit. The overall sustainability improvement brought by the application of technology set Nos. 1, 2, 5, 8, and 9 does not meet the requirement (i.e., 0.62). Technology set No. 3 can only enhance the economic sustainability to 0.48. The performance of social sustainability with the application of technology set No. 11 is 0.59, which is under the limit of 0.61. Therefore only technology set Nos. 4, 6, 7, and 10 meet all the requirements for sustainability improvement. Table 10.11 shows the analysis

Table 10.10 Results of Sustainability Improvement With Respect to Different Technology Options

No.	Selected Technology	E	V	L	S	B ($\times 10^5$ $)	I_{eff} ($\times 10^{-8}$)
1	T(1)	0.56	0.68	0.59	0.61	9.45	9.60
2	T(2)	0.45	0.60	0.50	0.52	6.55	0.10
3	T(3)	0.48	0.75	0.63	0.63	7.75	13.94
4	T(4)	0.62	0.72	0.62	0.65	12.55	10.59
5	T(1,2)	0.49	0.71	0.55	0.59	15.2	4.68
6	T(1,3)	0.51	0.80	0.61	0.65	16.34	7.88
7	T(1,4)	0.60	0.73	0.60	0.65	20.9	5.99
8	T(2,3)	0.47	0.76	0.56	0.61	13.59	6.49
9	T(2,4)	0.51	0.71	0.58	0.61	18.15	4.72
10	T(3,4)	0.53	0.82	0.63	0.67	19.29	7.75
11	T(1,2,3)	0.50	0.80	0.59	0.64	21.38	5.79
12	T(1,2,4)	0.53	0.73	0.58	0.62	25.69	3.92
13	T(1,3,4)	0.56	0.82	0.63	0.68	26.78	6.01
14	T(2,3,4)	0.52	0.81	0.59	0.65	24.17	5.52
15	T(1,2,3,4)	0.55	0.85	0.63	0.69	30.86	5.37

Table 10.11 Results of Sustainability Decision-Making Analysis

No.	Selected Technology	E	V	L	S	B ($\times 10^5$ $)	I_{eff} ($\times 10^{-8}$)
4	T(4)	0.62	0.72	0.62	0.65	12.55	10.59
6	T(1,3)	0.51	0.80	0.61	0.65	16.34	7.88
7	T(1,4)	0.60	0.73	0.60	0.65	20.9	5.99
10	T(3,4)	0.53	0.82	0.63	0.67	19.29	7.75

result (the technology sets use the same index number for consistency). The final results are then prioritized under three different orders. If the company wants to achieve the maximum improvement of sustainability, then the technology set No. 10 is the top choice while set Nos. 4, 6, and 7 can reach same sustainability performance. If the company prefers the lowest investment, then the order changes to Nos. 4, 6, 10, and then 7. If the investment efficiency is the priority, then the recommended technology sets are in the order of Nos. 4, 6, 10, and then 7.

CONCLUDING REMARKS

The electroplating industry has been greatly challenged because of economic globalization. Many plating companies maintain a low-profit business, and the technologies used are mostly very traditional. The industry is still among the most polluted in manufacturing industries. How to ensure this industry's sustainable development has been a major concern of the industry over the past two decades. It becomes imperatively important that the industry should conduct comprehensive sustainability assessment on virtually all plating companies and then develop short-to-long-term strategies for stage-wise sustainability performance improvement.

In this chapter, we introduced a sustainability metrics system for sustainability performance evaluation. Although this metrics system is currently very basic, it could serve as the starting point in this direction. The framework for assessment shows a systematic yet simple method to derive assessment results for plating lines as well as technologies potentially useful for sustainability performance improvement. A technology selection procedure facilitates identification of technology alternatives based on a company's development goal, strategies, and budget availability. A case study on an industrial process, although simplified, has demonstrated the efficacy of the introduced sustainability performance improvement methodology.

ACKNOWLEDGMENT

This work is support in part by NSF (award nos. 1434277 and 1140000).

REFERENCES

[1] Chase L. Electroplated thermoplastic automotive grille having improved flexibility. U.S. Patent No. 5487575. 1996.
[2] Zhang J, Yang Z, An M, Tu Z, Li M. A new process of electroplating on titanium and titanium alloy for aerospace. Trans Inst Met Finish 1996;74(1):25—7.
[3] Gong J, Luo K, Huang Y. Dynamic modeling and simulation for environmentally benign cleaning and rinsing. Plat Surf Finish 1997;84(11):63—70.
[4] Schlesinger M, Paunovic M. Modern electroplating. Hoboken (NJ): John Wiley & Sons; 2011.
[5] Luo K, Huang Y. Intelligent decision support for waste minimization in electroplating plants. Eng Appl Artif Intell 1997;10(4):321—33.
[6] Luo K, Gong J, Huang Y. Modeling for sludge estimation and reduction. Plat Surf Finish 1998;85(10):59—64.
[7] Yang Y, Lou H, Huang Y. Optimal design of a water reuse system in an electroplating plant. Plat Surf Finish 1999;86(4):80—4.
[8] Yang Y, Lou H, Huang Y. Synthesis of an optimal wastewater reuse network. Waste Manage 2000;20(4):311—9.
[9] Lou H, Huang Y. Profitable pollution prevention: concept, fundamentals & development. Plat Surf Finish 2000;87(11):59—66.
[10] Xiao J, Huang Y. Technology integration for sustainable manufacturing: an applied study on integrated profitable pollution prevention in surface finishing systems. Ind Eng Chem Res 2012;51(35):11434—44.
[11] Piluso C, Huang Y. Collaborative profitable pollution prevention: an approach for the sustainable development of complex industrial zones with uncertain information. Clean Technol Environ Policy 2009;11(3):307—22.
[12] Costa D, Diniz JC, Pagan R. Sustainability metrics for coal power generation in Australia. Process Saf Environ 2006;84(2):143—9.
[13] Sikdar S. Sustainable development and sustainability metrics. AIChE J 2003;49(8):1928—32.
[14] Clift R. Sustainable development and its implications for chemical engineering. Chem Eng Sci 2006;61(13):4179—87.
[15] López M, Garcia A, Rodriguez L. Sustainable development and corporate performance: a study based on the Dow Jones sustainability index. J Bus Ethics 2007;75(3):285—300.

[16] Saling P, Kicherer A, Dittrich-Krämer B, Wittlinger R, Zombik W, Schmidt I, et al. Eco-efficiency analysis by BASF: the method. Int J Life Cycle Assess 2002;7(4):203−18.

[17] Shonnard D, Kicherer A, Saling P. Industrial applications using BASF eco-efficiency analysis: perspectives on green engineering principles. Environ Sci Technol 2003;37(23):5340−8.

[18] Landsiedel R, Saling P. Assessment of toxicological risks for life cycle assessment and eco-efficiency analysis. Int J Life Cycle Assess 2002;7(5):261−8.

[19] Liu Z, Huang Y. Sustainable distributed biodiesel manufacturing under uncertainty: an interval-parameter-programming-based approach. Chem Eng Sci 2013;93:429−44.

[20] Liu Z, Huang Y. Technology evaluation and decision making for sustainability enhancement of industrial systems under uncertainty. AIChE J 2012;58(6):1841−52.

[21] Zhou Q, Lou H, Huang Y. Design of a switchable water allocation network based on process dynamics. Ind Eng Chem Res 2001;40(22):4866−73.

[22] Xu Q, Telukdarie A, Lou H, Huang Y. Integrated electroplating system modeling and simulation for near zero discharge of chemicals and metals. Ind Eng Chem Res 2005;44(7):2156−64.

[23] Kuntay I, Uygun K, Xu Q, Huang Y. Environmentally conscious hoist scheduling in material handling processes. Chem Eng Commun 2006;193:273−93.

[24] Xu Q, Huang YL. Graph-assisted optimal cyclic hoist scheduling for environmentally benign electroplating. Ind Eng Chem Res 2004;43:8307−16.

CHAPTER ELEVEN

Strategic Sustainable Assessment of Retrofit Design for Process Performance Evaluation

A. Carvalho
CEG-IST, University of Lisbon, Lisbon, Portugal

INTRODUCTION

Globalization associated with global economic crisis is imposing multiple burdens to business in general. Companies are forced to take actions to improve their business models toward a sustainable strategy. The World Commission on Environment and Development, through the well-known Brundtland Report, set forth the importance of new coordinated strategies among all the involved stakeholders, which put into practice the sustainable development [1]. This situation is imposing huge challenges to supply chain managers, especially at the process level, since this type of entity (factories) involves high costs and is responsible for significant environmental and social impacts in supply chains [2]. Retrofit design is a key action to achieve sustainable processes and in a holistic view to achieve sustainable supply chains. To propose sustainable retrofit designs, methods and tools that measure, compare, and assess sustainability in retrofit design are critical [3]. Over the past decades several metrics, fitting retrofit design assessment, have been proposed [4,5]. Nevertheless, sustainability still lacks clarity, especially on the standardization of what to measure and how to measure sustainability in retrofit designs. The proliferation of these metrics has raised several questions on the most suitable areas of assessment for retrofit design depending on the proposed design. The aforementioned problem indicates that literature is lacking on the presentation of guidelines that help decision makers to define the boundary of their analysis (e.g., supply chain or process level) and to select the most appropriate areas of assessment. The objective of this chapter is to present an overview of the current state of the art on the areas of process performance improvements that should be covered in the three pillars of sustainability assessment (economic, environmental, and social [6]), and to present a generic framework to guide decision makers in their sustainable retrofit design analysis, contributing to fill the identified research gaps. The framework indicates the required level of detail for the analysis, proposing the boundary of the study and indicating the most suitable areas for improvements.

Sustainability in the Design, Synthesis and Analysis of Chemical Engineering Processes

249

This chapter is organized as follows. In section: State of the Art an overview of the state of the art on sustainable retrofit design tools and metrics is presented. In section: Framework for Assessment of Retrofit Design Alternatives the strategic sustainable assessment of retrofit design (SARD) framework is presented with a step-by-step description. A case study is presented in section: Case Study: β-Galactosidase Production to highlight the application of the framework. Finally, conclusions are drawn in section: Conclusions.

STATE OF THE ART

Section: Sustainability in Retrofit Design underpins the importance of sustainability in the emerging theme of sustainable in retrofit design. Then a literature review on the areas of assessment required for retrofit design, which cover the three pillars of sustainability (sections: Economic Pillar, Environmental Pillar, and Social Pillar), will be presented.

Sustainability in Retrofit Design

Retrofit design has been defined by Guinand [6a] as follows:

Process retrofitting is the redesign of an operating chemical process to find new configuration and operating parameters that will adapt the plant to changing conditions to maintain its optimal performance.

Sustainability has shifted the paradigm in process design and consequently at retrofit actions. Clearly, retrofit design is critical to achieve viable changes toward a better process state in terms of process sustainability [7]. The introduction of sustainability aspects in process design implies that new process design alternatives should be designed, attaining optimal performance, not only at the economic level, but also at the environmental and social levels. Considering sustainable aspects in the early steps of retrofit design makes process redesign more efficient [8,9]. Retrofit design involves a series of actions, which can be summarized into four steps: (1) identify the bottlenecks in the process; (2) select the most relevant bottlenecks for improvements; (3) suggest new design alternatives that eliminate the identified bottlenecks; and (4) assess and select new design alternatives [10].

For the first step, Rong et al. [11] identified six types of bottlenecks that are primarily found in the industry: (1) scale—problems related to operation conditions used for a given size of the equipment; (2) energy consumption—problems related to utility consumptions; (3) raw material consumption—problems related to the inefficient reaction and/or separation operations; (4) environmental impact—problems related to emissions; (5) safety—problems related to the use of compounds and operations that involve risk in the safety of the process; and (6) feedstocks—problems related to market conditions. Regarding the second and third steps of the retrofit design process (Fig. 11.1), scientists in this field have developed several methodologies and tools that select the most relevant bottlenecks and suggest new sustainable design alternatives [4].

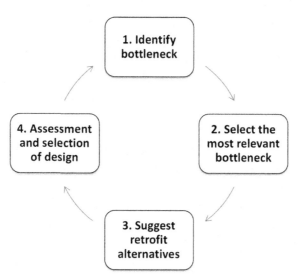

Figure 11.1 Retrofit design steps.

Azapagic et al. [12] proposed a methodology that integrates sustainability aspects into process design. The methodology guides the user through different design stages, suggesting the most appropriate technical, economic, environmental, and social criteria depending on the design stage. Bumann et al. [13] presented a methodology based in path flow indicators, which identify process alternatives with an improved performance. This methodology includes the unit occupancy time and the multiobjective process assessment to assess sustainable retrofitting actions.

Jayswal et al. [14] presented a methodology grounded on a sustainability root cause analysis, which employs the combination of a Pareto analysis and the fishbone diagram. This methodology helps decision makers to identify bottlenecks. Torres et al. [15] presented an environmental evaluation procedure for process design. The environmental performance of a process is employed and used as an objective function in combination with monetary and social concerns to achieve sustainable designs. Banimostafa et al. [16] presented a method based on principal component analysis, which helps the decision maker in the selection between various synthesis paths for the production of chemical compounds during the early phases of process design. Zheng et al. [5] presented a methodology to assist reaction pathway selection taking into consideration sustainability issues. This methodology helps decision makers establishing retrofit design alternatives. Later, Carvalho et al. [17] presented *SustainPro*, which is a tool for the screening and identification of the most critical bottlenecks in the processes, suggesting new sustainable design alternatives. *SustainPro* requires as input data the reference design/operational data of the process, such as the mass flows, the energy flows, and the prices of all components within the system. *SustainPro* will first perform a flowsheet decomposition, identifying all component mass and energy paths. Then it will calculate a set of mass and energy indicators that trace the paths of the components' "mass flows" and "energy flows" as they enter (or are generated) and leave (or are consumed) the process. With this information the critical points of the process will be determined, followed by a sensitivity analysis to identify the critical points, which have the potential for realizing the most sustainable improvements in the process. Through this sensitivity analysis the target indicators (that is the ones to be changed) will be selected and once target values are established, a local sensitivity analysis will be performed to determine the design and the operational parameters that influence those targets. Finally, *SustainPro* involves the matching of the desired target through changes in design/operational variables by using a set of algorithms for process synthesis that allow the determination of the new flowsheet/operational alternative.

For the last step of retrofit design (Fig. 11.1) several tools have been developed in the past decades to assess sustainability. For instance, "sustainability evaluator" is an impact assessment tool, intended to evaluate sustainability in processes, considering selected metrics that address economic, environmental, and health and safety concerns [18,19]. GREENSCOPE is a tool developed by the US Environmental Protection Agency (EPA), which calculates sustainability metrics to assess processes and helps decision makers in their assessment [3]. Kalakul et al. [9] proposed a framework, which integrates several tools, namely, *SustainPro*, ECON (software for economic analysis), and LCSoft (software for the evaluation of environmental impact of design of chemical—biochemical processes), proposing a global approach that covers all the steps of retrofit design.

The previous methodologies and tools cover nonstandardized sustainability metrics. With the myriad metrics it is difficult to select the most appropriate areas of improvement; therefore in the next section the most relevant areas of improvement of the three pillars of sustainability will be scrutinized and systematized.

Economic Pillar

Nowadays the economic pillar of sustainability represents the organizations' competitiveness in the demanding markets. Managers and engineers should be able to ensure the reliability of business through its profitability, so that business can move forward [20]. Traditionally, engineers working in chemical process design have focused their efforts in assessing retrofit design in terms of their economic viability [21], which is without doubt a prerequisite for any business financial viability [22]. A business can only be concerned with social and environmental aspects if it is economically viable. Therefore both academia and practitioners have proposed several economic indicators and metrics, covering different aspects of their processes [23]. The economic indicators should be the primary source of analysis in retrofit design, since they indicate the reliability of the proposed retrofit designs. The retrofit design alternatives can span from some minor changes at the operational level to changes that require new equipment or even new compounds with different supply chains. The retrofit design alternatives will require different degrees of process adaptations; this implies that retrofit designs should not be evaluated at the same level of detail. Assessing design alternatives is usually based on the microeconomics indicators, lacking the analysis of the big picture, which incorporates the macro-level economic indicators and the whole supply chain [12]. Established on an extensive literature review, Fig. 11.2 presents a systematization of the most important economic assessment areas that should be covered in retrofit design's evaluation.

Fig. 11.2 presents areas of improvement considering two boundaries for the system: supply chain assessment and process assessment. The economic indicators have been classified in these two levels, because: (1) some retrofit actions influence the different stakeholders and therefore the economic aspects of the supply chain should be considered; (2) some retrofit actions might have little influence on the external stakeholders and therefore the economic evaluation can be focused on the economic performance of the company.

One example of a retrofit design action, which might influence the external stakeholders, is the replacement of a chemical compound in the process. The new compound

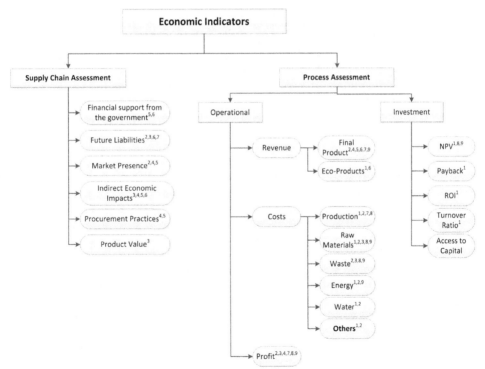

[1]Ruiz-Mercado et al. (2011), [2]WBCSD (2000), [3]Powell (2010), [4]Al-Sharrah (2010), [5]GRI (2004), [6]Tanzil (2006), [7]Azapagic et al. (2006), [8]Shokrian et al. (2014), and [9]shadiya (2013)

Figure 11.2 Economic assessment areas to evaluate retrofit design alternatives.

might be acquired from new suppliers, the production process of that compound might be different, or the transportation mode might have to be replaced, among many other restructures that might occur in the supply chain. Therefore the whole life cycle of the product should be contemplated. To account for the role of the supply chain in society, a set of assessment areas must be considered in the economic life cycle assessment, including the different stakeholders (Fig. 11.2). *Financial support from the government* should evaluate the financial assistance given from the government in terms of subsidies, tax credits, and access to capital. Different financial support can be given from national to local governments; therefore these incentives might have a significant influence in the supply chain (e.g., selection of suppliers, transportation mode, etc.). *Future liabilities* assess liabilities that might arise for a specific retrofit design alternative, including costs of externalities, environmental liabilities, insurance across stakeholders, fines, legal fees, and business interruptions. Different design alternatives might imply different liabilities at the process level and at the stakeholders' level. *Market presence* assesses aspects such as market share, market mix, market growth, wages, and hiring of local people. In this area of improvement all the costs incurred with the market presence and the attitude of the supply chain stakeholders' toward the society are considered. *Indirect economic impacts* account for all the costs that are not directly related to the value-added supply chain activities, for instance, economic development in high poverty areas, externalities, enhancing skills among the community, availability of services, reputation, customer acceptance, and loyalty. These indicators enlarge the scope of the analysis and allow more concise decisions. *Procurement practices* consider the analysis of the suppliers' selection and vertical integration options. Depending on the procurement decisions, hiring local suppliers might be an option for some design alternatives, while for other alternatives the local suppliers are not available. These aspects should be considered when a sustainable supply chain is at stake. Finally, *Product value* covers aspects related to the product in the market, this includes the perception of the product into the market, the price of the product, and the quality and safety of the final product. Retrofit design alternatives might lead to higher competitive advantages through an improved product in the market or in an opposite way; they might compromise the reliability and the viability of the product in the eyes of the consumer. This aspect is a major avenue towards a sustainable supply chain.

For the retrofit action, which does not affect the external stakeholders, the process assessment area is the most suitable boundary for sustainability analysis (Fig. 11.2). In the process assessment, the areas of improvement are the traditional economic indicators applied to retrofit design analysis. The operational results as well as the investment analysis of the proposed retrofit designs are accounted for. The operational results are the traditional indicators applied by engineers in retrofit design assessment [12]. They should be always contemplated, since after the implementation of a retrofit design those costs are going to be incurred at least for the next 5 years (minimum lifetime for a project). Regarding the operational results, several authors point out different aspects that should be covered in retrofit design assessment; nevertheless, there are three key aspects that any retrofit analysis should include: (1) revenue; (2) operating costs; and (3) profit. The aforementioned aspects can be analyzed in more detail, according to the proposed retrofit design. For instance, if a design alternative involves energy reduction, it might be important to contemplate specifically the energy costs, so that alternatives are compared in a structured way. In addition, the profit of the process can be calculated with different levels of detail, from the gross margin, the earnings before interest and taxes, to the profit after taxes. The detail of the calculation is specified by the decision maker. Fig. 11.2 presents a series of specific areas under the main three aspects, which are case dependent.

The investment analysis includes the traditional project management indicators, namely, net present value (NPV), payback time, return on investment (ROI), turnover ratio, and access to capital. If these indicators are not favorable, the companies' board will never approve the proposed retrofit design project, which makes these indicators mandatory for any retrofit design assessment involving investments.

Some tools can be used in order to calculate the economic indicators. For instance, ECON [29] is an economic analysis software. The input for the software is the dimensions of all process equipment and the prices of raw materials and final products, then the software estimates costs and revenues; presenting as an output the traditional economic indicators for process investment analysis, for example, the total capital investment cost, operating cost, ROI, and NPV.

Environmental Pillar

"The environmental dimension of sustainability concerns an organization's impacts on living and non-living natural systems, including ecosystems, land, air, and water" [25, p. 52]. Companies have become aware of environmental issues because of the significant impact of production on the environment. Moreover, stakeholders and governments are forcing companies to restructure their production processes, so that environmental impacts are reduced [30]. Several methods have been presented in the literature to assess the environmental retrofit design [31]. The life cycle assessment (LCA) approach has been put into practice through the development of several methods. LCA methodology enables the assessment of the environmental impacts of a product, service, or process over its whole life cycle and its general requirements, goal, and scope are well defined by the International Organization for Standardization [32]. LCA attempts to quantify all inputs and outputs needed to all entities of the supply chain [33], identifying and quantifying the energy and materials used, and wastes released to the environment of those inputs and outputs [34]. Carvalho et al. [31] presented an extensive literature review on environmental impact methods (25 methods), available to assess retrofit design alternatives. The authors suggest grouping the reviewed methods into three groups according to the boundary of the system: (1) single issue; (2) process; and (3) LCA intended (Table 11.1).

Table 11.1 Classification of Environmental Impact Assessment Methods

Single Issue	LCA Intended	Process
CED [35]	CML [36]	EFRAT [37]
CExD [38]	Eco-indicator 99 [39]	EPI [40]
Ecological Footprint [41]	Ecological Scarcity 2006 [42]	MBEI [15]
EDP [43] and Koellner and Scholz [43a]	EDIP 2003 [44]	WAR [45]
IPCC2007 [46]	Impact 2002+ [47]	
USES-LCA [48]	LCA Nets [49]	
USEtox [50]	LIME [51]	
	LUCAS [52]	
	MEEuP [53]	
	ReCiPe [53a]	
	TRACI [54]	
	EPS2000 [55]	

The single issue methods allow users to assess the environmental impact from a single issue perspective. For example, the Intergovernmental Panel on Climate Change (IPCC) 2007 [46] only assesses global warming and the USEtox [50] investigates the impacts of toxicity (ecosystems and human health). Process methods were presented in the literature as methods where the boundary is defined at the chemical processes. For instance, the waste reduction (WAR) algorithm [45] is the most important method in this category, since it is the most widely applied in the chemical process' assessment. Another example is the environmental fate and risk assessment tool (EFRAT) method [37], which is intended to assess gate-to-gate processes. Finally, the LCA intended methods were developed to assess environmental impacts in an LCA context, covering the assessment of all entities in the supply chain. The methods shown in Table 11.1 present several impact categories with similar meaning and sometimes with the same denomination. Therefore Carvalho et al. [31] grouped these impact categories into categorical groups (groups obtained by aggregation of all impact categories that are alias, that have similar meaning, or that complement each other in assessing a certain effect). The 25 methods accounted for 167 different impact categories that were grouped into 64 impact categorical groups. The categorical groups represent the environmental aspects, which can be assessed through the application of the aforementioned methods. Several categorical groups assess aspects related to the same area of improvement, therefore these aspects were further aggregated into three classes, ecological, resources, and human health. The ecological and the human health classes evaluate the impact of the process' emissions in the ecosystems and human heath, while the resource class evaluates the extraction of raw materials. These classes are independent and they are not overlapped. The methods presented in Table 11.1 cover the categorical groups summarized in Table 11.2.

Table 11.2 Environmental Assessment Areas to Evaluate Retrofit Design Alternatives Impact Categorical Group

Resources	Ecological	Human Health	Others
Abiotic resources	Acidification	Carcinogenic	Damage to built structures
Base cation capacity	Acidification/ eutrophication (combined)	Causalities	Noise
Biotic resources	Air pollution	Human health	Product-specific emissions
Crop production capacity	Bioaccumulation	Human health—other effects	
Element reserves	Damage to flora	Human toxicity	
Energy	Ecotoxicity	Ionizing radiation	
Fish and meat production capacity	Eutrophication	Life expectancy	
Gravel extraction	Extinction of species	Morbidity	
Land use	Fish toxicity	Noncarcinogenic	
Mineral resources	Global warming	Nuisance	
Natural resources	Hazard substances	Respiratory effects	
Nonrenewable energy	Heavy metals		
Nonrenewable, fossil	Hydrosphere pollution		
Nonrenewable, metals	Impairment of soil fertility		
Nonrenewable, nuclear	Malodorous air		
Nonrenewable, primary forest	Oxygen consumption		
Production capacity for water	Ozone depletion		
Raw material consumption	Particulate matter		
Recycling effect	Persistent organic pollutants		
Renewable energy	Photochemical oxidation		
Renewable, biomass	Volatile organic compounds		
Renewable, geothermal	Waste		
Renewable, solar	Waste heat		
Renewable, wind			
Resources consumption			
Water			
Wood production capacity			

Adapted from Carvalho A, Mendes AN, Mimoso AF, Matos HA. From a literature review to a framework for environmental impact assessment index. J Clean Prod 2014;64:36—62.

To wrap up, the single issue methods are the simplest methods, covering only one aspect and avoiding tradeoffs, since the detail of the assessment is low. Process intended methods cover some aspects presented in Table 11.2 and they perform a gate-to-gate analysis (further details on the impact of categories should be consulted in the original references of the methods, presented in Table 11.1). The input data are all the inlet and outlet streams of the process. The most complex methods are the LCA methods, where the boundary can be extended to other entities of the supply chain. These methods cover different aspects and they require high-intensity data collection, thus they are more time-consuming methods. In order to select the most appropriate method the work presented by Carvalho et al. [31] should be consulted. The authors present a table where all the methods are described and the correspondent impact categories are identified for each method. There are several commercial software packages available to employ the aforementioned environmental methods, namely, *SimaPro* and *Gabi*, among others. The main drawback of these software packages is the license price, which might not be affordable to all companies, especially for small and medium enterprises. However, there are already some tools available for free, such as the WAR graphical user interface (WAR GUI) or LCSoft. WAR GUI has been developed by US EPA and allows the WAR algorithm analysis to be performed in a user-friendly way [56]. LCSoft [57] is an LCA software, which is exclusively developed for evaluation of chemical–biochemical processes. LCSoft presents three main features: (1) a knowledge management tool, which organizes the life cycle inventory data; (2) a calculation factor estimation tool, where missing characterization factors are estimated; and (3) calculation of LCA, where it calculates the carbon footprint and other environmental impact categories.

Social Pillar

"Social sustainability is a quality of societies. It signifies the nature–society relationships, mediated by work, as well as relationships within the society" [58, p. 72]. Social indicators are mainly related to qualitative and semiqualitative criteria, which are built on human intuitive and affected by the decision maker's knowledge and experience [14]. Social aspects of sustainability involve a variety of societal impacts, which are difficult to quantify. Consequently, several social metrics have been developed, assessing different impacts on societies [58a]. Chemical processes encompass critical social aspects, including materials and process safety, employees' accidents, and toxicity potential, among others [14]. Meckenstock et al. [1] identified that health and safety issues are the most relevant social aspects to be considered at the process level. All retrofit design alternatives require an assessment in terms of health and safety issues, thus it is of paramount importance to account for the legal issues and companies' image in terms of safety [14]. Several indices have been presented in the literature for evaluating processes. For instance, Khan and Abbasi [59] proposed the Hazard Identification and

Ranking System, Heikkila [60] presented the Inherent Safety Index, the Inherently Safer Design Index was presented by Gupta and Edwards [61], and more recently the Integrated Inherent Safety Index was developed by Khan and Amyotte [62]. The SUSTAINABILITY EVALUATOR presents a set of indicators to assess social sustainability through the evaluation of safety and health aspects [18]. However, retrofit design can go far beyond the health and safety issues at the plant. After considering the health and safety issues at the process level, labor conditions are the most relevant social aspect to take into consideration. Azapagic et al. [12] presented a set of social indicators covering the traditional health and safety issues; nonetheless, they add some other indicators related to the employees well-being. IChemE [63] considers the labor practices and decent work, a cornerstone for process design assessment. Therefore a set of metrics should be included to cover this aspect, namely, the number of employees hired for that process, benefits/promotions for the employees, types and rates of injuries, and training hours [63,25].

Nevertheless, as stated before, some retrofit actions might affect the whole supply chain, forcing the assessment of the whole social life cycle, evaluating the interaction of the process with the different stakeholders. Benoît et al. [63a, p. 158] defined social life cycle assessment as follows: "social life cycle assessment is a systematic process using best available science to collect best available data on and report about social impacts (positive and negative) in product life cycles from extraction to disposal." Simões [64] proposed 16 midpoint categories to assess social aspects in supply chains. Some midpoints are related to the labor conditions and health and safety. The remaining aspects cover business impact on the community, corruption in business, fair business operations, stakeholders' participation, basic human rights, human rights implementation, consumer health and safety, product management, and customer satisfaction. Fig. 11.3 summarizes the social areas of improvement to consider in retrofit design.

FRAMEWORK FOR ASSESSMENT OF RETROFIT DESIGN ALTERNATIVES

A framework called SARD, intended to guide the decision makers in their retrofit design assessment, is presented in this section. Fig. 11.4 presents the flow diagram of SARD's steps.

SARD's steps will be described in detail in the following points.

Step 1: Identify Bottleneck

The aim of this step is to characterize the problem in the study by identifying the type of bottleneck that should be improved in the retrofit deign. Decision makers should detect the bottleneck and classify it according to Rong et al. [11]: (1) scale; (2) energy consumption; (3) raw material consumption; (4) environmental impact; (5) safety; and

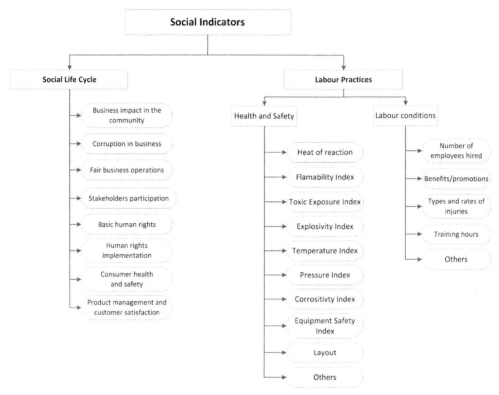

Figure 11.3 Social assessment areas to evaluate retrofit design alternatives.

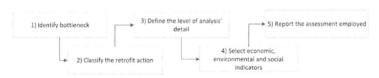

Figure 11.4 SARD framework for assessment of retrofit design alternatives.

(6) bottleneck of feedstock. This step is critical to underpin the possible retrofit design alternatives. Any methodology for bottleneck identification and for retrofit design alternatives proposal can be employed (see some options in section: Sustainability in Retrofit Design, earlier). The output of this step is the bottleneck identification.

Step 2: Classify the Retrofit Action

From the identified bottlenecks (step 1), new design alternatives should be proposed. The decision makers need to be aware of to what extent the retrofit actions are influencing the system. Depending on the retrofit actions undertaken, consequences of those actions might be situated at the operational level, with some small adaptations of the process, or might influence the whole supply chain, for example, with the replacement of raw materials. SARD framework proposes the classification of the retrofit actions boundary into three groups: (1) fine-tune operations; (2) investments in the process; and (3) supply chain changes. Table 11.3 lists some examples of retrofit actions that can be taken to eliminate the bottleneck and classify those actions according to their implications in the general system.

It is important to mention that scale bottlenecks are not usually related to supply chain changes and therefore their assessment does not require the most complex level of assessment. Regarding the feedstock bottlenecks, they are directly related to market changes, thus they are usually complex problems that influence the whole supply chain.

Summarizing, in this step new retrofit actions should be proposed and classified according to the classification presented in Table 11.3.

Table 11.3 Retrofit Action Classification Based on the Type of Bottleneck

| | **Retrofit Actions Classification** | | |
Bottleneck	Fine-Tune Operations	Investments in the Process	Supply Chain Changes
Scale	• Small adjustments at the operation conditions	• Replacement of equipment • New operations	–
Energy consumption	• Small adjustments at the operation conditions	• Replacement of equipment • New equipment with same utilities	• New sources of energy
Raw material consumption	• Small adjustments at the reaction • Recycling without extra investment	• Change reactor • Recycling requiring extra investment	• Change of catalyst • New raw materials • New solvents
Environmental impact	• Small adjustments at the operation conditions	• New treatment units • New operations	• New compounds
Safety	• Small adjustments at the operation conditions • New layout	• New equipment	• New compounds
Feedstock	–	• New operations for recovery of subproducts with market demand	• New raw materials • New products to sell

Step 3: Define the Level of Analysis' Detail

After classifying the retrofit design actions (step 2), the decision maker should determine the most suitable level of analysis' detail for assessing the retrofit designs. For that purpose, the SARD framework proposes a three-layer evaluation for the assessment of retrofit designs.

- **Level I: Quick sustainable operation decisions**: This assessment employs the most basic analysis of the retrofit design assessment. It requires less data and time to analyze the retrofit alternatives; however, it aggregates data and it does not include several aspects that could be important for the analysis. It will avoid possible tradeoffs that occur in more complex analyses; nonetheless, these tradeoffs might be important to achieve more sustained decisions.

- **Level II: Sustainable operation decisions**: The analysis performed at this level employs more precise methods, which focus on the implications of retrofit designs at the process level. This level gives a grounded process analysis and deeply studies the sustainability of the proposed retrofit actions. This analysis is more intensive data, since it requires more input information, therefore the time for the analysis will be longer than for Level I. Possible tradeoffs and critical decisions will be exposed through this analysis. Companies undertaking this type of analysis are aware of the importance of sustainability and they want to consider sustainability in their daily decisions.

- **Level III: Sustainable supply chain decisions**: The highest level of analysis considers the retrofit designs integrated in the whole system. This level of analysis represents a holistic view and it is intended for companies aiming to achieve sustainable supply chains. This analysis requires time and it is data intensive. It covers several aspects of the process and assesses the relationship between stakeholders. Decisions will include conflicting aspects, which should be weighted, so that a final decision can be achieved.

Figs. 11.5—11.7 present a guideline for economic, environmental, and social assessment in retrofit design. The decision makers might use these figures to identify the most appropriate detail of analysis, depending on the retrofit design actions. The user can use Figs. 11.5—11.7 in two distinct ways: (1) for companies without a sustainable strategy in their primary agenda, they should first identify the retrofit design action and classify it according to the retrofit action classification proposed in Table 11.3 (step 2 of SARD). Based on that classification, read in the figures the most appropriate level of detail for the sustainability analysis (Level I to III), and identify the most suitable areas of improvement required for the identified level of assessment (read in the top of the bars); (2) for companies with a strong emphasize on the development of a sustainable strategy, the user might decide the detail of the retrofit design assessment (Level I to III) based on the strategic practices of the company, independently from the retrofit design action. Based

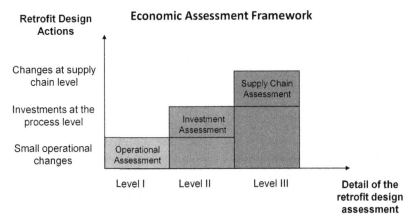

Figure 11.5 Characterization of the required level of detail for assessing the economic pillar in retrofit design decisions.

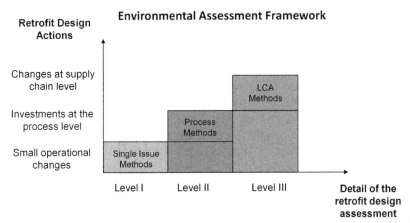

Figure 11.6 Characterization of the required level of detail for assessing the environmental pillar in retrofit design decisions.

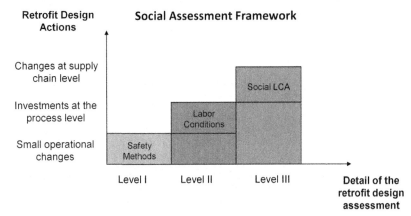

Figure 11.7 Characterization of the required level of detail for assessing the social pillar in retrofit design decisions.

on the assessment level decision, Figs. 11.5—11.7 indicate the most suitable areas of improvement to be assessed in a specific level of analysis for the three pillars of sustainability.

From this framework the decision maker has a guide to the best practices in terms of retrofit design assessment, but at the end, based on the available constraints, he should be the one deciding the level of analysis. For example, the decision maker identifies from the plots that the most suitable assessment level would be a Level III to all sustainability pillars; however, he might opt to do a Level III economic analysis, since he knows that shareholders prioritize this pillar and a Level II analysis for environmental and social aspects, to show that until a certain extent the selected design alternative is more sustainable. At this stage the practitioner should define the best strategy for the retrofit design assessment considering all possible constraints of the company.

Summarizing, the decision maker in this step has the role of deciding the level of the sustainability analysis performed to evaluate the retrofit design suggested in step 2. The current methodology helps to guide him in that decision.

Step 4: Select Economic, Environmental, and Social Indicators

After a conscious decision on the level of detail for the retrofit design assessment (step 3), the most suitable areas for improvement should be selected. In the sections Economic Pillar, Environment Pillar, and Social Pillar, earlier, a guide through the most appropriate areas of assessment and methods is presented. It is important to highlight that areas of improvement, which should be assessed for a sustainable retrofit action, are selected depending on the retrofit action, which will be considered. For instance, if the retrofit action is related to energy reduction, more specific indicators covering energy costs should be contemplated in the analysis. It is important to mention that Level III analysis includes the analysis of the previous levels. For instance, if a third-level economic analysis should be conducted, the supply chain areas of assessment listed in Fig. 11.2 should be assessed, as well as the process assessment area, where the investment and the operational areas are assessed. The specific indicators to assess those areas of improvement are described in the aforementioned literature review.

Step 5: Report the Assessment Employed

In the previous step the decision maker has determined the levels of detail for the analysis. The degree of detail in the assessment of retrofit design might be different for the three pillars of sustainability, spanning from Level I to Level III. A company's board and external stakeholders should be aware of the followed strategy, in terms of the three sustainability pillars, taken in the retrofit process decisions. The SARD framework proposes a radar format to inform and monitor the company about their employed strategies, regarding to sustainable retrofit design. Retrofit analysis can be classified into three strategic levels as represented in a radar format in Fig. 11.8.

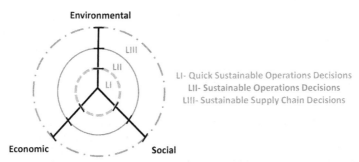

Figure 11.8 Strategic radar for reporting sustainable retrofit design decisions.

Engineers usually lack reporting these decisions to the shareholders; however, reporting is a key aspect for the success of the company. For instance, decisions can be taken applying the Level I approach, pointing out one retrofit design alternative as the most sustainable. However, if Level II is applied, a new alternative might be slightly worse in economic terms, but it might be more sustainable in terms of environment and social aspects. If the company opts for the first alternative, in the short term the results will be better; however, with the new legislations and government policies, which increasingly appear regarding environment and social responsibility issues, the company would have more benefit in the long run with the second alternative.

CASE STUDY: β-GALACTOSIDASE PRODUCTION

This case study involves the production process of β-galactosidase (β-gal), an intracellular enzyme produced by *Escherichia coli*. The case study has been presented by Carvalho et al. [17]. β-Gal allows people to digest milk or milk products, having a critical role in the production of lactose-free milk products. The flowsheet for the β-gal production process is shown in Fig. 11.9.

The β-gal production flowsheet can be divided into three sections. (1) Fermentation, where *E. coli* cells are used to produce the β-gal through a fermentation process. The fermentation process consists of four operations: charge, reaction, discharge, and clean. (2) Primary recovery, where first a cell harvesting operation to reduce the volume of the broth and to remove extracellular impurities takes place. Then an operation for cell disruption is performed in a high-pressure homogenizer, followed by a homogenization and a centrifuge to remove most of the cell debris. A dead-end polishing filter removes the remaining cell debris. The resulting protein solution is concentrated by an ultrafilter. (3) Purification, where the product stream is purified by an ion exchange chromatography column. Then it is concentrated by a second ultrafiltration unit and polished by a gel filtration unit.

The SARD framework will be applied to this case study in order to illustrate its application.

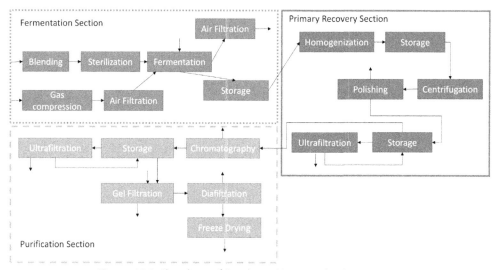

Figure 11.9 Flowsheet of β-galactosidase production process.

Step 1: Identify Bottleneck

In this step for the bottleneck identification, *SustainPro* [17] has been applied. Other methodologies could have been used to identify process bottlenecks (see section: Sustainability in Retrofit Design). As explained in that section, *SustainPro* follows a step-by-step procedure to identify the main bottlenecks. *SustainPro* has decomposed the flowsheet into 251 mass open-paths, 17 mass closed-paths, 36 energy open-paths, and one energy closed-paths. Based on a set of indicators, which assess value, waste, energy consumption, accumulation, and time of the different paths, and employing a sequence of sensitivity analysis, the bottlenecks, which have the highest potential for improvement, were selected. For this particular case, two paths, where water was being wasted, presented the lowest value of material value-added indicator, which means a capital loss in those paths. Then a bottleneck of scale regarding the flow rate of the chromatographic column was identified, because it presented the highest value of operational time factor (OTF). Therefore the flow rate should increase to reduce processing time. For the detailed analysis of *SustainPro* applied to this case study, please see the work presented by Carvalho et al. [17].

Step 2: Classify the Retrofit Action

SustainPro has been applied to generate possible new design alternatives, based on the identified bottleneck. Any other methodology for synthesis design could be employed. The water leaving the ultrafiltration has properties, which allow it to be recycled directly to the blending operation. However, the water wasted in chromatographic column contains proteins and therefore it cannot be recycled directly to the blending operation. These proteins require difficult separation processes in order to purify the water. Consequently, it would not be economically viable to purify this water and recycle it, so this water will be sent for treatment. In order to reduce the operation time a higher flow should pass through a chromatographic column (eliminate OTF bottleneck). A retrofit design considering the recovery of water has been proposed and the acquisition of a new ion-exchange chromatographic column for β-gal purification has been considered. SARD assumes that the user will apply the most convenient methodology for retrofit design generation. After that, following SARD, Table 11.3 should be used in order to classify the retrofit action into one of three classes (fine-tune operations; investment in the process; supply chain changes). The retrofit action related to the bottleneck of scale is classified as an *investment in the process*, because of replacement of equipment. The bottleneck of raw material consumption is considered a retrofit action of *fine-tune operations*, where a recycle with small adjustments in the process is considered (see Table 11.3 for classification). After classifying the retrofit design actions, the next step of SARD should be followed in order to decide the detail of the analysis (Level I to III).

Step 3: Define the Level of Analysis' Detail

Following the SARD framework the detail of the sustainability analysis should be decided. SARD helps the user in that decision, through the analysis of Figs. 11.5–11.7. After deciding on the level of detail for the sustainability analysis, Fig. 11.2, Table 11.1, and Fig. 11.3 should be analyzed in order to identify the areas of improvement that will be assessed in an analysis of a given level. The bottleneck of scale, which is in the retrofit action class of *investment in the process*, should be assessed at a Level II (Figs. 11.5–11.7). Therefore economic aspects should include an *investment assessment*, which includes the traditional project assessment metrics and the operational metrics (Fig. 11.2). In terms of environmental aspects, environmental process methods are suggested (Table 11.1) and labor conditions evaluation is proposed for social assessment (Fig. 11.3). The bottleneck of raw material consumption requires the simplest level of analysis, with economic performance being measured through the traditional operational indicators (Fig. 11.2), the environmental aspects being assessed through single issue methods (Table 11.1), and social aspects being measured through safety methods (Fig. 11.3).

The decision maker has now to evaluate the tradeoffs between the time spent in the analysis and the reward obtained from the selected analysis. For the bottleneck of scale the decision maker is obliged to conduct an investment analysis, otherwise any company will accept a new acquisition without knowing about it is viability. Regarding the environmental and social aspects, the replacement of the equipment will influence these pillars only slightly. The addition of the chromatographic column will only reduce time, so effluents, raw material consumption, and worker's conditions would remain the same.

Therefore the decision maker can opt for a Level II economic analysis and a Level I environmental and social analysis. With respect to the bottleneck of raw material consumption, Level I analysis is recommended to all pillars.

Since both retrofit actions can be considered together, the decision maker can opt to perform a Level II economic analysis, since it is required for the bottleneck of scale and it fits the raw material consumption bottleneck analysis. For the environmental and social pillars, the Level I will be conducted, since it is enough for a sustained assessment of both retrofit designs.

Step 4: Select Economic, Environmental, and Social Indicators

To apply Level II, operational and investment indicators are suggested (Fig. 11.2). ECON is a project evaluation software, which calculates all these indicators. ECON output is presented in Fig. 11.10, showing all the economic indicators. The user can perform the economic analysis applying any other tool or simply by doing the calculations personally. Here ECON appears as just an option to employ faster calculations of the economic analysis.

The high net return value indicates full recovery of the investment. Additionally, it is possible to verify that high profits will be achieved in the future with the β-gal production. The rate of return is also high and the payback time is very short, meaning that in less than 1 year the full investment is recovered and therefore the risk for this investment is low. This implies that the new retrofit designs are viable. Regarding environmental aspects, Level I analysis points out for single issue methods. The most common single issue method is the IPCC method [31], which assesses the global warming potential through the carbon footprint analysis. Therefore this method has been selected to assess the proposed retrofit design alternative. LCSoft is a software for environmental assessment, which has been applied to calculate the carbon footprint. Any other tool could have been used to calculate the environmental impact. The results are presented in Fig. 11.11.

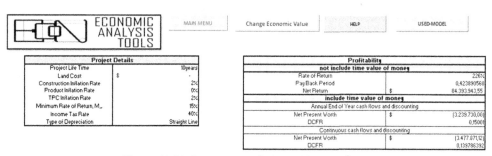

Figure 11.10 Economic analysis—ECON interface.

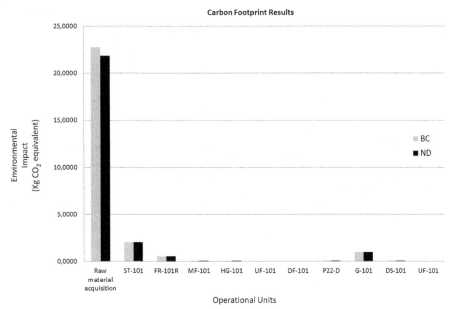

Figure 11.11 Carbon footprint—LCSoft results.

Table 11.4 Summary of the Environmental and Social Indicators

Indicators	Base Case	New Design	Improvement
Net water consumed per unit mass of product (kg/kg)	885.04	599.23	32%
Net water consumed per unit value added (kg/$)	1.69×10^{-05}	1.14×10^{-05}	32%
Carbon footprint (kg CO_2 eq.)	23	22	4%
Safety index (points)	28	28	0%

From Fig. 11.11 it is possible to verify the improvement in terms of carbon emission from the base case design to the retrofit design proposed (black bars lower than gray bars). However, this retrofit design involved the recovery of water, so metrics related to water impact should be also assessed. The IChemE metrics have been selected to assess the water impact of the new retrofit design [63]. For the social assessment the health and safety issues were considered, through the application of the Inherent Safety Index defined by Heikkila [60]. The results obtained for the environmental and social assessments are summarized in Table 11.4.

From Table 11.4 it can be verified that the new retrofit design considerably improves the environmental aspects, but it does not have influence on the social indicators.

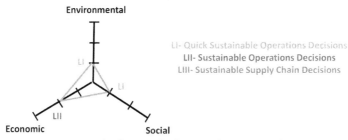

Figure 11.12 Strategic radar for retrofit design's decisions in the β-gal case study.

Step 5: Report the Assessment Employed

The final step of the framework is to report through a radar chart the level of detail employed in the analysis of the retrofit design alternatives. Fig. 11.12 illustrates the radar for this specific case study.

In this case, a Level I analysis has been performed for environmental and social aspects and a Level II analysis has been performed for economic aspects. This radar chart indicates that this company is not following a sustainable strategy, but it is already considering at a basic level the influence of sustainability issues in the retrofit design.

CONCLUSIONS

This chapter summarizes the different areas of improvement that should be assessed in retrofit design assessment. Retrofit design involves the elimination of different bottlenecks striving to achieve more sustainable processes. Nonetheless, the amplitude of the retrofit action varies, influencing different stakeholders. For the three pillars of sustainability, three layer boundaries were considered in the system: operational, process, and supply chain level. For each system the most important areas of improvement were identified and systematized considering the three pillars of sustainability. From this analysis a framework, SARD, was proposed. SARD guides the decision makers in their retrofit analysis, evaluating the designs in terms of their sustainability. SARD incorporates several critical aspects, creating a consistent roadmap for practitioners in their retrofit design decisions. These aspects can be summarized as: (1) the retrofit design alternatives are classified in terms of bottleneck type, clearly underpinning the problem in the study; (2) the bottlenecks are classified according to their amplitude of influence in the internal and external stakeholders; (3) the required level of detail for a valuable analysis is pointed out through SARD; (4) the areas of improvement intended to be assessed are listed according to the boundary of the system; and (5) the retrofit design decisions are reported to the shareholders, so that strategic retrofit design decisions can be aligned with the corporation's strategy. SARD is a valuable framework to practitioners working on the process engineering field and on charge for retrofit designs. Several authors have pointed out the existence of myriad sustainable indicators and the need for systematization frameworks; SARD is intended to fulfill this gap.

REFERENCES

[1] Meckenstock J, Barbosa-Póvoa AP, Carvalho A. The wicked character of sustainable supply chain management: evidence from sustainability reports. Bus Strategy Environ 2014. http://dx.doi.org/10.1002/bse.1872.

[2] Enríquez AH, Tanco M, Kim J-K. Simulation-based process design and integration for the sustainable retrofit of chemical processes. Ind Eng Chem Res 2011;50:12067–79.

[3] Ruiz-Mercado GJ, Gonzalez MA, Smith RL. Sustainability indicators for chemical processes: III. Biodiesel case study. Ind Eng Chem Res 2013;52:6747–60.

[4] Bakshi BR. Methods and tools for sustainable process design. Curr Opin Chem Eng 2014;6:69–74.

[5] Zheng K, Lou HH, Gangadharan P, Kanchi K. Incorporating sustainability into the conceptual design of chemical process-reaction routes selection. Ind Eng Chem Res 2012;51:9300–9.

[6] Elkington J. Cannibals with forks. The triple bottom line of 21st century business. New Society Publishers; 1998. p. 407.

[6a] Guinand EA. Optimization and network sensitivity analysis for process retrofitting [Ph.D. thesis]. Boston: Massachusetts Institute of Technology (MIT), Chemical Engineering Department; 2001.

[7] Nourai F, Rashtchian D, Shayegan J. An integrated framework of process and environmental models, and EHS constraints for retrofit targeting. Comput Chem Eng 2001;25:745–55.

[8] Diwekar U, Shastri Y. Design for environment: a state-of-the-art review. Clean Technol Environ Policy 2011;13:227–40.

[9] Kalakul S, Malakul P, Siemanond K, Gani R. Integration of life cycle assessment software with tools for economic and sustainability analyses and process simulation for sustainable process design. J Clean Prod 2014;71:98–109.

[10] Carvalho A. Design of sustainable processes: systematic generation & evaluation of alternatives [Ph.D. thesis]. Portugal: Instituto Superior Técnico; 2009.

[11] Rong BG, Yu HF, Kraslawski A, Nystrom L. Study on the methodology for retrofitting chemical processes. Chem Eng Technol 2000;23(6):479–84.

[12] Azapagic A, Millington A, Collett A. A methodology for integrating sustainability considerations into process design. Chem Eng Res Des 2006;84(A6):439–52.

[13] Bumann AA, Papadokonstantakis S, Fischer U, Hungerbühler K. Investigating the use of path flow indicators as optimization drivers in batch process retrofitting. Comput Chem Eng 2011;35:2767–85.

[14] Jayswal A, Li X, Zanwar A, Lou HH, Huang Y. A sustainability root cause analysis methodology and its application. Comput Chem Eng 2011;35:2786–98.

[15] Torres CM, Gadalla MA, Mateo-Sanz JM, Esteller LJ. Evaluation tool for the environmental design of chemical processes. Ind Eng Chem Res 2011;50:13466–74.

[16] Banimostafa A, Papadokonstantakis S, Hungerbühler K. Evaluation of EHS hazard and sustainability metrics during early process design stages using principal component analysis. Process Saf Environ Prot 2012;90:8–26.

[17] Carvalho A, Matos HA, Gani R. SustainPro – a tool for systematic process analysis, generation and evaluation of sustainable design alternatives. Comput Chem Eng 2013;50:8–27.

[18] Shadiya OO, High KA. Designing processes of chemical products for sustainability: incorporating optimization and the sustainability evaluator. Environ Prog Sustain Energy 2013;32(32):762–76.

[19] Shadiya OO, High KA. Sustainability evaluator: tool for evaluating process sustainability. Environ Prog Sustain Energy 2013;32(3):749–61.

[20] Doane D, MacGillivray A. Economic sustainability: the business of staying in business. The Sigma Project; 2001.

[21] Ouattara A, Pibouleau L, Azzaro-Pantel C, Domenecha S, Baudetb P, Yaoc B. Economic and environmental strategies for process design. Comput Chem Eng 2012;36:174–88.

[22] Stamford L, Azapagic A. Sustainability indicators for the assessment of nuclear power. Energy 2011;36(10):6037–57.

[23] Ruiz-Mercado GJ, Smith RL, Gonzalez MA. Sustainability indicators for chemical processes. I. Taxonomy. Ind Eng Chem Res 2011;51:2309–28.

[24] Al-Sharrah G, Elkamel A, Almanssoor A. Sustainability indicators for decision-making and optimisation in the process industry: the case of the petrochemical industry. Chem Eng Sci 2010;65:1452–61.

[25] GRI-Global Report Initiative. G4 sustainability reporting guidelines – reporting principles and standard disclosures. 2004.

[26] Powell JB. In: Harmsen J, Powell JB, editors. Sustainability metrics, indicators and indices for the process industries, sustainable development in the process industries: cases and impact. John Wiley & Sons, Inc.; 2010.

[27] Tanzil D, Beloff BR. Assessing impacts: overview on sustainability indicators and metrics. Environ Qual Manag 2006. http://dx.doi.org/10.1002/tqem.20101.

[28] Verfaillie HA, Bidwell R. Measuring eco-efficiency – a guide to reporting company performance. WBCSD—World Business Council for Sustainable Development; 2000.

[29] Saengwirun P. Cost calculations and economic analysis [M.Sc thesis]. Bangkok, Thailand: The Petroleum and Petrochemical College, Chulalongkorn University; 2011.

[30] Seuring S. A review of modeling approaches for sustainable supply chain management. Decis Support Syst 2012;54(4):1513–20.

[31] Carvalho A, Mendes AN, Mimoso AF, Matos HA. From a literature review to a framework for environmental impact assessment index. J Clean Prod 2014;64:36–62.

[32] ISO14040. Environmental management – life cycle assessment – principles and framework. International Organization of Standardization; 2006.

[33] Ashby A, Leat M, Hudson-Smith M. Making connections: a review of supply chain management and sustainability literature. Supply Chain Manag An Int J 2012;17(5):497–516.

[34] Miettinen P, Hamalainen RP. How to benefit from decision analysis in environmental life cycle assessment (LCA). Eur J Operational Res 1997;102(2):279–94.

[35] Goedkoop M, Oele M, de Schryver A, Vieira M, Hegger S. SimaPro database manual. Methods library. PRé Consultants; 2010.

[36] Guinée JB, Gorrée M, Heijungs R, Huppes G, Kleijn R, Koning A, et al. Life cycle assessment – an operational guide to the ISO standards. Centre of Environmental Science – Leiden University (CML); 2001.

[37] Shonnard DR, Hiew DS. Comparative environmental assessments of VOC recovery and recycle design alternatives for a gaseous waste stream. Environ Sci Technol 2001;34(24):5222–8.

[38] Boesch ME, Hellweg S, Huijbregts MAJ, Frischknecht R. Applying cumulative exergy demand (CExD) indicators to the ecoinvent database. Int J Life Cycle Assess 2007;12(3):181–90.

[39] Goedkoop M, Spriensma R. The eco-indicator 99: a damage oriented method for life cycle impact assessment. Methodology Report. Pré Consultant; 2001.

[40] Ramzan N, Degenkolbe S, Witt W. Evaluating and improving environmental performance of HC's recovery system: a case study of distillation unit. Chem Eng J 2008;140(1–3):201–13.

[41] Wackernagel M, Rees WE. Our écological footprint: reducing human impact on the earth. New Society Publishers; 1996.

[42] Frischknecht R, Jungbluth N, Althaus H-J, Bauer C, Doka G, Dones R, et al. Implementation of life cycle impact assessment methods. Ecoinvent report No. 3, v2.0. Dübendorf: Swiss Centre for Life Cycle Inventories; 2007.

[43] Koellner T, Scholz RW. Assessment of land use impacts on the natural environment – part 1: an analytical framework for pure land occupation and land use change. Int J Life Cycle Assess 2007;12(1):16–23.

[43a] Koellner T, Scholz RW. Assessment of land use impacts on the natural environment – part 2: generic characterization factors for local species diversity in central Europe. Int J Life Cycle Assess 2008;13(1):32–48.

[44] Hauschild M, Potting J. Spatial differentiation in life cycle impact assessment − the EDIP2003 methodology. Environmental Protection Agency − Danish Ministry of the Environment; 2005.

[45] Young DM, Cabezas H. Designing sustainable processes with simulation: the waste reduction (WAR) algorithm. Comput Chem Eng 1999;23(10):1477−91.

[46] Solomon S, Qin D, Manning M, Marquis M, Averyst K, Tignor MMB, et al. Climate change 2007: the physical science basis, intergovernmental panel on climate change. 2007.

[47] Humbert S, Margni M, Charles R, Torres Salazar OM, Quiros AL, Jolliet O. Toxicity assessment of the main pesticides used in Costa Rica. Agric Ecosyst Environ 2007;118(1−4):183−90.

[48] Van Zelm R, Huijbregts MAJ, van de Meent D. USES-LCA 2.0−a global nested multi-media fate, exposure, and effects model. Int J Life Cycle Assess 2009;14(3):282−4.

[49] Sampattagul S, Kato S, Kiatsiriroat T, Maruyama N, Widiyanto A. LCA-NETS tool for environmental design of natural gas-fired power generation systems in Thailand. In: 2003 3rd international symposium on environmentally conscious design and inverse manufacturing − Ecodesign '03; 2003. p. 141−6.

[50] Rosenbaum RK, Bachmann TM, Gold LS, Huijbregts MAJ, Jolliet O, Juraske R, et al. USEtox-the UNEP-SETAC toxicity model: recommended characterisation factors for human toxicity and freshwater ecotoxicity in life cycle impact assessment. Int J Life Cycle Assess 2008;13(7):532−46.

[51] Itsubo N, Inaba A. A new LCIA method: LIME has been completed. Int J Life Cycle Assess 2003;8(5):1.

[52] Toffoletto L, Bulle C, Godin J, Reid C, Deschenes L. LUCAS − a new LCIA method used for a Canadian-specific context. Int J Life Cycle Assess 2007;12(2):93−102.

[53] MEEUP. Methodology report, final. VHK − Van Holsteijn en Kemma BV; 2005.

[53a] Goedkoop M, Heijungs R, Huijbregts M, Schryver AD, Struijs J, Van Zelm R. A life cycle impact assessment method which comprises harmonised category indicators at the midpoint and the endpoint level. In: Report I: characterisation. ReCiPe 2008; 2009.

[54] Bare J. TRACI 2.0: the tool for the reduction and assessment of chemical and other environmental impacts 2.0. Clean Technol Environ Policy 2011;13(5):687−96.

[55] Steen B. A systematic approach to environmental priority strategies in product development (EPS). Version 2000−models and data of the default method. CML − Centre for Environmental Assessment of Products and Material Systems; 1999.

[56] US EPA. WAR GUI. 2008. http://www.epa.gov/nrmrl/std/war/sim_war.htm.

[57] Piyarak S. Development of software for life cycle assessment. Bangkok, Thailand: The Petroleum and Petrochemical College, Chulalongkorn University; 2012 [M.S. thesis].

[58] Littig B, Griessler E. Social sustainability: a catchword between political pragmatism and social theory. Int J Sustain Dev 2005;8:65−79.

[58a] Shokrian M, High KH, Sheffert Z. Screening of process alternatives based on sustainability metrics: comparison of two decision-making approaches. Int J Sust Eng 2014. http://dx.doi.org/10.1080/19397038.2014.958601.

[59] Khan FI, Abbasi SA. Multivariate hazard identification and ranking system. Process Saf Prog 1998;17(3):157−70.

[60] Heikkila AM. Inherent safety in process plant design: an index-based approach. Espoo, Finland: VTT Publications 384, Technical Research Centre of Finland; 1999.

[61] Gupta JP, Edwards DW. A simple graphical method for measuring inherent safety. J Hazard Mater 2003;104:15−30.

[62] Khan FI, Amyotte PR. Integrated inherent safety index (I2SI): a tool for inherent safety evaluation. Process Saf Prog 2004;23(2):136−48.

[63] IChemE. Sustainable development progress metrics. UK: IChemE Sustainable Development Working Group, IChemE Rugby; 2002.

[63a] Benoît C, Norris GA, Valdivia S, Ciroth A, Moberg A, Bos U, Prakash S, et al. The guidelines for social life cycle assessment of products: just in time! Int J Life Cycle Assess 2010;15(2):156−63.

[64] Simões M. Social key performance indicators − assessment in supply chains [Master thesis]. Lisboa, Portugal: Instituto Superior Técnico; 2014.

CHAPTER TWELVE

Chemical Engineering and Biogeochemical Cycles: A Techno-Ecological Approach to Industry Sustainability

S. Singh
Purdue University, West Lafayette, IN, United States

B.R. Bakshi
The Ohio State University, Columbus, OH, United States

MOTIVATION
Chemical Industry and Biogeochemical Cycles

The role of the chemical industry in the present world is to provide desired products from raw feedstock materials. Several natural feedstocks in their original state are not useful as products and must be processed through chemical processes to make them fit for consumption. Therefore chemical industries provide man-made processes to change the form and composition of resources to the desired form for consumption. The phenomenon of conversion of chemical compounds to various stable states is also present in natural systems, which form the basis of biogeochemical cycles. Biogeochemical cycles are defined as "a pathway by which a chemical substance moves through both biotic (biosphere) and abiotic (lithosphere, atmosphere and hydrosphere) compartments of Earth" [1]. Thus biogeochemical cycles are the key natural processes that provide resources for use on Earth, which is a valuable ecosystem service. Some of the key biogeochemical cycles are carbon, nitrogen, oxygen, phosphorus, sulfur, water, and rock. At least four of these play a very crucial role in providing feedstock resources for chemical industries (carbon, nitrogen, phosphorus, and sulfur). Fossil fuels that form the basis of all the petrochemical and chemical industries are products of years of biogeochemical cycle functioning. This establishes a core connection of chemical industries with Earth's natural processes that has always been present, but neglected. The chemical industries are inherently series of chemical changes that are perceived to be disconnected from these natural cycles because of time scale differences, but impact the natural cycles because of rapid increase in demand of specific flows. Thus the chemical sector will be affected by disruption of these natural processes in the long run. The recent trend of shift to more biobased products, which are considered renewable on the time scale of human life, provides a stronger incentive to realize the basic dependence of chemical industries on these biogeochemical cycles. This push comes from understanding that several biogeochemical cycles such as carbon and nitrogen cycles have been hugely disrupted by human activities [2–4]. In one study it was projected that by 2030, 35% of global chemical production may be biobased [5]. Before large-scale adoption of biobased chemical production systems, it is important to understand potential challenges arising because of dependence on biogeochemical cycles for provision of biomass raw materials.

Chemical Industry and the Carbon Cycle

The natural carbon cycle is basically an oxidation-reduction process where the carbon in oxidized form, i.e., CO_2, is converted to fixed or reduced form of carbon in organic compounds. Thus the carbon cycle circulates the element between the atmospheric pool and other natural system pools such as plants, soil, aquatic ecosystems, and fossil reserves [4]. Photosynthesis is the main process that is involved in the conversion of oxidized carbon to organic carbon. Therefore ecosystems play a very crucial role in this natural carbon cycle by providing the pathway for reduction of carbon to organic forms such as in biomass. The natural processes of carbon cycling have been hugely altered by anthropogenic impacts because of human intervention in ecosystem structure, change in soil reserves caused by land use practices [6], and depletion of fossil reserves at an unprecedented rate. Chemical industries play a major role in this alteration of the carbon cycle because of huge dependence on feedstocks derived from fossil reserves or biomass in recent years. Hydrocarbons are the basic chemical molecules that are ubiquitous and essential for production of the majority of chemical products. Most of these chemical compounds are directly derived from petrochemicals or the processing of products obtained from petrochemicals. Thus chemical industries are not only reliant on fossil fuels for energy but also for precursors of several chemical compounds. For example, natural gas is purified into the natural gas delivered to consumers and produces by-products such as ethane, natural gas liquids, propane, butanes, and natural gasoline. Similarly, crude oil is converted into several downstream products by fractionation, which is then used to synthesize different chemical products. Recent years have seen an increasing shift toward biobased feedstocks and value-added products [7] that have shifted the reliance from long time-scale carbon-based fossil reserves to shorter turnaround of soil carbon profile. Thus carbon forms the most fundamental block for chemical industry processing and these chemical industries have a huge role to play in carbon cycling. On the one hand, the sustainable management of the carbon biogeochemical cycle requires special attention from chemical industries. On the other hand, sustainability of the chemical manufacturing sector itself depends heavily on the natural carbon cycle. As an example, the global human emissions of carbon are about 9 GtC/year (or 33,000 MT-CO_2/year) [8] and the emissions related to the chemical industry in the United Kingdom alone for the year 2010 were 110 MT-CO_2-eq. [5]. In the United States, plastics are made from hydrocarbon gas liquids and natural gas [9]. About 412 billion cubic feet (bcf) of natural gas was used for the plastics and resins industry in 2010 within the United States [9]. Out of this, 13 bcf (equivalent to 1.93E-04 GtC) of natural gas was used as feedstock, whereas 399 bcf (equivalent to 5.93E-03 GtC) of natural gas was used for energy in the plastics and resins industry. When compared to total anthropogenic emissions of 9 GtC, this is about 0.06% for the United States alone in 2010. The percentage of carbon being used in feedstock as natural gas is very low as compared to the flows of the carbon cycle (about 0.002% of 9 GtC of anthropogenic emissions), but does illustrate the connections of chemical industries with carbon cycling in the long term. As the chemical industry shifts to more biobased feedstock, the connections with various flows of the carbon cycle will also shift, such as dependence on a faster carbon cycle of land-based emissions. This may lead to comparable flows of the carbon cycle related to chemical industries.

Chemical Industry and the Nitrogen Cycle

Nitrogen is a basic element that is present in almost every life supporting chemical in one or the other form. Some examples include proteins, enzymes, polymers, fibers, fertilizers, etc. The Haber–Bosch process can be easily cited as one of the most revolutionary chemical processes invented in the history of the human race. The Haber–Bosch process involves reaction of nitrogen gas (N_2) and hydrogen gas (H_2) to produce ammonia (NH_3).

$$N_2 + 3H_2 \rightarrow 2NH_3 \ (\Delta H = -92.4 \ kJ/mol)$$

Prior to the advent of the Haber–Bosch process most of the required nitrogen-based raw materials were provided by natural cycling of nitrogen. The nitrogen cycle in nature involves the process of nitrogen fixation, which is the biological conversion of inert N_2 in air to reduce other forms of nitrogen that are reactive and therefore can be converted to several products [4]. The nitrogen cycle in nature consists of the following steps:

1. Nitrogen fixation: This step involves conversion of N_2 to NH_3 by biological processes such as microbial activity or atmospheric lightning.
2. Ammonification: This process converts organic nitrogen to ammonium, which can then be converted to other forms of nitrogen that can be assimilated by plants or organisms. The process of ammonification is facilitated by bacteria or fungi.
3. Nitrification: The next step in the natural nitrogen cycle is nitrification. In this process the soil living bacteria and other nitrifying bacteria convert ammonia or ammonium to nitrite, which is followed by oxidation of nitrite to nitrate.

4. Denitrification: This process is the closing loop of the nitrogen cycle. In this process nitrate is reduced to ultimately produce molecular N_2; however, the process also produces a by-product of N_2O gas, which is a potent greenhouse gas.

The industrial process of Haber–Bosch basically emulates the "nitrogen fixation" process to provide the resources for agricultural production at a faster pace to meet the rising demand. The industrial production of ammonia therefore provided a boost to production of nitrogen-based products. Ammonia is the basic building block for numerous chemical compounds downstream, such as conversion to polymers, fertilizers, etc. (Table 12.1).

One of the most in-demand downstream products of nitrogen from the chemical industry is fertilizer for agricultural production. Fig. 12.1 shows the various pathways for different fertilizer products starting from ammonia [12].

These interactions of chemical industrial production with biogeochemical cycles are not normally captured in any industrial sustainability analysis. Therefore a systems approach that integrates the ecosystem flows such as biogeochemical cycles with industrial production is necessary. This motivation led to development of the ecologically based life cycle analysis (Eco-LCA) approach that can account for the role of ecosystem goods and services in supporting industrial production [13]. The Eco-LCA framework was further adapted by Singh and Bakshi [14,15] to understand the role of biogeochemical cycles of carbon and nitrogen in industrial production. In the following sections the Life Cycle Analysis (LCA) approach is described along with

Table 12.1 Main Chemical Compounds Produced From Ammonia as a Basic Reactant

Basic Compounds	Processes	Catalysts	End Products
Ammonia and carbon dioxide	Heating at 450K and 200 atm pressure	—	Urea
Ammonia and oxygen	Ostwald process	Pt	Nitric acid
Ammonia and nitric acid	Neutralization process at specific temperature and pressure [10]	None	Ammonium nitrate
Ammonia and HCl	Engelcor process—mixing of dilute HCl with NH_3 at 353K, reduced pressure, and excess NH_3	None	Ammonium chloride
Ammonia and sulfuric acid	Exothermic reaction in saturators [10].	None	Ammonium Sulfate
Ammonia and phosphoric acid	Reaction in stirred tank reactor (PhoSAI [at atm pressure] and Minifos process [at 0.21 Mpa]) [11]	None	Monoammonium phosphate powder [12]

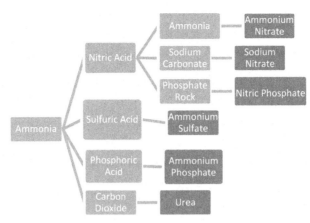

Figure 12.1 Downstream products derived from ammonia for the fertilizer industry.

the Eco-LCA approach and application to chemical industry interaction with biogeochemical cycles.

LIFE CYCLE ANALYSIS FOR CHEMICAL INDUSTRY INTERACTION WITH CARBON AND NITROGEN CYCLES

LCA is an analytical approach that accounts for full life cycle scale environmental impact of a product or process. The ISO 14040 [16] sets the standard for life cycle assessment guiding the principles and framework. According to ISO, LCA involves four basic steps: (1) goal and scope definition, (2) inventory analysis, (3) impact assessment, and (4) interpretation. Based on the scope of study, LCA can be categorized as process LCA, economic input–output LCA (EIO-LCA), or hybrid LCA. Process LCA draws the system boundaries based on the most important processes that need to be included in the LCA study. The selection of processes that are included is usually based on the size of mass flows or the size of environmental impact. Thus a cut-off is used to decide the boundary of the process LCA study. The system boundaries for studying impacts can be cradle to gate or cradle to grave. The cradle-to-gate analysis studies the impact from extraction of raw materials to the point where product is ready for use, whereas the cradle-to-grave analysis studies the impact from raw material extraction to the final disposal of product. Because of the use of cut-off criteria there may be underestimation of the impact estimated by the specific process LCA study. This is termed as "truncation" error. To overcome this limitation of process LCA, EIO-LCA was proposed [17]. The strength of EIO-LCA is that it does not need selection of boundaries since it includes the whole economy as the system involved in production for the LCA boundary. Thus it overcomes the truncation error challenge; however, it loses the fine-scale resolution of processes caused by the economic sectors being at coarse scale, which aggregates impact from several processes. This limitation of EIO-LCA is overcome by using hybrid LCA. A hybrid LCA combines EIO-LCA with fine-scale data for the important processes to improve resolution, thus combining the strengths of both process LCA and EIO-LCA. Eco-LCA is similar to the EIO-LCA framework; however, it utilizes a supply-side model [13,18] for calculating the life cycle impact as discussed later. The supply-side model allows for accounting of life cycle consumption of resources being utilized in the economy for various productions.

Eco-LCA Inventory

Eco-LCA accounts for the role of ecological systems in supporting different production systems. A recently developed Eco-LCA inventory that integrates the flow of the natural carbon and nitrogen cycle therefore provides insights into dependence of various production pathways or products on these natural cycles [13,14]. The inventory developed utilizes the EIO model given in Eq. [12.1]:

$$R = \widehat{X}^{-1}\left(I - G^T\right)^{-1} V_{ph} \qquad\qquad [12.1]$$

In Eq. [12.1], $(I - G^T)^{-1}$ is called the Ghosh inverse and V_{ph} is the vector that contains information about the resources entering the economy or emissions from the economy. Additional details about the underlying theory for Eco-LCA are available in [12]. In the case of the biogeochemical cycles these are the flows related to extraction of resources or emissions to the natural systems. X is the total economic throughput for the economic sectors. Thus the vector R provides the resources/emissions per unit economic output from sectors. The life cycle impact of a demand from a particular sector on resource consumption or emissions is obtained by multiplying R with final demand (f) in economic units.

Direct and Indirect Impact/Dependence on Carbon and Nitrogen Cycles

The direct and indirect dependence on various flows of carbon and nitrogen cycles (Tables 12.2 and 12.3) is calculated by using a Taylor series expansion of Eq. [12.1]. Direct impact is defined as the resources and emissions that are associated with the activities that are in immediate association to support production from the chosen sector for which final demand is given. Indirect impact is defined as the emissions and resources associated with the activities that are indirectly related to the chosen sector through those activities that were directly involved. The direct impact intensity is calculated as $\widehat{X}^{-1}(I + G^T)V_{ph}$ and the indirect impact intensity is calculated by $\widehat{X}^{-1}\left((G^T)^2 + (G^T)^3 + (G^T)^4 + ...\right)V_{ph}.$

Table 12.2 Carbon Cycle Flows in the Eco-LCA Inventory

Flows	Description	Sectors Directly Related to Flows
Carbon Sequestration		
CO_2—farmland	Carbon sink occurring on farmlands during biomass growth	Oilseed farming, grain farming, vegetable and melon farming, tree nut farming, fruit farming, greenhouse, nursery, and floriculture production, tobacco farming, cotton farming, sugar cane and sugar beet farming, all other crop farming
CO_2—ranchland	Carbon sink occurring on ranchlands involved in cattle grazing and cattle farming	Dairy cattle and milk production and cattle ranching and farming
CO_2—forestland	Carbon sink on the forestlands	Forest nurseries, forest productions, and timber tracts
Soil sink (cropland remaining cropland)	Carbon sink in soil of all croplands that had been cropland for last 20 years	Oilseed farming, grain farming, vegetable and melon farming, tree nut farming, fruit farming, greenhouse, nursery, and floriculture production, tobacco farming, cotton farming, sugar cane and sugar beet farming, all other crop farming
Soil sink (land converted to grassland)	Carbon sink in all grasslands that had been in another land use at any point during the previous 20 years	Oilseed farming, grain farming, vegetable and melon farming, tree nut farming, fruit farming, greenhouse, nursery, and floriculture production, tobacco farming, cotton farming, sugar cane and sugar beet farming, all other crop farming, dairy cattle and milk production, cattle ranching and farming
Urban trees sink	Carbon sequestration because of trees in urban areas	Nonresidential commercial and health care structures, nonresidential manufacturing structures, other nonresidential structures, residential permanent site single and multifamily structures, other residential structures, nonresidential maintenance and repair, residential maintenance and repair

(*Continued*)

Table 12.2 Carbon Cycle Flows in the Eco-LCA Inventory—cont'd

Flows	Description	Sectors Directly Related to Flows
Yard trimming and food scrap stocks in landfills (sink)	Carbon storage occurring because of yard trimmings and food scrap in landfills	Waste management and remediation services
Soil sink (grassland, remaining grassland)	CO_2 sequestration in soil on grasslands that had been grassland for the previous 20 years	Dairy cattle and milk production and cattle ranching and farming
Carbon Emissions		
Fossil fuel emissions Liming emissions	CO_2 emissions caused by fossil fuel burning	All respective sectors based on fuel consumption Oilseed farming, grain farming, vegetable and melon farming, tree nut farming, fruit farming, greenhouse, nursery, and floriculture production, tobacco farming, cotton farming, sugar cane and sugar beet farming, all other crop farming
Emissions by land converted to cropland	Soil carbon losses or emissions because of land converted to cropland that had different use for previous 20 years	Oilseed farming, grain farming, vegetable and melon farming, tree nut farming, fruit farming, greenhouse, nursery, and floriculture production, tobacco farming, cotton farming, sugar cane and sugar beet farming, all other crop farming
Emissions by urea fertilization	CO_2 emissions because of application of urea on land (emission of CO_2 that was fixed during industrial production)	Oilseed farming, grain farming, vegetable and melon farming, tree nut farming, fruit farming, greenhouse, nursery, and floriculture production, tobacco farming, cotton farming, sugar cane and sugar beet farming, all other crop farming, forest nurseries, forest production and timber tracts, support activities for agriculture and forestry

For details of Carbon dependence on whole economy (see Ref. [15]).

Table 12.3 Nitrogen Cycle Flows in the Eco-LCA Inventory

Flows	Description	Sectors Directly Related to Flow
Nitrogen mobilization	All flows that convert inert nitrogen to reactive nitrogen. Includes natural processes such as nitrogen fixation in soybean, microbial fixation, and artificial processes such as the Haber–Bosch process	Related agricultural sectors such as oilseed farming, etc. and fertilizer manufacturing sector
Nitrogen product	Flows that are associated with conversion of nitrogen product ammonia to various other products in the economy	Respective sectors such as plastics manufacturing, explosives manufacturing, etc.
Nitrogen emissions	Emission flows both from fossil fuel combustion and land use and land use change	Respective sectors for fossil fuel combustion. Agricultural sectors mostly for land use and land use change emissions

For detailed description of flows in nitrogen profile refer to [14].

CHEMICAL INDUSTRY PROFILE FOR CARBON
Chemical Industry Profile for Carbon Sequestration

Fig. 12.2 shows the carbon sequestration profile for chemical sectors in the year 2002. It is to be noted here that "carbon sequestration" is considered as an ecosystem service because of various flows of the carbon cycle that provide the sequestration service. This carbon sequestration is not the same as long-term "sequestration," which is created artificially, but these are the carbon cycle flows that occur in the ecosystem and provide the ecosystem service with carbon sequestration. Also it is not obvious that the productivity of chemical sectors depends on the carbon sequestration services; however, the Eco-LCA inventory establishes that connection by including carbon sequestration flows in the life cycle inventory. As shown in Fig. 12.2 the dominant flows of carbon sequestration are CO_2 sink in forestland and CO_2 sink in farmland. The sector that has highest intensity of CO_2 sequestration is the "Pulp mills (322110)" sector (Table 12.4). The highest contribution to this sector's high intensity of carbon sequestration is from the CO_2 sink in forestland. This is because of this sector's productivity directly and indirectly related to the "forestry and logging" industry that is responsible for the CO_2 sink in forestland. In the absence of forestry activities and sink in forestland, the pulp mills

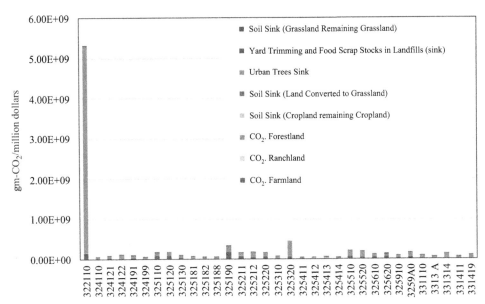

Figure 12.2 Carbon sequestration profile for chemical sectors (refer to Table 12.4 for sector names).

activity will be severely affected, thus highlighting the connections of chemical industries to natural system flows. Similarly, the sector with the highest contribution for CO_2 sink in farmland is the "Other basic organic chemical manufacturing (325190)" sector. This sector mainly utilizes biobased raw feedstocks to manufacture chemicals, thus relying on the CO_2 sequestration in farmlands. On an alternate perspective this sector's activity has a role in promoting carbon sink through activities on farmlands. However, the carbon sequestration activity is restricted by natural systems' ability, which makes carbon sequestration an ecosystem service. Also, since the biomass is utilized for production, it is not a long-term carbon sequestration.

Fig. 12.3 shows the carbon sequestration profile for the chemical sectors after removing the dominant flows of carbon sequestration in forestland and farmland. There is an increasing realization that other anthropogenic activities such as land conversion to grassland, soil sink in grasslands, or croplands and urban trees plantations also have led to changes in carbon cycle flows. The Eco-LCA inventory also captures the carbon flows in these activities and provides a database to study the life cycle impacts of various sectors on these flows. As shown in Fig. 12.3, the dominant chemical sector for impact on these flows is "Pulp mills (322110)." The next highest impact comes from "Other basic organic chemical manufacturing (325190)," which is related to the dependence of this sector on biobased feedstock that leads to the activities affecting these carbon sinks. Other high-impact sectors are "Soap and cleaning compound manufacturing (325610)" and "Toilet preparation manufacturing (325620)," which are involved with manufacturing soaps, detergents, cosmetics, etc. The category of carbon sink that has

Table 12.4 Chemical Sectors With NAICS Code for Figs. 12.2–12.8

NAICS Code	Sector
322110	Pulp mills
324110	Petroleum refineries
324121	Asphalt paving mixture and block manufacturing
324122	Asphalt shingle and coating materials manufacturing
324191	Petroleum lubricating oil and grease manufacturing
324199	All other petroleum and coal products manufacturing
325110	Petrochemical manufacturing
325120	Industrial gas manufacturing
325130	Synthetic dye and pigment manufacturing
325181	Alkalies and chlorine manufacturing
325182	Carbon black manufacturing
325188	All other basic inorganic chemical manufacturing
325190	Other basic organic chemical manufacturing
325211	Plastics material and resin manufacturing
325212	Synthetic rubber manufacturing
325220	Artificial and synthetic fibers and filaments manufacturing
325320	Pesticide and other agricultural chemical manufacturing
325411	Medicinal and botanical manufacturing
325412	Pharmaceutical preparation manufacturing
325413	In-vitro diagnostic substance manufacturing
325414	Biological product (except diagnostic) manufacturing
325510	Paint and coating manufacturing
325520	Adhesive manufacturing
325610	Soap and cleaning compound manufacturing
325620	Toilet preparation manufacturing
325910	Printing ink manufacturing
3259A0	All other chemical product and preparation manufacturing
331110	Iron and steel mills and ferroalloy manufacturing
33131A	Alumina refining and primary aluminum production
331314	Secondary smelting and alloying of aluminum
331411	Primary smelting and refining of copper
331419	Primary smelting and refining of nonferrous metal (except copper and aluminum)
325310	Fertilizer manufacturing

highest contribution for these sectors is the CO_2 sequestration in ranchland, which is because of the dependence on these sectors on the animal-derived feedstock such as animal fat. Similarly, the contribution of several other chemical product manufacturing sectors on different carbon sink flows is felt indirectly through use of intermediate products. The Eco-LCA inventory provides insights into several of these indirect impacts or dependence of chemical sectors on carbon cycle flows. These insights enable assessment of the sustainability of chemical sectors in the context of natural cycle disruption.

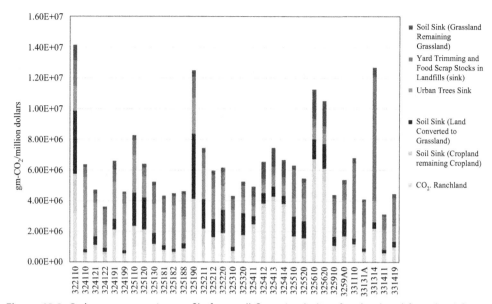

Figure 12.3 Carbon sequestration profile for small flows (excluding farmland and forestland flows).

Chemical Industry Profile for Carbon Emissions

The carbon emissions intensity (g-CO$_2$/million $) profile is shown in Fig. 12.4. The intensity profile including fossil fuel emissions (Fig. 12.4) shows that the top five sectors with highest impact are: "Alumina refining and primary aluminum production (33131A)," "Alkalies and Chlorine Manufacturing (325181)," "Industrial gas

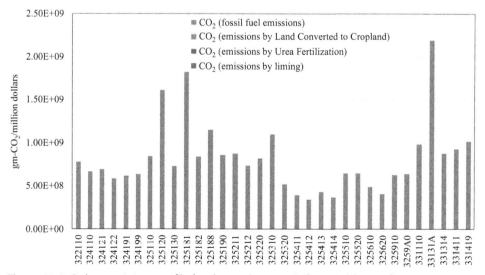

Figure 12.4 Carbon emissions profile for chemical sectors (refer to Table 12.4 for sector names).

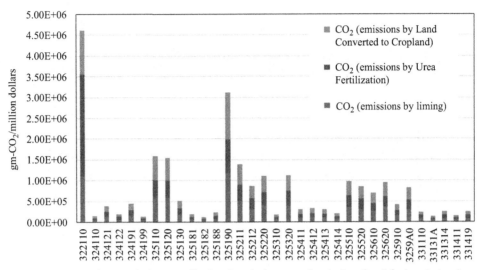

Figure 12.5 Carbon emissions profile for chemical sectors (excluding fossil fuel emissions).

manufacturing (325120)," "Fertilizers manufacturing (325310)," and "All other basic inorganic chemical manufacturing (325188)." Fig. 12.5 does not include the fossil fuel CO_2 emissions to highlight the connection of chemical sectors with other flows of CO_2 emissions related to land change and waste management activities. The chemical sectors with the highest impact on other categories of CO_2 emissions are "Pulp mills (322110)" and "Other basic organic chemical manufacturing (325190)," which have a direct impact on the land use and urea–related emissions because of dependence on biobased feedstock.

CHEMICAL INDUSTRY PROFILE FOR NITROGEN
Nitrogen Mobilization Profile for Chemical Sectors

Fig. 12.6 shows the life cycle flows for nitrogen mobilization in selected chemical sectors per million $ of output from those sectors. In Fig. 12.6 the "Nitrogen fertilizer manufacturing (325310)" sector was removed since that sector dominates the impact on total nitrogen mobilization by order of magnitude. The intent was to highlight the role of other significant chemical sectors on nitrogen mobilization. As is clear from Fig. 12.6, the "Pulp mills (322110)" sector has the highest impact on nitrogen mobilization flows as represented by high nitrogen mobilization intensity. Other top five chemical sectors with high nitrogen mobilization intensity are: "Other basic organic chemical manufacturing (325190)," "Petrochemical manufacturing (325110)," "All other basic inorganic chemical manufacturing (325188)," "Plastics material and resin manufacturing (325211)," and "All other chemical product and preparation manufacturing (3259A0)." Among the categories of nitrogen mobilization, the industrial nitrogen fixation as ammonia dominates the flow followed by "Nitrogen from atmospheric deposition."

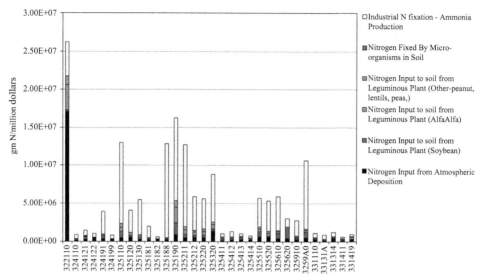

Figure 12.6 Nitrogen mobilization profile for chemical industries.

The reason for high atmospheric deposition of nitrogen because of pulp mills is caused by the indirect dependence on logging activities that have huge land requirement, which was used as a proxy for allocating total nitrogen deposition. The dependence on nitrogen mobilization by leguminous plant such as alfalfa and soybean is highest in "Other basic organic chemical manufacturing (325190)" because of the obvious dependence of this sector on biomass. The increasing shift toward biomass-derived molecules is expected to increase this flow in future, thus increasing the stress on the nitrogen cycle further.

Nitrogen Product Profile for Chemical Sectors

The flows of nitrogen product imply the conversion of mobilized nitrogen into various products downstream. Chemical industries are the key agents in the conversion of the mobilized nitrogen to various products, thus indirectly effecting the natural nitrogen cycle by key users of reactive nitrogen in the economy. Fig. 12.7 shows nitrogen product intensity for various chemical sectors. The highest intensity for nitrogen products is shown by the "All other chemical production and preparation manufacturing sector (3259A0)" followed by "Plastics material and resin manufacturing (325211)," "Artificial and synthetic fibers and filaments manufacturing (325220)," "Synthetic rubber manufacturing (325212)," and "Other basic inorganic chemical manufacturing (325188)." The dominating flow of nitrogen as products in intensity is nitrogen in explosives. The next highest intensities of nitrogen product categories that are being used in chemical sectors are for the "Plastics and synthetics" and "Animal feed" manufacturing. In the current scenario, most of these products are manufactured using ammonia that is produced by the Haber—Bosch process utilizing natural gas as feedstock.

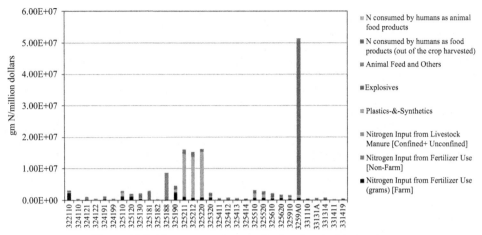

Figure 12.7 Nitrogen product profile for chemical industries.

As there is an increasing shift toward greener manufacturing processes there will be greater reliance on biomass-based products that will shift the nitrogen mobilization profile of chemical sectors.

Nitrogen Emissions Profile for Chemical Sectors

The nitrogen emissions profile of chemical sectors includes land, air, and water emissions because of energy combustion and indirect impact on upstream nitrogen emissions caused by consumption of nitrogen-based products. The land emissions imply nitrogen waste to land such as sludge, etc., air emissions imply nitrogen emissions to air such as NO_x and N_2O, etc., and the water emissions imply nitrogen emissions to water such as organic and inorganic nitrogen going to water as run-off. This profile is shown in Fig. 12.8, which does not contain emissions from the plant site at these manufacturing sectors except for the energy use nitrogen emissions. The sector with highest emissions intensity is "Alumina refining and primary aluminum manufacturing (33131A)" followed by "Alkalies and chlorine manufacturing (325181)," "Industrial gas manufacturing (325120)," "Petroleum refineries (324110)," and "Asphalt paving mixture and block manufacturing (324121)." It is clear from Fig. 12.8 that the nitrogen emissions profile for chemical sectors is dominated by air emissions. The few sectors where water nitrogen emissions show up are the sectors with indirect connections to biobased products such as in "Pulp mills manufacturing (322110)" and "Other basic organic chemical manufacturing (325190)." It is clear that the chemical sectors that show high impact on the nitrogen cycle while focusing on emissions are different to the sectors that show high impact through nitrogen mobilization and nitrogen product use. This emphasizes the necessity of assessing sustainability of the chemical sector through studying all flows related to nitrogen cycle disruption and that these sectors have heavy

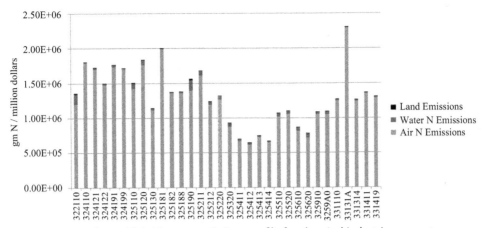

Figure 12.8 Nitrogen emissions profile for chemical industries.

dependence. The current focus on emissions reduction only captures the partial impact of chemical sectors on the nitrogen cycle, which is inadequate for managing the natural cycles.

TECHNO-ECOLOGICAL APPROACH AND CHEMICAL INDUSTRY SUSTAINABILITY

In previous sections "Chemical Industry Profile for Carbon" and "Chemical Industry Profile for Nitrogen" it was demonstrated that chemical sectors have dependence on natural biogeochemical cycles through direct or indirect interactions. However, in traditional analysis, chemical industries are considered to be more reliant on fossil fuels, as is the case currently. The petrochemical sector accounted for almost 6% of oil consumption, and about 10% of global mineral oil consumption was by petrochemical sectors [19]. In an effort to reduce fossil consumption, there has been an increasing push toward biobased feedstock in chemical industries. Biobased polymers are very rapidly replacing the fossil-based polymers founded on similar functional properties. Many of these are likely to be commercially available in the next 10 years, as suggested by chemical engineering news [7]. Succinic, fumaric, and malic acids are being commercialized currently and are used in downstream products such as solvents, polyesters, polyurethanes, nylon, fabrics, inks, etc. The routine industrial production of polyesters, nylon, etc. is usually based on the petroleum feedstocks. For example, polyester is produced from purified terephthalic acid, which is obtained by oxidation of *p*-xylene. *p*-Xylene is obtained from catalytic reforming of petroleum naphtha. Thus shifting to a different feedstock provides reduction of toxic chemicals along with switching to a renewable feedstock. Table 12.5 shows some of the promising shifts in using renewable feedstock for producing downstream products that were originally derived from

Table 12.5 Biomass-Derived Compounds Identified to Replace Petroleum-Derived Chemicals [7]

Chemical Products	Original Chemical Route/ Products Replaced	Biobased Production Route	Downstream Products	Key Uses and Products
Succinic acid	Hydrogenation of maleic acid, oxidation of 1,4-butanediol and carbonylation of ethylene glycol	Bacterial fermentation of glucose	1,4-Butanediol, tetrahydrofuran	Solvents, polyesters, polyurethanes, nylon, food and beverage acidity control, fabrics, etc.
2,5-Furandicarboxylic acid	Can substitute for terephthalic acid in the production of polyesters and other polymers that have aromatic moiety	Chemical dehydration of glucose, oxidation of 5-hydroxymethylfurfural	2,5- Dihydroxymethylfuran, 2,5-bis(aminomethyl) tetrahydrofuran	Polyethylene terephthalate, polyamides such as nylon, plastic bottles and containers, fabrics, carpet fiber
3-Hydroxypropionic Acid (3–HP)	This compound is used in the production of acrylates that were originally produced from acrylic acid that was obtained from propene. Propene was obtained as a by-product of ethylene and gasoline production	Bacterial fermentation of glycerol or glucose. 3-HP is a commercial monomer derived from corn	1,3-Propanediol, acrylic acid, methyl acrylate, acrylamide	Acrylate polymers, carpet fibers, paints and adhesives, polytrimethylene terephthalate, contact lenses, polymers for diapers
Glycerol	Synthetic glycerol was produced from propylene by chlorination or chlorine-free process	Chemical or enzymatic transesterification of vegetable oils	Propylene glycol, ethylene glycol, 1,3–propanediol, glyceric acid, lactic acid, acetol, acrolein, epichlorohydrin	Polyesters, butanol, soaps, cosmetics, foods and beverages, antifreeze/ deicing fluids, coatings, carpet fiber, pharmaceuticals
Sorbitol	Sorbitol is used in the production of alkanes, ethylene glycol, etc., which are originally produced from natural gas	Hydrogenation of glucose from corn syrup, bacterial fermentation (under development)	Isosorbide, propylene glycol, ethylene glycol, glycerol, lactic acid, alkanes	Sweeteners, mouthwash and toothpaste, sugar-free chewing gum, cough drops and medicines, antifreeze/ deicing fluids, water treatment

(Continued)

Table 12.5 Biomass-Derived Compounds Identified to Replace Petroleum-Derived Chemicals [7]—cont'd

Chemical Products	Original Chemical Route/Products Replaced	Biobased Production Route	Downstream Products	Key Uses and Products
Xylitol		Hydrogenation of xylose, extraction from lignocellulose, bacterial fermentation (under development and under research)	Propylene glycol, ethylene glycol, glycerol, xylaric acid, furfural	Sweeteners, sugar-free chewing gum, cough drops and medicines, antifreeze/deicing fluids, new polyesters

The following biobased molecules are less likely to become commercial biobased products in the next 10 years (2025)

Chemical Products	Original Chemical Route/Products Replaced	Biobased Production Route	Downstream Products	Key Uses and Products
Levulinic acid		Acid-catalyzed dehydration of sugars	2-Methyltetrahydrofuran, valeroacetone, 1,4-pentanediol, acetylacrylic acid, diphenolic acid, caprolactam, adiponitrile, pyrrolidones	Fuel ingredients, solvents, acrylate polymers, bisphenol A-free polycarbonates, polyesters, polyamides, pharmaceuticals, herbicides, plastic bottles and containers
Itaconic acid		Fungal fermentation of glucose	4-Methyl-γ-butyrolactone, 3-methyltetrahydrofuran, pyrrolidones	Styrene–butadiene copolymers, polytaconic acid, rubber, plastics, paper and architectural coatings
3-Hydroxybutyrolactone		Multistep chemical synthesis from starch	3-Hydroxytetrahydrofuran, acrylate-lactone, 3-aminotetrahydrofuran	Solvents, synthetic intermediates for pharmaceuticals, polyurethane fiber analogs, new polymers
Glutamic acid		Bacterial fermentation of glucose	1,5-Pentanediol, glutaric acid, 5-amino-1-butanol	Polyesters, nylon analogs, glutamate flavor enhancers, fabrics, plastics
Glucaric acid		Oxidation of starch or glucose by nitric acid or bleach	Lactones, polyhydroxypolyamides, adipic acid	Solvents, nylon analogs, branched polyesters, fabrics, plastics, detergents
Aspartic acid		Enzymatic amination of fumaric acid, fermentation route (under development)	2-Amino-1,4-butanediol, 3-aminotetrahydrofuran, aspartic anhydride, amino-γ-butyrolactone	Aspartame, poly-aspartame, sweeteners, chelating agents for water treatment, superabsorbent, polymers for diapers

petroleum feedstock. None of these shifts have accounted for dependence on the ecosystem or biogeochemical cycle flows, which are also rapidly declining. Thus, as shown in previous sections, an integrated approach should be followed to quantify the dependence of these shifts on natural system flow. This will enable an understanding of whether these shifts are truly sustainable.

Similar to the shift in chemical manufacturing toward renewable feedstocks, there is also a shift toward biobased fuels over fossil fuels. However, the life cycle studies are reporting opposing views about the benefits of switching to biobased fuels. Further, understanding the connection of biobased fuels with disruption of biogeochemical cycles because of direct and indirect connections using Eco-LCA inventory shows that biobased production may shift the burden between the carbon and nitrogen cycles. Thus decisions for adopting biobased products that focus only on the carbon cycle may reduce the carbon footprint, but may increase the nitrogen footprint [20]. This again emphasizes that the sustainability of proposed alternative fuel/feedstock should be quantified using an integrated holistic approach instead of unidimensional metrics. Thus redesign of the chemical industry for sustainability should be performed using a techno-ecological approach. One step in this direction is incorporating the impact of the chemical sector on biogeochemical cycle flows on a life cycle scale. In future, more design shifts for sustainable chemical production should also include the dynamics of natural systems and ecosystem changes. This will ensure a truly holistic design framework for sustainable chemical industries by moving the design to match the planetary limits both at regional and global scale [21].

REFERENCES

[1] Biogeochemical cycle, Wikipedia, http://en.wikipedia.org/wiki/Biogeochemical_cycle [accessed February 2015].
[2] Ayres RU, Schlesinger WH, Scolow RH. Human impacts on the carbon and nitrogen cycles [chapter in Industrial Ecology and Global Change]. Cambridge University Press; 1997.
[3] Vitousek PM, Aber JD, Howarth RW, Likens GE, Matson PA, Schindler DW, Schlesinger WH, Tilman DG. Human alteration of the global nitrogen cycle: sources and consequences. Ecol Appl 1997;7(3):737−50.
[4] Schlesinger WH. Biogeochemistry: an analysis of global change. Academic Press; 1997.
[5] Gilbert P, Roeder M, Thirnley P. The chemical industry in the UK − market and climate change challenges. Tyndall Manchester: University of Manchester; May 2013. Technical Reports, http://www.tyndall.ac.uk/sites/default/files/chemical_industry_in_the_uk_-_final.pdf [Report accessed July 2015].
[6] Pongratz J, Reick CH, Raddatz T, Claussen M. Effects of anthropogenic land cover change on the carbon cycle in the last millennium. Glob Biogeochem Cycles 2009;23.
[7] Bomgardner MM. Biobased polymers. Chem Eng News October 27, 2014;92(43):10−4.
[8] UCAR, National Center for Atmospheric Research http://scied.ucar.edu/imagecontent/carbon-cycle-diagram-doe-numbers [accessed 10.07.15].
[9] US−Energy Information Administration, http://www.eia.gov/tools/faqs/faq.cfm?id=34&t=6 [accessed 20.07.15].
[10] Zapp Karl H, et al. Ammonium compounds. In: Ullmann's encyclopedia of industrial chemistry. KGaA: Wiley-VCH Verlag GmbH & Co; 2014.

[11] Kongshaug G, et al. Phosphate fertilizers. In: Ullmann's encyclopedia of industrial chemistry. KGaA: Wiley-VCH Verlag GmbH & Co; 2014.

[12] The Fertilizer Institute, http://www.tfi.org/industry-resources/fertilizer-economics/us-fertilizer-production.

[13] Zhang Y, Baral A, Bakshi BR. Accounting for ecosystem services in life cycle assessment, Part II: toward an ecologically based LCA. Envi Sci Techol 2010;44(7):2624–31.

[14] Singh S, Bakshi BR. Accounting for the biogeochemical cycle of nitrogen in input-output life cycle assessment. Environ Sci Technol 2013;47(16):9388–96.

[15] Singh S, Bakshi BR. Accounting for emissions and sinks from the biogeochemical cycle of carbon in the US economic input-output model. J Ind Ecol 2014;18(6):818–28.

[16] ISO 14040 for LCA, http://www.iso.org/iso/catalogue_detail?csnumber=37456.

[17] Hendrickson CT, Lave LB, Matthews HS. Environmental life cycle assessment of goods and services : an input-output approach. Resour Future 2006.

[18] Ghosh A. Input-output approach in an allocation system. Economica 1958;25:58–64.

[19] Renewable raw materials: new feedstocks for the chemical industry, 1st ed. In: Ulber R, Sell D, Hirth T. Wiley-VCH Verlag GmbH & Co. KGaA; © 2011.

[20] Gibbemeyer EL. Incorporating the demand for and capacity of ecosystem services in analysis and design [Ph. D. thesis]. The Ohio State University; 2014.

[21] Rockstrom J, et al. A safe operating space for humanity. Nature 2009;461(7263):472–5.

Challenges for Model-Based Life Cycle Inventories and Impact Assessment in Early to Basic Process Design Stages

S. Papadokonstantakis
Chalmers University of Technology, Gothenburg, Sweden

P. Karka
National Technical University of Athens, Athens, Greece

Y. Kikuchi
The University of Tokyo, Tokyo, Japan

A. Kokossis
National Technical University of Athens, Athens, Greece

INTRODUCTION

Decision making in process synthesis typically follows a hierarchical structure [1]. It usually starts from product specifications and proceeds to selection of the synthesis path, identification and screening of process design alternatives, equipment design, and selection of operational strategies, to name just a few main steps before operability and risk analysis studies refine or challenge the aforementioned decisions and complete the basic stages of the design life cycle. The decision-making procedure entails technical and financial criteria to be considered together with ecological and social factors to screen diverse process alternatives from a multicriteria point of view. This multicriteria assessment promotes sustainability aspects in process design, creating also challenges and opportunities [2] for new process systems engineering approaches of high industrial relevance [3].

In this context, substantial efforts have been made to provide robust tools that can deal with the challenges of designing sustainable processes. Taking into account the diversity of the sustainability objectives and that the hierarchy in decision making usually implies different levels of prior knowledge utilization, data availability, and therefore modeling detail, it is inevitable that various methods and tools have been developed, ranging from simple indicators to holistic frameworks for an overall process assessment. Some examples of various sustainability aspects applied to different process design stages include sustainability indicators for industry [4], technology screening [5], life cycle assessment (LCA) approaches [6,7], hierarchical decision making [8] in grassroots process

---

design [9,10,11], and process retrofitting [12,13]. Ruiz-Mercado et al. [14] have provided a taxonomy of sustainability indicators according to the GREENSCOPE methodology [15] and have further discussed the data needs for the calculation of the respective indicators [16].

Sustainability Frameworks for Process Design

A rather generic framework promoting sustainability concepts in the design of chemical processes is the one based on the 12 green chemistry [17,18] and 12 green engineering principles [19] proposed by Anastas and coworkers and applied in a series of publications over the last 15 years [20]. Most of the principles of green chemistry (e.g., waste prevention rather than treatment, atom economy in chemical synthesis routes, use of inherently safer chemicals and conditions, use of renewable feedstocks, design for safer products during their use, and product degradation at the end of their function) refer to decisions made within the scope of the process synthesis problem formulation and early phases of process design, highlighting the importance of this task. Some of these principles are also referred to as green engineering principles. These are complemented by principles for later stages of basic process design (i.e., design of a basic process flowsheet), such as the design of upstream and downstream separation processes targeting material and energy efficiency (e.g., through process integration), avoiding material diversity resulting in multicomponent systems (i.e., to promote disassembly and value retention), and avoiding unnecessary capacity (e.g., oversizing of equipment).

Another, more specialized, framework of academic research and industrial practice for designing sustainable chemical processes refers to process risk analysis and the relevant safety, health, and environmental hazard assessment. The evaluation of process hazards and risks is critical for designing sustainable chemical processes, since no process can be perceived as sustainable if it has the potential to lead to severe accidents with serious impacts for the workers, the near-by population, and the environment. As a matter of fact, some of the aforementioned principles of green chemistry and engineering [18,19] obviously refer to reducing process hazards and risks. Prominent frameworks in the area of process risk analysis include, but are not limited to, the hazard and operability analysis [21], fault tree analysis [22], and failure mode and effect analysis [23]. A review of risk analysis methodologies in industrial plants can be found elsewhere [24]. However, such methods often require detailed plant specifications and process layout, expert knowledge, and are sometimes time-consuming, making them inappropriate for early phases of process design (e.g., for screening and selection of less hazardous chemical synthesis paths and their respective basic process flowsheets), where the amount of data is limited and the goal is often to screen a wide domain of possible process alternatives. On the other hand, some of the proposed hazard assessment frameworks are using indices derived by empirical scales of "dangerous properties" (e.g., toxicity, flammability, reactivity, degradation), amounts and conditions under which the various chemicals are used in the process. These frameworks are applicable to early design stages because of the simplicity of the respective calculations and the fact that no detailed process information is required. Reviews of the most popular hazard assessment methods can be found elsewhere [25,26], and perhaps the most interesting conclusion when comparing them is that the merit that distinguishes each method is its level of simplicity according to the early design phase where it can be applied. However, this simplicity is often achieved through subjective scaling and weighting schemes used in the definition of the index-based approaches, unclear index resolution used for ranking the process alternatives, and possibly in the expense of neglected effects in the overall assessment [27,28].

A further specialized framework focuses on the topic of waste generation from both prevention and waste treatment perspectives. No matter if this goal is achieved by innovative process design or by end-of-pipe technologies, complying with legal emission limits is clearly one of the first priorities of the chemical industry as far as environmental impact is concerned and it is often perceived as such by society too. The importance of this challenge is already included in multiple principles of the generic green chemistry and engineering framework. In general the integration of the waste problem in the development of chemical processes is relatively recent [29], including, for example, methodologies like the "waste reduction" algorithm [30,31], the synthesis of mass-exchange networks [32], and expert systems like ENVOP [33]. However, despite the clear benefits of these waste minimization approaches in process design, waste generation is often not negligible. Thus, end-of-pipe technologies are still needed and waste treatment design efforts typically try to achieve low treatment costs while respecting the legislation limits for emissions to air, water, and soil. In this direction, mathematical models have been used to describe the waste treatment processes and the overall waste treatment system [34–36], including also recycling opportunities and considering economic, energy, and environmental impact objectives [37–39]. In these studies, typically simplified waste processing models were used to calculate utility consumptions and emissions as a linear function of the waste stream compositions and flows and a superstructure design was optimized for long-term operation and investment planning under uncertainty conditions.

Life Cycle Assessment Framework

The frameworks mentioned previously represent some main trends for sustainability assessment of chemical processes and range from stating generic principles to addressing specific aspects of sustainability. On the other hand, if the focus is only on the environmental impact of normal process operation and production, LCA is a suitable framework. As a typical system-oriented approach, it has been demonstrated that LCA can not only highlight the relative importance of diverse environmental impacts from a "cradle-to-grave" perspective but can also allocate the crucial system interdependencies leading to impact tradeoffs and/or "win–win" situations [40,41]. In this context, LCA can effectively assist decision makers to have an overall system-based perception of environmental impacts in process design beyond the traditional plant emissions-oriented one. LCA achieves these goals following a systematic procedure well defined in ISO norms [42,43], comprising the steps of goal and scope definition (i.e., system boundaries, functional unit, and width/depth of the analysis), inventory analysis (i.e., resource requirements, resulting emissions, and assessment of data quality), impact assessment (i.e., on humans, property, and environment), and interpretation (i.e., relative importance of impacts, sensitivity analysis, and optimization potentials).

Already from this short overview of the LCA framework, it is evident that there are many methodological aspects involved in performing an LCA study. As an example, in the chemical industry, typical decisions for the goal and scope definition include functional units (e.g., "production of 1 kg of chemical compound-x," "use of 1 kg of raw material-x," "production or of 1 kg of final chemical product-x"), retrofitting an existing process or developing grassroots designs, and answering questions of strategic importance for a company, such as the evaluation of the supply network, the product portfolio, and the geographical and temporal investment planning. Usually the factors playing a role at this stage of decision making include the initiator and target group of the study, the subject of the study (e.g., in the case of comparing different production options of the same chemical compound, the use phase of this compound is not important and a "cradle-to-gate" approach is adequate, while in the case of comparing different products with similar functions or options for utilizing available raw material, the use phase of the products should be included in the LCA study and a "cradle-to-grave" approach is necessary), and the levels of available technology detail and diversification. Typically, the LCA practitioner will come back to these questions after the core stage of LCA, namely, the calculation of life cycle inventories (LCIs).

Importance of Life Cycle Inventories

There are many challenges in the calculation of LCIs, most of which are related to the aforementioned decisions in the goal and scope definition stage. For instance, if the subject of the study refers to assessment of process retrofitting scenarios, then the process to be retrofitted defines the base case scenario, the scope of technologies, and the level of detail to be considered. However, retrofitting a process rarely means that the whole value chain is to be retrofitted; a part of the process typically lies within the interest of the company that has ordered the LCA study. These boundaries exist also when new processes and products need to be designed, since the company business and know-how typically define such boundaries, although the degrees of freedom are generally more in the grassroots design projects. It is therefore rather common and good practice [44] to:

- Define the functional unit in relation to the product (or service) that the initiator or target group of the LCA study is interested in;

- Use as a starting point for the calculation of LCIs the "nearest" processes to the functional unit and expand both upstream and downstream (e.g., end-of-pipe waste treatment, product use phase if relevant with the LCA study);
- Flag unfamiliar and/or neglected processes;
- Define the crossing points of three types of boundaries, those between the system under study and the environment, the system under study and other interrelated systems, and the relevant and unfamiliar, irrelevant, or neglected processes;
- Start at a low process resolution (i.e., combine processes into single process building blocks), since quite often only aggregated data are available (e.g., for waste treatment processes, energy utilities, and other auxiliaries production).

As mentioned previously, it is quite common to rethink and restate the goal and scope definition of the LCA study based on the outcome of the LCIs calculation stage. Moreover, the life cycle impact assessment (LCIA) stage that follows heavily depends on the calculated LCIs, which are transformed through established classification and characterization schemes (i.e., the choice of environmental aspects such as the potential for abiotic depletion, energy depletion, global warming, ozone depletion, ecotoxicity and human toxicity, acidification, photochemical oxidant creation, and nitrification) into relevant environmental impacts [45] and even further into midpoint scores and single environmental indices [46,47]. There may be again an iterative procedure between LCIs calculation and the LCIA stage in terms of focusing on the quantification of specific energy, material, or emission flows to reduce uncertainties. For instance, if the LCA target is to quantify the global warming potential of the production of a chemical compound, the relevant direct and indirect gas emissions (e.g., carbon dioxide, dichlorodifluoromethane, dinitrogen oxide, methane, trichloromethane, etc.) associated with the production process and the raw material and auxiliaries used, respectively, are of much higher importance than used or emitted toxic substances such as heavy metals, benzene, etc.

Although the LCIs calculation lies in the core of every LCA study, it is by far less standardized compared to the rest of the LCA stages. This is even more evident at early to basic design stages of new processes and products. Here a combination of process design, simulation, and flowsheeting techniques, and data collection and estimation procedures (e.g., based on company, commercial and public databases, state-of-the-art models, and heuristics) is required, which typically does not follow a commonly accepted protocol, and therefore makes it cumbersome to combine different sources of information within the system under study and other interrelated systems. This causes additional problems for LCA studies in multiproduct or multifunctional systems. These systems are characterized by one or more process subdivision points, where a new process path begins for the production of one or more products. Ideally, these multifunctional systems should be divided into subprocesses connected to specific products, up to the point where a comprehensive understanding of the subdivision processes allows a first principles-based allocation of the respective mass and energy flows. It is evident that applying this principle may have immense implications in the depth of the LCIs calculation procedure, and vice versa the feasibility of collecting the necessary information for the LCIs calculation may dictate a different procedure of allocating the environmental impacts in multifunctional processes (e.g., by partitioning at subdivision points of lower resolution on the basis of generic principles, such as heat content or market value of the coproducts, or by expanding the system and changing the functional unit to include all relevant coproducts, or by giving credit to main products for avoided environmental impacts of coproducts) [48,49].

To the points raised previously regarding the importance of LCIs calculation, it should be added that similar type of information, if not the LCIs themselves, is needed for the generic to specialized sustainability frameworks briefly mentioned at the beginning of this paragraph. For instance, Barton et al. [50] described a way to model unit operations for flowsheeting of waste management systems to get the necessary LCIs and their interdependencies with waste characteristics. This led to the development of several models for municipal waste LCA tools, and comparative studies [51] have also been performed to quantitatively analyze their results [52] and identify the differences in the model assumptions [53]. The safety, health, and environmental hazard assessment frameworks often comprise LCIs relevant for the ecotoxicity and human toxicity potential. Besides that, they typically share the common calculation ground of material flow analysis. This common computational background is also true for many of the environmental indicators in the GREENSCOPE taxonomy [14] and the green chemistry and engineering principles.

LCI ASPECTS IN EARLY TO BASIC PROCESS DESIGN STAGES

The term "early to basic stages of process design" typically means those primary design stages after the product and its specifications have been clearly defined. For instance, for a chemical product, these early design stages comprise the selection of the chemical synthesis path, process flowsheet layout, and operating parameters, including upstream and downstream processes for raw material pretreatment, product recovery, and recycle of valuable material. They are typically characterized by limited data availability and accuracy, but they also involve more degrees of freedom in the sense that a large number of process options is usually screened in these stages. Sometimes the problem can be even coupled, hierarchically or through advanced integration, with the design of the supply chain networks and the waste management systems. Moreover, if the environmental fate of the produced chemical after its use is also considered in the assessment, a full "cradle-to-grave" analysis can be performed within the scope of LCA. The target is to identify the best performing process options from a multicriteria point of view, as well as the critical "tradeoffs" and "win−win" situations between the considered objectives. Depending on the modeling detail, the system boundaries, and the considered frameworks for process assessment, the design problem can be addressed by methods ranging from heuristics or rules of thumb to rigorous mathematical programming techniques [54].

LCI Data Gaps and Process Design Decisions

The LCIs clearly depend on process design decisions. Similarly, when access to LCIs of existing processes is available, it would be beneficial to have some insights about design decisions that have been made during the development of the process. A generic example (Fig. 13.1) illustrates some of the main concepts related to the decisions made in early to basic process design stages. In this generic example a mixture of chemical compounds (A, B, C, and D) produced by previous processes (i.e., P_{i-1} in Fig. 13.1) is to be separated (i.e., in process P_i in Fig. 13.1), into the primary (i.e., in higher amounts in the original mixture), lighter, pure products A and B, and the secondary (i.e., in lower amounts in the original mixture), heavier, pure products C and D. The primary product A is then fed to a subsequent process (i.e., P_{i+1} in Fig. 13.1) and so on, until a final product Pr. Assuming that the LCIs for product Pr are to be estimated, there are a number of options:

- If the product Pr is registered in LCA databases (e.g., Ecoinvent [45]), a comprehensive list of LCIs and LCA metrics can be typically found, as well as background information for the considered processes, assumptions, etc. In this case, these can be used as reference values to be compared with new, innovative processes for the

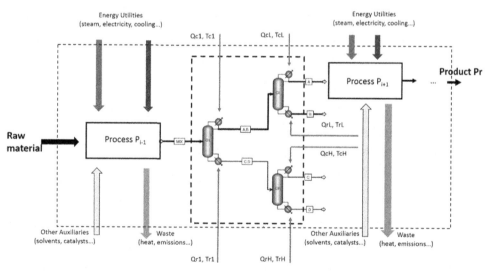

Figure 13.1 Generic example of process decomposition for extracting cradle-to-gate life cycle inventories.

production of Pr. For instance, this can be the case for the design of biobased production of chemicals and fuels, where reference values are needed (e.g., the fossil fuel-related ones) to infer the "greenness" of the new processes.

- If the product is not registered in LCA databases, estimations from similar processes can be used or tools calculating LCA metrics on the basis of the molecular structure of the product Pr. The Finechem tool [55] is a unique, state-of-the-art tool of this kind. A more detailed application of this approach is presented in Case Study 1 for the solvent-based CO_2 capture process.

- If the process can be decomposed into subprocesses (e.g., processes P_{i-1}, P_i, etc. in Fig. 13.1), LCA databases for LCIs and/or LCA metrics can be used for the intermediate products (e.g., "MIX" in Fig. 13.1) and model the remaining process blocks (e.g., processes P_i, P_{i+1}, etc. in Fig. 13.1). The modeling of these remaining blocks may be based on a range of indicators, heuristics, short-cut models, or detailed process models. For instance, Bumann et al. [56] has evaluated a proxy indicator for the estimation of the gate-to-gate energy consumption in the early design stages and has demonstrated its applicability in the case of organic solvents; the estimation error of this proxy indicator-based model has been shown to be similar with other calculations in these early stages (e.g., capital cost estimations). Besides filling LCIs data gaps for LCA studies, this type of proxy indicator has been also integrated into more generic process design frameworks [9,57]. As far as short-cut models are concerned for filling LCIs data gaps, the work of Köhler et al. [58] and Seyler et al. [59] provides generic multiinput allocation models for LCIs of wastewater treatment and waste incineration processes, respectively.

Focusing on the last case for the modeling of identified subprocesses, as mentioned earlier, process design decisions can be highly important. If the design of P_i refers to a grassroots design with no further limitations, then the process layout presented in Fig. 13.1 is only one of the possible layouts for separating "MIX." In principle, a different sequence of separations may be feasible, namely, to separate only product A at the top of the first column, product B at the top of a second column, etc. Moreover, other types of separations can be considered; for instance, based on the thermodynamic properties of the compounds in the mixture, recovery of the compound D in solid state through crystallization out of the original mixture could be an option. The pioneering work of Jackson et al. [60] is an example of a systematic method that summarizes basic thermodynamic properties to be considered in screening options for the design of separation sequences.

Another process design decision is the degree of process integration. This refers to both heat and mass integration within a process block as well as to "process to process" integration and "total site analysis," if the process is to be realized in an industrial site or an industrial park sharing infrastructure, utilities, etc., between the processes [61,62]. Obviously, integration contributes to decreasing the overall LCIs (although tradeoffs may also appear when, for instance, material recovery involves energy-intensive processes), and therefore also the ones allocated to each of the integrated processes. In the simple example of Fig. 13.1, even if only the process P_i is considered, there is, in principle, some heat integration potential; the reboiler duty of the light components distillation (QrL) can be, at least partially, covered by the condenser duty of the heavy components column (QcH), since the respective temperatures allow such a heat transfer (TcH > TrL).

Another design tradeoff that may affect the resulting LCIs is that of operating and capital costs, depending on decisions about the operating parameters and the equipment sizes (e.g., the additional size of a plug-flow reactor to increase the conversion of the reaction, bigger distillation columns instead of reboiler/condenser duties, smaller temperature differences in heat integration in the cost of bigger heat exchangers). In the example of Fig. 13.1 for the process P_i, it is logical to assume that the recoveries of the A, B, C, and D components are set at very high values for all columns. This means that each column can be designed relatively independently (i.e., selecting a reflux ratio or number of theoretical column stages), although some degrees of freedom may be covered by the heat integration potential of QrL/QcH, which clearly depends on the selected reflux ratios for the CL and CH columns. Therefore for a given "process scale" (i.e., input flow rate of "MIX") there is an economic optimum for each column, representing the tradeoff between operating and capital cost. The operating cost comprises in this specific example the energy utilities of the three columns (Qr1, QrL, QrH, Qc1, QcL, QcH) and is often dominated by the reboiler duties, if there is no need for cooling at lower than ambient temperature, operating at vacuum or elevated pressure, use of an extraction medium, etc. In this simple case the LCA metrics will also be linearly proportional to the operating costs and both will be linearly proportional to the process scale. However, this is not true for the capital costs. Smaller process scales will shift the economic optimum in favor of decreasing the capital costs more than the operating costs. Since in the majority of LCA studies the LCIs related with the construction of the plant are neglected (partially because of no available data and/or relatively smaller environmental impacts compared to their economic implications), the consequence of a smaller process scale will be higher environmental impacts per functional unit, assuming that the process is indeed optimized accordingly.

Importance of the Process Scale

When the environmental impacts of a process are compared with the capacity of the environment or with the impacts of other processes from different production sectors, the role of the process scale lies at the center of the discussion. For instance, in a study for the production of fine chemicals it has been demonstrated that the respective environmental impacts per kilogram of product can be more than two orders of magnitudes higher than those of bulk chemicals [63]; however, bulk chemical production volumes can be more than four orders of magnitude higher. Another example refers to the utilization of biomass for replacing fossil-based production of chemicals and fuels. In this case it should be kept in mind that, currently, less than 10% of fossil-based production is directed toward chemicals. Similarly, capturing CO_2 from power plant emissions refers to a different (i.e., higher) scale compared to acid gas cleaning and CO_2 recovery in the process industry or the techno-economically viable CO_2 utilization options (e.g., through power to gas or power to methanol technologies) with respect to the demand side for closing the carbon cycle.

When the impact of process scale is viewed from the planetary boundaries perspective, the inherent multicriteria nature of any sustainability assessment is indispensable. Even when only environmental LCA impacts are accounted for, studies have shown that certain boundaries have been crossed or are very close to the limit (i.e., with respect to climate change, biodiversity loss, and nitrogen and phosphorous cycles), while others are still reasonably well safeguarded (i.e., stratospheric ozone depletion, ocean acidification, and freshwater use) [64]. It is therefore possible that different production sectors may have an impact on different planetary boundaries; some of which may be within or already outside their safe operating space. For instance, studies have indicated the severe impacts of plastic debris on marine organisms [65]. Thus, from a cradle-to-grave LCA perspective, fossil-based plastics production may have a more direct or at least a different kind of effect in terms of biodiversity compared to fossil-based fuel production, which is certainly in higher production scales.

A somewhat different issue, related also to the impact of the process scale on LCIs, refers to the case of multifunctional processes. In this case, there are various approaches for allocating LCIs and therefore environmental impacts on the multioutput products, depending also on whether an attributional or consequential LCA is the goal of the study [48,49]. Generally, the following allocation methods appear in practical applications [66]:

- All LCIs and environmental impacts are allocated to the main product, which is often called determining product, because its demand determines the production capacity (i.e., process scale). The coproducts are free of environmental burden, or even contribute to it as waste streams.
- All LCIs and environmental impacts are allocated to the main product, but the co-products substitute other equivalent products and this credit is ascribed to the main product.
- The LCIs are allocated to the various products of the multifunctional process based on a characteristic property (e.g., heat content, exergy, economic value). This allocation can be done either for the whole process or on a given process subdivision point. For instance, in process P_i of Fig. 13.1, either the overall LCIs (Qr1, QrL, QrH, Qc1, QcL, QcH) are allocated to A, B, C, and D based on their economic value or C1 is identified as a subdivision point and Qr1, Qc1 are allocated to products A, B and C, D based on their economic value, while QrL and QcL are allocated only to A, B and QrH, QcH are allocated only to C, D.
- Recently, hybrid methods have also been proposed, where the allocation factors are based on the type of product that is substituted in the market.

As can be seen, in multifunctional processes the key question is the utilization of the secondary flows, either as valuable coproducts or waste streams for further treatment. The role of the process scale in this case becomes more relevant when some of the coproducts are produced in significantly smaller amounts than the main product; as the process scale decreases, the main product (i.e., assuming it exists) can obviously still be utilized or sold, while the coproducts may become valuable chemicals to useful waste streams as energy carriers (e.g., for industrial incineration plants) or waste streams for further treatment. In the example of process P_i in Fig. 13.1, if products A, B are available in much higher amounts compared to C, D in the "MIX" stream, in smaller process scales it may not be any more sensible to recover the pure C and D through the distillation in CH; instead the stream C, D at the bottom of the column C1 may be considered as fuel, depending on the energy content of C, D, or waste water stream if one of C, D is actually water.

CASE STUDIES

In the following sections three case studies are presented where, among other LCA aspects, some of the LCI issues discussed previously are highlighted. Case Study 1 refers to solvent-based postcombustion CO_2 capture processes, Case Study 2 refers to lignocellulosic biorefineries, and Case Study 3 refers to the poly(methyl methacrylate) recycling process.

CASE STUDY 1: LCA ASPECTS OF SOLVENT SELECTION POSTCOMBUSTION CO_2 CAPTURE

Motivation

The aim of this case study is first to demonstrate the lack of life cycle inventory data in the case of established process technologies finding new applications, and thus to highlight the importance of generic short-cut models for filling in the respective data gaps. The case of solvent-based postcombustion CO_2 capture constitutes such a technology when, for instance, its application to retrofitting power plants is concerned. The motivation for studying this process system is generally driven by environmental concerns for climate change, making LCA a suitable methodology to provide relevant answers. However, in the quest for not only technologically feasible but also economically viable and environmentally favorable options, process optimization is required, which typically starts from screening a wide range of conventional and unconventional options (e.g., starting from the solvent selection at very early design stages), for which severe data gaps exist.

Process System and Scope of the LCA

Solvent-based CO_2 capture has been applied for many years in the production scale and conditions of oil refineries for acid gas cleaning; the conditions for both absorption and solvent regeneration are relatively easy to meet and the process can be easily retrofitted into existing plants. However, applying it to the scale of power plants is causing some concerns, since it has been estimated that a cost penalty of over 40% is introduced to the operation of the plant, 70% of which is because of the thermal separation process for solvent regeneration. Other important aspects include the capacity of the solvent for CO_2 capture, the relevant chemisorption kinetics, the toxicity of the solvent, its degradation potential and the associated production impacts, the corrosion of equipment caused by the solvent itself under the CO_2 capture conditions, and the implications of the solvent selection on the process flowsheet design, control, and integration with the flue gas emitting plant [67].

In principle, all these aspects related with the solvent selection will have an impact on the outcome of the LCA metrics of the various alternatives. Fig. 13.2 presents a generic flowsheet for the solvent-based postcombustion CO_2 capture processes, illustrating also the main concerns from an LCA point of view. If the LCA scope is to screen solvents for postcombustion CO_2 capture without performing process simulations (i.e., in a very early phase of process design, where perhaps the number of solvent molecules to be screened is immense, for instance, in computer-aided molecular design [CAMD] of solvent molecules [68]), the potential solvents should be characterized based on "properties" that would indicate their expected performance in the capture process (e.g., the standard flowsheet of Fig. 13.2). These "properties" can be thermodynamic in nature (e.g., solubility parameters between CO_2 and the solvent, solvent heat of vaporization,

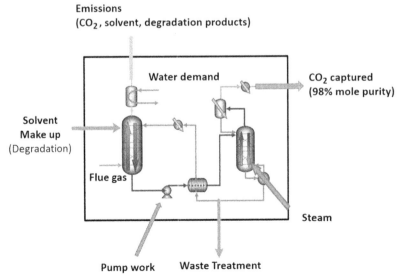

Figure 13.2 Generic flowsheet for the solvent-based postcombustion CO_2 capture processes including the main life cycle assessment-related issues.

liquid heat capacity, density, viscosity, vapor pressure, etc.) in order to consider effects associated with the potential of the solvent to capture CO_2, the process fugitive losses, the heating and cooling loads, the mass transfer in the separation columns, the pumping work, etc. The "properties" can also be related with the solvent oxidative and thermal degradation potential [69,70] and therefore the solvent make-up, the resulting degradation products, and the required purge stream to avoid their accumulation. The purge stream will introduce additional solvent losses and there is typically a solvent reclaimer process (i.e., in Fig. 13.2 the reclaimer process is included in the generally stated waste treatment process, since the purge may also be treated as waste) that can recover more than 90% of the solvent at the expense of additional energy. Finally, at this early stage of process design it would be important to have LCA metrics for the solvent production process to consider the cradle-to-gate perspective.

Life Cycle Inventories: Short-Cut Models for Filling in Data Gaps

While there are available databases (e.g., Ecoinvent) for obtaining "cradle-to-process gate" data of a substantial number of chemicals including solvents, these should not be considered as the sole source of information, especially for innovative processes or chemicals. Fig. 13.3 presents an illustrative example for a number of conventional solvents that have been proposed for CO_2 capture, a very small percentage of which is available in the Ecoinvent database. In this case the alternative would be to use a tool such as FineChem to fill in the cradle-to-gate gaps for the solvent production or to construct flowsheets for the production of each solvent and derive from there the necessary LCIs. It is clear that the second choice, although perhaps more accurate, may be inefficient in

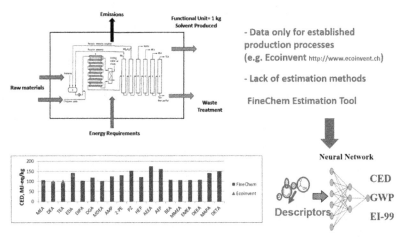

Figure 13.3 Life cycle data availability for conventional solvents that have been proposed for CO_2 capture. *CED*, cumulative energy demand; *GWP*, global warming potential; *EI-99*, EcoIndicator 99.

early design phases where potentially a large number of alternatives (i.e., in this case solvent molecules) have to be screened. For instance, even with today's computational means it is still inefficient, if not infeasible, to include a flowsheet generation and assessment procedure into a more generic optimization algorithm (e.g., CAMD), especially when extending the boundaries to a cradle-to-gate scope.

It is important to note that this short-cut method through molecular structure-based models for LCIA estimation is not a panacea. A first reason is that the FineChem tool is trained with specific classes of molecules and extrapolating to other classes is not recommended. A second reason is that FineChem only refers to conventional, fossil-based technologies; thus it cannot be used when innovative, biobased production paths are investigated. A third reason is that, although the FineChem tool is a state-of-the-art model, it only provides certain LCA metrics, such as cumulative energy demand, global warming potential, and EcoIndicator 99, and not specific LCIs, such as water footprint or emissions, as it is, for instance, proposed by the environmental performance strategy map [71]. Another reason is that the FineChem molecular structure is not using established group contribution methods, such as those applied for the calculation of physical properties. Therefore when FineChem needs to be included in a CAMD procedure, where group contribution methods are typically used, additional algorithmic approaches have to be designed to ensure compatibility; these algorithmic approaches are not straightforward, because quite often the transformation from one molecular description to the other is not an injective function. Finally, after a first screening of alternatives, the focus is obviously constrained on a limited number of preferred options, for which more accurate LCA calculations are necessary.

CASE STUDY 2: LCA ASPECTS IN THE DESIGN OF LIGNOCELLULOSIC BIOREFINERIES

Motivation

The aim of this case study is to illustrate the environmental impact assessment through a life cycle approach of lignocellulosic biorefineries, which are representative examples of multifunctional systems comprising various production lines. This example highlights the need to provide environmental performance metrics by allocating the total biorefinery impacts to the biorefinery final products. This kind of allocation is particularly useful when a certain final product of interest can also be produced by alternative ways (i.e., other biomass- and fossil-based pathways) and, therefore, a comparison between these alternatives for the functional unit of 1 kg of product is required. In other words, in this case the focus is not on the utilization of 1 kg of biomass where the total biorefinery environmental performance would be of interest, but on a specific production path within the biorefinery production network, considering of course the joint or coproduction nature of the system. This allocation procedure (here the term "allocation" is used in its more general definition, namely, including the system expansion approaches) is usually a critical issue in LCA studies without, still, a straightforward commonly accepted answer and practice.

Thus this case study focuses on key methodological steps for handling a multifunctional system and the assumptions made when conducting the LCIs. Additionally, it discusses the effect of the type of the LCA studies, namely, prospective and retrospective, on the selection of the allocation methods and therefore on the LCA metrics. According to some researchers [72], attributional LCA requires market-oriented allocation while consequential applications require system expansion. However, it has also been reported that various allocation methods have been applied in practical applications, depending on the goal and scope of the study and the LCA practitioner preference and expert knowledge [66].

Process System and Scope of the LCA

Biomass can be converted to fuels and chemicals through thermochemical and (bio) chemical routes. Gasification is considered as one of the most promising thermochemical processes producing syngas and other light gases. This process can be combined with technologies that produce methanol or dimethyl ether. These can be considered as final products or may be further converted to gasoline or olefins. Syngas can also be converted to mixed alcohols or substitute natural gas. A conceptual representation of such a biorefinery system is shown in Fig. 13.4. Its boundaries include three interacting processes, which convert biomass (wood chips) into a range of intermediate or final products. For example, syngas, which is considered as an intermediate product, can be further converted through downstream processes into the final upgraded

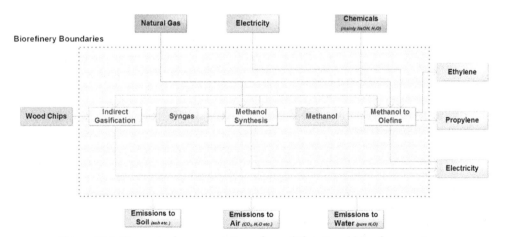

Figure 13.4 Conceptual representation of a multifunctional biorefinery system.

products depending on the desired product portfolio. This study intends to evaluate the environmental impacts of olefins (propylene and ethylene) as final products coproduced with electricity along the production routes. Electricity is produced either from heat recovery and exploitation of high-pressure steam in steam turbines or from fuel-spent gas in gas turbines; fuel-spent gas is typically produced from chemical reactions and subsequent separation stages (e.g., mixture of H_2 and CH_4 produced in the methanol to olefins process).

Life Cycle Inventories: Allocation Through Detailed Process Flowsheeting

In this case study the mass and energy flow data are given at unit process level, facilitating a detailed allocation approach and the critical subdivision points (i.e., the points where allocation can be applied inside the system boundaries) are known. The functional unit is defined as 1 kg of ethylene. The methodological steps followed in this study are in accordance with the ISO 14040 norms. When applying the sub-stitution method, the first step is to define the determining (main) product based on criteria such as mass flow, market value, or other specific rules [72]. This main product determines the scale of the production system as well as the production rate of all the dependent coproducts. It should be noted that in general the "methanol to olefins" conversion is an example of a combined production process where the relative proportions of ethylene and propylene can be adjusted by changing the reactor operating conditions, giving a propylene to ethylene ratio in the range 1.2—1.8 to meet the growing demand for propylene [73]. In this case study, however, ethylene is considered to be the determining product and it is assumed that the ethylene to propylene mass ratio is roughly 1:1, where the maximum yield is ach-ieved. Then the way of utilization is defined for the dependent coproducts (i.e., H_2 and C_2H_4, higher alcohols, energy recovery from hot streams). For instance, co-products can be treated as waste streams or displace other equivalent products; in both cases the main product is accredited the environmental burden of the waste treatment or the avoided environmental impact of product displacement according to the respective Ecoinvent database values [45].

Life Cycle Inventories: Model-Based Calculations for Impact Assessment

Various data categories are necessary to perform LCA using the Simapro software [74]; these comprise, for instance, the yield of the total process, the allocation factors, input processes from the technosphere, which represent materials and energy used in the study system, emissions to soil, water bodies, and air, and information about waste treatment facilities for the several waste streams. The key parameters and selected inventory data of the unit processes included in the biorefinery boundaries are summarized in Table 13.1. The available information includes the yield of each process (i.e., the conversion rate of one intermediate product to another), the production ratio of the dependent coproducts relative to the main product as well as information for handling some coproducts (e.g., a mixed stream with higher alcohols directed to the incineration process because of a low degree of purification instead of being subjected to further separation levels). It should be noted that because of confidentiality reasons only a small part of information and inventory data is provided as an example. Additionally, because of simplifications in the presentation of the study process, some waste streams and chemical and energy auxiliaries have been omitted in Table 13.1.

The models for the waste treatment units are taken from previous LCA studies based on industrial data that describe incineration and wastewater treatment facilities [58,59]. Streams with organic load and water content lower than 50% w/w are assumed to be treated in incineration units, whereas streams with water content greater than 50% w/w are treated in wastewater treatment plant units. Streams with small inorganic load (<1% w/w) are considered to be treated in municipal waste treatment facilities using the model "Treatment, sewage, to wastewater treatment, class 1/CH" in the Ecoinvent V.2 database. Processes describing the production of chemicals, energy, and water used in the study system have also been taken from the Ecoinvent database V.2 [75]. Table 13.2 includes properties, such as market prices and other physical properties, which have been used to calculate the partitioning coefficients, before entering data in the Simapro software 8.0.2 to perform the LCIA calculations.

For the case study system, the subdivision points have to be defined as those points inside the biorefinery boundaries where more than one product is produced (i.e., indirect gasification, methanol synthesis, methanol to olefins). As explained in the previous paragraph, this is possible because of the system insights given in Fig. 13.4. Table 13.3 shows the values of the partitioning factors for each coproduct and subdivision point and the displaced processes used in the three allocation approaches. The selection of the process that is displaced because of the production of a coproduct has a significant effect on the results of the LCA. In this assessment, natural gas-derived electricity (e.g., in state-of-the-art gas turbines) is assumed in the displacement procedure in order to take into account the fossil-based best available technology having a greenhouse gas (GHG) emission factor of 0.24 kg CO_2 eq./kWh. If coal-derived electricity was considered as the displaced process (i.e., having a GHG emission factor of 1.08 kg CO_2 eq./kWh on average for the countries belonging to the Union for the Coordination of the Transmission of Electricity), the GHG emission savings of the biorefinery systems would be greater.

Based on the LCIs, the LCA metrics were calculated using the RECIPE method [46]. Biorefineries constitute emerging technological options based on diverse production activities. Assessing their performance from an environmental sustainability

Table 13.1 Key Parameters and Selected Data for the Biorefinery System of Case Study 2

Unit Processes		Value	Comments
Biomass production	Wood chips		Wood chips, mixed, from industry, $u = 40\%$, at plant/RER
	Density of wood chips	536 kg/m^3	Source: Process flowsheet
Gasification	Syngas yield	0.933 kg syngas/1 kg wood chips	Source: Process flowsheet
	Electricity ratio	0.207 kWh/1 kg syngas	Source: Process flowsheet
	Steam equivalent	0.124 kg/1 kg syngas	Amount of steam producing the equivalent amount of electricity. Source: Process flowsheet
Methanol synthesis	Methanol yield	0.591 kg methanol/1 kg syngas	Source: Process flowsheet
	Electricity ratio	0.127 kWh/1 kg methanol	Source: Process flowsheet
	Steam equivalent	1.370 kg/1 kg methanol	Amount of steam producing the equivalent amount of electricity. Source: Process flowsheet
Methanol to olefins	Ethylene yield	0.158 kg ethylene/1 kg methanol	Source: Process flowsheet
	Propylene ratio	0.964 kg propylene/1 kg ethylene	Source: Process flowsheet
	Electricity ratio	1.238 kWh/1 kg ethylene	Assumed that it is produced from the combustion of fuel gas (1.458 kWh/1 kg ethylene) in gas turbine based on the model "natural gas, burned in gas turbine, GLO" (using the conversion factor: 1.178 MJ of input fossil energy produces 1 MJ electricity)
	Higher alcohols ratio	0.411 kg higher alcohols/1 kg ethylene	Treated as waste to incineration unit based on the model presented by Seyler et al. [59]

In Ecoinvent 2.2, GLO means "global" and represents activities which are considered to be an average valid for all countries in the world. RER represents Europe.

Table 13.2 Properties for Coproducts of the Biorefinery System in Case Study 2

Product Properties		Value		Data Source/Description
Syngas	Price	0.150	€/m³	Ecoinvent 3, market for synthetic gas (GLO)
	Density	1.15	kg/m³	Ecoinvent 2, synthetic gas, production mix, at plant/CH
	Thermal content	5.20	MJ/Nm³	Ecoinvent 2, synthetic gas, production mix, at plant/CH
Steam	Thermal content	2.75	MJ/kg	Ecoinvent 3, steam production in chemical industry
Electricity	Price	0.0977	€/kWh	Ecoinvent 3, electricity production, natural gas, combined cycle power plant. Comment: for electricity produced in steam turbine
	Price	0.0057	€/kWh	Ecoinvent 3, natural gas, burned in gas turbine, for compressor station. Comment: for electricity produced from fuel gases of processes in combustion engine
Ethylene	Price	0.753	€/kg	Ecoinvent 3, ethylene production, average, RER
	Thermal content	47.2	MJ/kg	http://chemistry.tutorvista.com/organic-chemistry/addition-reaction.html
Propylene	Price	0.753	€/kg	Ecoinvent 3, propylene production, average, RER
	Thermal content	45.8	MJ/kg	http://chemistry.tutorvista.com/organic-chemistry/addition-reaction.html

In Ecoinvent 2.2, GLO means "global" and represents activities which are considered to be an average valid for all countries in the world. RER represents Europe and CH represents Switzerland.

point of view comprises a wide range of impacts apart from those caused by energy use and GHG emissions [76–79]. RECIPE is a newly developed method following the concept of existing methods, such as CML [80] and EcoIndicator 99 [47], which provide a wide range of indicators for various environmental impacts such as climate change, eutrophication, acidification, etc. RECIPE is advantageous as it combines the principles of midpoint- and endpoint-oriented approaches. At the midpoint level, 18 impact categories cover a wide range of common concern impacts modeling the cause—effect chain caused by the release of substances and the consumption of resources. The endpoint approach covers the environmental impacts expressed in terms of areas of protection, such as human health, ecosystems quality, and resource depletion, and can be adapted to address different types of goals and analyses.

Table 13.3 Values for Partitioning Coefficients and Displaced Products Used in the Different Allocation Methods in Case Study 2

Method	Key	Subdivision Points							
		Gasification		Methanol Synthesis		Methanol to Olefins			
		Syngas	Electricity	Methanol	Electricity	Ethylene	Propylene	Electricity
Allocation	Heating value	0.93	0.07	0.84	0.16	0.49	0.46	0.05
Allocation	Market price	0.86	0.14	0.99	0.01	0.51	0.49	0.005
Substitution		Main product	Natural gas, burned in gas turbine/CH	Main product	Natural gas, burned in gas turbine/CH	Main product	Propylene, at plant/RER	Natural gas, burned in gas turbine/CH

In Ecoinvent 2.2, RER represents Europe and CH represents Switzerland.

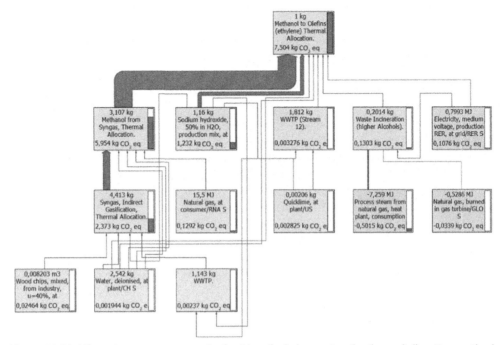

Figure 13.5A Life cycle assessment metrics for 1 kg of ethylene using the thermal allocation method. *WWTP*, wastewater treatment plant.

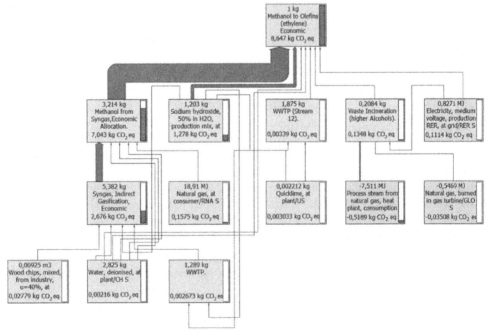

Figure 13.5B Life cycle assessment metrics for 1 kg of ethylene using the economic allocation method. *WWTP*, wastewater treatment plant.

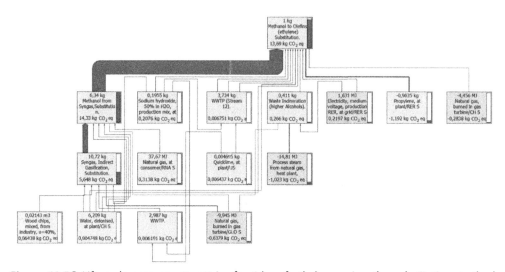

Figure 13.5C Life cycle assessment metrics for 1 kg of ethylene using the substitution method. *WWTP*, wastewater treatment plant.

Table 13.4 Summary of the Results for the Midpoint Impact Categories of the RECIPE Method for the Biorefinery System of Case Study 2 According to Different Allocation Methods and Comparison With the Ecoinvent Data

Impact Category	Unit	Methanol to Olefins (Ethylene) Substitution	Methanol to Olefins (Ethylene) Economic Allocation	Methanol to Olefins (Ethylene) Thermal Allocation	Ethylene, Average, at Plant/ RER
Climate change	kg CO_2 eq.	13.7	8.6	7.5	1.2
Ozone depletion	kg CFC-11 eq.	−8.8E-08	8.6E-08	8.2E-08	3.0E-10
Human toxicity	kg 1,4-DB eq.	79.2	77.5	69.9	0.52
Photochemical oxidant formation	kg NMVOC	−0.0043	0.0040	0.0036	0.0048
Particulate matter formation	kg PM10 eq.	0.0027	0.0039	0.0035	0.0012
Ionizing radiation	kg U235 eq.	0.38	1.07	1.03	0.00
Terrestrial acidification	kg SO_2 eq.	0.016	0.016	0.014	0.004
Freshwater eutrophication	kg P eq.	0.0005	0.0014	0.0014	0.0000
Marine eutrophication	kg N eq.	0.0007	0.0008	0.0007	0.0001
Terrestrial ecotoxicity	kg 1,4-DB eq.	0.0011	0.0027	0.0025	0.0000
Freshwater ecotoxicity	kg 1,4-DB eq.	0.045	0.041	0.037	0.000
Marine ecotoxicity	kg 1,4-DB eq.	26.6	42.1	39.45	0.4
Agricultural land occupation	m^2a	1.2	0.5	0.5	0.0
Urban land occupation	m^2a	0.015	0.013	0.012	0.000
Natural land transformation	m^2	−2.0E-05	2.2E-04	2.1E-04	8.8E-08
Water depletion	m^3	0.017	0.025	0.024	0.002
Metal depletion	kg Fe eq.	0.020	0.106	0.102	0.001
Fossil depletion	kg oil eq.	−1.2	0.7	0.6	1.6

In Ecoinvent 2.2, RER represents Europe.

The process network was created in Simapro software 8.0.2 and the ethylene production results for the climate change midpoint impact category are presented in Figs. 13.5A—C for the three allocation scenarios. Table 13.4 presents an overview of the results for the midpoint impact categories of the RECIPE method, whereas the respective results for the production of ethylene taken from the Ecoinvent database are also included for the sake of comparison. The tradeoff between the fossil depletion impact category and the rest of the categories is clear when comparing the biobased ethylene production to the conventional, fossil-based one.

CASE STUDY 3: POLY(METHYL METHACRYLATE) RECYCLING PROCESS

Motivation

The aim of this case study is to demonstrate the inventory data required to assess, from a life cycle perspective, the performance of recycling systems for a multifunctional material that is found in many different end products of everyday use. This case study extends the typical boundaries of a chemical production system. Although process modeling is again used for comparing process alternatives and filling in data gaps for LCA (i.e., which lies again at the center of interest in the presented case study), additional information is required to design and assess recycling processes; global market data for estimation of the waste material generated and its collectability are critical for estimating the process size, which in turn may affect the efficiency of the selected technologies for the material recovery.

A material of this kind is poly(methyl methacrylate) (PMMA). PMMA is a special plastic because of its significantly high transparency, resistance to weather, and technical properties. Regarding transparency, the transmission is higher than 90%, which is the highest level among plastics. PMMA is used in products requiring high-level transparency and resistance, for instance, automobile lamp covers, lighting equipment, optical fibers, and light guide panels (LGPs) of liquid crystal displays (LCDs). With the increased popularity of LCD products, it is likely that the use and stored amount of PMMA have also increased, while the current shipment of PMMA makes up about 1—2% of the total shipments of all plastics in Japan [81] and Europe [82]. Although the amount of waste PMMA included in such appliances will also increase in the future, there are currently no effective recycling systems for PMMA.

Process System and Scope of the LCA

The systematic analysis on PMMA recycling systems has been previously presented by Kikuchi et al. [83,84]. Fig. 13.6 shows the life cycle of PMMA-containing products. Three types of PMMA materials based on their categorization in Japanese statistical data [81] are specified, namely, high-molecular-weight sheet (h-sheet), low-molecular-

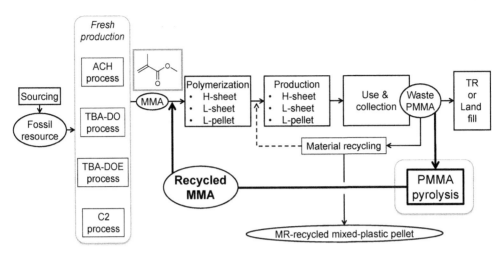

Figure 13.6 Life cycle of poly(methyl methacrylate) (PMMA)-containing products. *MMA*, Methyl methacrylate; *ACH*, acetone cyanohydrin method; *TBA-DO*, tertiary butyl alcohol (isobutylene) direct oxidation method; *TBA-DOE*, tertiary butyl alcohol (isobutylene) direct oxidative esterification method; *C2*, ethylene method; *TR*, thermal recovery; *MR*, material recycling. *Modified from Kikuchi Y, Hirao M, Ookubo T, Sasaki A. Design of recycling system for poly(methyl methacrylate) (PMMA). Part 1: recycling scenario analysis. Int J Life Cycle Assess 2014;19(1):120–9.*

weight sheet (l-sheet), and low-molecular-weight pellet (l-pellet). The sheet-type PMMA provides the raw materials for LGP of LCD products such as television sets, PC monitors, laptops, and mobile phones. The l-pellets can be molded and formed into designed shapes. L-sheet is one of the products from l-pellets. In this regard, however, h-sheet is produced by direct polymerization of the MMA monomer and cannot be produced by molding, because high-molecular-weight PMMA cannot be melted to form a shape for products.

After PMMA products are used, they must be collected on individual pathways based on the condition of waste products containing PMMA. Existing recycling laws in Japan enable the collection of some kinds of PMMA-containing products, such as home electronic appliances. Most of the PMMA contained in LCD panels may be collected and treated in recycling plants. An investigation of an appliances recycling plant in Japan revealed that LGPs in LCDs were gathered and accumulated as by-products from the manual separation of valuable parts and metals in LCD products. A limited amount of PMMA can be collected as waste electrical and electronic equipment (WEEE), while other molding products may be mixed with other plastics as combustible or plastic wastes. Based on these conditions of PMMA products, recycling PMMA in the form of PMMA pellets may not be effective. This is because most of material recycling has difficulties avoiding contamination of impurities in the recycled pellet from the optical perspective; for instance, the PMMA used for LGP requires the highest possible level of transparency. Therefore monomer recycling has a larger potential to recycle PMMA with lower environmental impacts than material recycling [83].

Life Cycle Inventories: Background Information and Process Simulation

To expand and examine the possibility of PMMA pyrolysis, scenario analysis based on LCA is necessary, because the plastic recycling system has diversified styles resulting from the multiple uses of plastics. PMMA monomer recycling processes should be used as a recycling method of PMMA to reduce the environmental impacts. Because the valuable parts of PMMA-containing waste are segregated by hand in some countries, for example, Japan, PMMA parts are also disassembled at the recycling process. At that time, however, small amounts of additives can result in impurities for PMMA. For example, titanium oxide attached to the surface of LGP, sealing chemicals for assembling parts, and metals included in conducting wires can be mixed with plastic wastes at collection and disassembling plants. Contaminating impurities in collected PMMA parts can cause unexpected reaction in PMMA pyrolysis.

In the block flow diagram shown in Fig. 13.7, the liquid recovery segment comprises a condenser and a mist separator. The temperature of effluent gas from the pyrolysis reactor is increased from 350 to 500°C at atmospheric pressure. After cooling to about room temperature, the obtained crude methyl methacrylate (MMA) monomer is

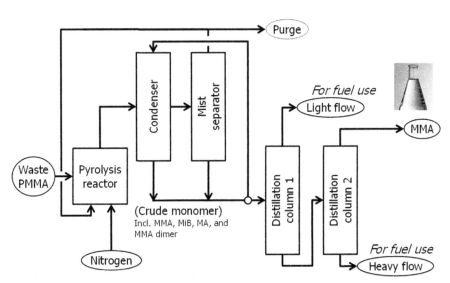

Figure 13.7 Block flow diagram of current poly(methyl methacrylate) (PMMA) monomer recycling process. *MIB*, Methyl isobutyrate; *MA*, methyl acrylate; *MMA dimer*, 1,4-cyclohexane dicarboxylic acid dimethyl ester [87]. *Modified from Kikuchi Y, Hirao M, Sugiyama H, Papadokonstantakis S, Hungerbuehler K, Ookubo T, et al. Design of recycling system for poly(methyl methacrylate) (PMMA). Part 2: process hazards and material flow analysis. Int J Life Cycle Assess 2014;19(2):307–19.*

condensed before its purification. The purification segment is composed of two distillation columns, where the first column removes low-boiling-point chemicals and the second column purifies the MMA monomer. All residues from the liquid recovery and purification segments are used as fuel in the heat recovery unit for heating sand used as a fluidized bed in the pyrolysis reactor. The contaminants may cause a reduction in the recovery ratio of MMA. Moreover, the energy efficiency of pyrolysis and purification of MMA are reduced by the increase of impurities in the feed for the pyrolysis reactor. To use the recycled MMA in the same way as fresh MMA synthesized from fossil resources, the purity of MMA at the top condensate of the second distillation column must satisfy a respective design specification. At the same time the recovery of MMA is the most important objective to increase the environmental efficiency of monomer recycling [83].

MMA includes a partly oxidized functional group in its molecular structure. To synthesize this structure, multiple reactors and separators are required in various synthesis routes [85]. As a result the environmental load for producing PMMA is higher than that of the other commodity resins such as polyethylene, polypropylene, or polyethylene terephthalate. As discussed in the literature [83], additional fuel use for recovering more recycled MMA may lead to more environmentally benign performance of monomer recycling compared to MMA synthesis. The process simulation for purification of MMA after pyrolysis plays a key role for designing the PMMA pyrolysis process. In this simulation the process inventory is needed for LCA (e.g., power and heat input to operate the process or the composition and flow rate of input and output streams of the pyrolysis process shown in Fig. 13.7). The purity of MMA in the product flow and the composition of waste PMMA can be constraints for the simulation of a recycling process from wastes to valuable products. Possible heat integration should be conducted based on pinch analysis for hot and cold streams. The saved environmental loads originating in the MMA recycling by pyrolysis and the additionally caused ones because of fossil fuel consumption should be carefully examined. MMA contained in the top condensate of the first column can be a fuel for heating sands used in the pyrolysis reactor. If more MMA is recovered from this condensate, more fuel is required for heating sands. The breakeven points can be specified by process simulation.

The purification of MMA has potentially several alternatives. The current process adopts multiple distillation columns to cut the lighter and heavier components in the stream after pyrolysis, while MMA and methyl isobutyrate, for example, are not easily separated because of their similar boiling points. On the other hand, the fresh production of MMA from crude oil also comprises a purification process, where MMA itself is used as a solvent in the extraction of MMA. Therefore an extraction process may be applicable for MMA purification and solvents may be also available if the PMMA pyrolysis process is located near the fresh MMA production processes such as the Mitsubishi Rayon Co. Ltd. Process integration including not only heat and power but also materials such as solvents in extraction should be addressed by process simulation for performing LCA on process alternatives.

Life Cycle Inventories: Global Market Data and Process Size

The design of recycling processes requires the forecasting of market conditions of target products. According to the global market conditions on LCD products, the shipment will keep increasing [86]. This will result in the generation of WEEE after the lifetime of LCD products in the world, as discussed by Kikuchi et al. [84]. PMMA may be an important resin for such highly transparent panels and should be effectively recycled after collecting them as WEEE. PMMA pyrolysis has a potential for contributing to the establishment of closed-loop recycling of PMMA. For this reason, appropriate estimation of generated and collectable waste PMMA is needed for designing the scale of the pyrolysis process. Lower estimation of waste PMMA may necessitate the incineration of waste PMMA without any recycling mechanism to avoid the creation of piled waste PMMA. Overestimation, on the other hand, may decrease the energy efficiency of the pyrolysis reactor and MMA purification separators.

CONCLUSIONS AND OUTLOOK

In this chapter, LCA is first placed within the scope of various sustainability assessment frameworks that facilitate decision making in process design. Then the most important methodological aspects of LCA are highlighted and the estimation of LCIs is identified as the core of the LCA results as well as of other sustainability frameworks; without exaggerating, the phrase "garbage in—garbage out" describes the relation of LCIs to the rest of the LCA stages.

Among the various challenges in calculating LCIs mentioned in this chapter, perhaps the most important one is that related to the early design stages and the respective data gaps. The resolution of the conceptual process design will define the possible process subdivision points, which in turn will affect the type and amount of cradle-to-gate data that must be collected or modeled as well as the allocation approaches in multifunctional processes. The decision about the scale of the process may also play a role with respect to process optimization for tradeoffs between capital and operating costs and utilization of the secondary process flows considering also the mass and heat integration potential. This will also affect the respective LCIs per kilogram of product, which may be different for different scales of the process.

Some of these aspects have been further analyzed in practical applications such as those described in the case studies of this chapter. In the first case study, it is demonstrated how screening potential solvents for postcombustion CO_2 capture requires short-cut models for estimating the cradle-to-gate environmental impact of solvent production to make up for the degraded solvent during the CO_2 capture process. When only a few solvents are to be compared, a database search for the most common ones and conceptual flowsheet development for the least common ones could be considered as an alternative to estimate the relevant cradle-to-gate LCIs and impacts. However, when the solvent selection stage includes the screening of a vast number of solvent molecules and the design of new molecules through CAMD approaches, then developing flowsheets for each molecule is highly inefficient, if not infeasible, even with today's computational means. In this case the importance of estimation tools such as those correlating the molecular structure to LCA metrics is indispensable. However, these tools are limited and more research effort has to be dedicated to this direction.

In the second case study the importance of various allocation methods is highlighted for reporting the LCA metrics of various products in multifunctional biorefinery processes. The importance of subdividing the multifunctional process into its main building blocks is discussed with respect to the calculation of allocation factors according to thermal and economic product properties and displacement of equivalent products. Based on this the life cycle impacts according to the midpoint categories of the RECIPE method are reported for ethylene production via thermochemical conversion of wood chips to syngas, methanol synthesis, and methanol to olefins processes. In this case study, another type of short-cut model is used to fill in waste treatment-related data gaps.

The challenges of model-based LCIs is also evident in the third case study for the design of PMMA recycling processes. Impurities may follow the PMMA-containing waste and affect the pyrolysis process. Additionally, various MMA purification options exist affecting the potential of heat and mass integration of the recycling process. The scale of the future recycling process will depend on market conditions of PMMA-containing products and, therefore, end-of-life PMMA waste; if the process scale is wrongly estimated, it will lead to the need for additional incineration processes or inefficient process operation, affecting the environmental performance of the recycling process.

Throughout this chapter the use of models (short-cut, flowsheeting, etc.) for LCIs calculation lies at the center of discussion. The use of models lies at the core of process systems engineering as an efficient alternative for avoiding costly lab-, pilot-, or plant-scale experimentation to fill in data gaps. In following this approach it is common knowledge that, in return, a decrease in accuracy will be tolerated. However, what is less often discussed is that contrary to conducting experiments or collecting data from running processes, the procedure of developing process models and flowsheets is far less standardized; in fact, even today, general good practices or heuristics may be available for process flowsheeting but no general protocol exists. Considering that the LCA framework is standardized according to ISO norms, when less standardized model-based LCIs are used in it, the value of the LCA studies may be decreased. For instance, the comparison between different studies using model-based LCIs may be misleading, because of the lack of protocols for flowsheet development to facilitate the transparency of the modeling effort. Questions regarding the degree to which the process is integrated, the alternative upstream and downstream separation options that have been considered, the impurities or secondary flows that have been neglected, and the maturity of the technologies that have been used (e.g., with respect to their investment costs) are often not easy to answer because of this lack of standardization. Therefore an extensive use and widespread acceptance of model-based LCIs will require the development of flowsheet protocols, and this should be recognized as an important target for exploiting the synergies between the systematic methods of process systems engineering and LCA.

REFERENCES

[1] Douglas JM. A hierarchical decision procedure for process synthesis. AlChE J 1985;31(3):353−62.
[2] Bakshi BR, Fiksel J. The quest for sustainability: challenges for process systems engineering. AlChE J 2003;49(6):1350−8.
[3] Klatt K-U, Marquardt W. Perspectives for process systems engineering - Personal views from academia and industry. Comput Chem Eng 2009;33(3):536−50.
[4] Azapagic A, Perdan S. Indicators of sustainable development for industry: a general framework. Process Saf Environ 2000;78(B4):243−61.
[5] Hoffmann VH, Hungerbühler K, McRae GJ. Multiobjective screening and evaluation of chemical process technologies. Ind Eng Chem Res 2001;40(21):4513−24.
[6] Kniel GE, Delmarco K, Petrie JG. Life cycle assessment applied to process design: environmental and economic analysis and optimization of a nitric acid plant. Environ Prog 1996;15(4):221−8.
[7] Saling P, Kicherer A, Dittrich-Kramer B, Wittlinger R, Zombik W, Schmidt I, et al. Eco-efficiency analysis by BASF: the method. Int J Life Cycle Assess 2002;7(4):203−18.
[8] Chen H, Shonnard DR. Systematic framework for environmentally conscious chemical process design: early and detailed design stages. Ind Eng Chem Res 2004;43(2):535−52.
[9] Sugiyama H, Fischer U, Hungerbühler K, Hirao M. Decision framework for chemical process design including different stages of environmental, health, and safety assessment. AlChE J 2008;54(4):1037−53.
[10] Torres CM, Gadalla MA, Mateo-Sanz JM, Esteller LJ. Evaluation tool for the environmental design of chemical processes. Ind Eng Chem Res 2011;50(23):13466−74.
[11] Tugnoli A, Santarelli F, Cozzani V. Implementation of sustainability drivers in the design of industrial chemical processes. AlChE J 2011;57(11):3063−84.
[12] Bumann AA, Papadokonstantakis S, Fischer U, Hungerbühler K. Investigating the use of path flow indicators as optimization drivers in batch process retrofitting. Comput Chem Eng 2011;35(12):2767−85.
[13] Carvalho A, Gani R, Matos H. Design of sustainable chemical processes: systematic retrofit analysis generation and evaluation of alternatives. Process Saf Environ 2008;86(5):328−46.
[14] Ruiz-Mercado GJ, Smith RL, Gonzalez MA. Sustainability indicators for chemical processes: I. Taxonomy. Ind Eng Chem Res 2012;51(5):2309−28.
[15] Gonzalez MA, Smith RL. A methodology to evaluate process sustainability. Environ Prog 2003;22(4):269−76.
[16] Ruiz-Mercado GJ, Smith RL, Gonzalez MA. Sustainability indicators for chemical processes: II. Data needs. Ind Eng Chem Res 2012;51(5):2329−53.
[17] Anastas P, Eghbali N. Green chemistry: principles and practice. Chem Soc Rev 2010;39(1):301−12.
[18] Anastas PT, Warner JC. Green chemistry: theory and practice. USA: Oxford University Press; 2000.
[19] Anastas PT, Zimmerman JB. Design through the 12 principles of green engineering. Environ Sci Technol 2003;37(5):94A−101A.
[20] Mulvihill MJ, Beach ES, Zimmerman JB, Anastas PT. Green chemistry and Green engineering: a framework for sustainable technology development. In: Gadgil A, Liverman DM, editors. Annual review of environment and resources. Annual review of environment and resources, vol. 36; 2011. p. 271−93.

[21] Imperial Chemical Industries. Hazard and operability studies: process safety report 2. London: ICI; 1974.

[22] Parmar JC, Lees FP. The propagation of faults in process plants - Hazard identification. Reliab Eng Syst Safe 1987;17(4):277—302.

[23] Lees FP. Loss prevention in process industries. 2nd ed. Oxford: Butterworth-Heinemann; 1996.

[24] Tixier J, Dusserre G, Salvi O, Gaston D. Review of 62 risk analysis methodologies of industrial plants. J Loss Prev Proc 2002;15(4):291—303.

[25] Adu IK, Sugiyama H, Fischer U, Hungerbühler K. Comparison of methods for assessing environmental, health and safety (EHS) hazards in early phases of chemical process design. Process Saf Environ 2008;86(2):77—93.

[26] Srinivasan R, Natarajan S. Developments in inherent safety: a review of the progress during 2001—2011 and opportunities ahead. Process Saf Environ 2012;90(5):389—403.

[27] Gupta JP, Edwards DW. A simple graphical method for measuring inherent safety. J Hazard Mater 2003;104(1—3):15—30.

[28] Srinivasan R, Nhan NT. A statistical approach for evaluating inherent benign-ness of chemical process routes in early design stages. Process Saf Environ 2008;86(3):163—74.

[29] Cano-Ruiz JA, McRae GJ. Environmentally conscious chemical process design. Annu Rev Energy Env 1998;23:499—536.

[30] Hilaly AK, Sikdar SK. Pollution balance: a new methodology for minimizing waste production in manufacturing processes. J Air Waste Manage Assoc 1994;44(11):1303—8.

[31] Hilaly AK, Sikdar SK. Pollution balance method and the demonstration of its application to minimizing waste in a biochemical process. Ind Eng Chem Res 1995;34(6):2051—9.

[32] El-Halwagi MM, Manousiouthakis V. Synthesis of mass exchange networks. AIChE J 1989;35(8):1233—44.

[33] Halim I, Srinivasan R. Systematic waste minimization in chemical processes. 1. Methodology. Ind Eng Chem Res 2002;41(2):196—207.

[34] Cavin L, Fischer U, Hungerbühler K. Multiobjective waste management under uncertainty considering waste mixing. Ind Eng Chem Res 2006;45(17):5944—54.

[35] Chakraborty A, Linninger AA. Plant-wide waste management. 1. Synthesis and multiobjective design. Ind Eng Chem Res 2002;41(18):4591—604.

[36] Roberge HD, Baetz BW. Optimization modeling for industrial waste reduction planning. Waste Manage Oxf 1994;14(1):35—48.

[37] Alidi AS. A multiobjective optimization model for the waste management of the petrochemical industry. Appl Math Model 1996;20(12):925—33.

[38] Hogland W, Stenis J. Assessment and system analysis of industrial waste management. Waste Manage Oxf 2000;20(7):537—43.

[39] Rerat C, Papadokonstantakis S, Hungerbuehler K. Integrated waste management in batch chemical industry based on multi-objective optimization. J Air Waste Manage Assoc 2013;63(3):349—66.

[40] Wernet G, Mutel C, Hellweg S, Hungerbuehler K. The environmental importance of energy use in chemical production. J Ind Ecol 2011;15(1):96—107.

[41] Kralisch D, Ott D, Gericke D. Rules and benefits of life cycle assessment in green chemical process and synthesis design: a tutorial review. Green Chem 2015;17(1):123—45.

[42] Standardisation ECf. ISO 14040, Environmental management — life cycle assessment — principles and framework. 2006.

[43] Standardisation ECf. ISO 14044, Environmental management — life cycle assessment — requirements and guidelines. 2006.

[44] Baumann H, Tillman AM. The Hitch Hiker's Guide to LCA. An orientation in life cycle assessment methodology and application. Lund: Studentlitteratur; 2004.

[45] Frischknecht R, Jungbluth N, Althaus HJ, Doka G, Dones R, Heck T, et al. The ecoinvent database: overview and methodological framework. Int J Life Cycle Assess 2005;10(1):3—9.

[46] Goedkoop M, Heijungs R, Huijbregts MAJ, De Shruyver A, Struijs J, Van Zelm R. ReCiPe 2008—a life cycle impact assessment method which comprises harmonised category indicators at the midpoint and the endpoint level. The Netherlands: Ministerie van Volkhuisvesting, Ruimtleijke Ordening en Milieubeheer; 2009.

[47] Goedkoop M, Spriensma R. The eco-indicator 99: a damage orientated method for life-cycle impact assessment. The Netherlands: Methodology Annex, Pre-Consultants; 2000.

[48] Suh S, Weidema B, Schmidt JH, Heijungs R. Generalized make and use framework for allocation in life cycle assessment. J Ind Ecol 2010;14(2):335−53.

[49] Weidema BP, Schmidt JH. Avoiding allocation in life cycle assessment revisited. J Ind Ecol 2010;14(2):192−5.

[50] Barton JR, Dalley D, Patel VS. Life cycle assessment for waste management. Waste Manage Oxf 1996;16(1−3):35−50.

[51] Morrissey AJ, Browne J. Waste management models and their application to sustainable waste management. Waste Manage Oxf 2004;24(3):297−308.

[52] Winkler J, Bilitewski B. Comparative evaluation of life cycle assessment models for solid waste management. Waste Manage Oxf 2007;27(8):1021−31.

[53] Gentil EC, Damgaard A, Hauschild M, Finnveden G, Eriksson O, Thorneloe S, et al. Models for waste life cycle assessment: review of technical assumptions. Waste Manage Oxf 2010;30(12): 2636−48.

[54] Grossmann IE, Guillen-Gosalbez G. Scope for the application of mathematical programming techniques in the synthesis and planning of sustainable processes. Comput Chem Eng 2010;34(9): 1365−76.

[55] Wernet G, Papadokonstantakis S, Hellweg S, Hungerbühler K. Bridging data gaps in environmental assessments: modeling impacts of fine and basic chemical production. Green Chem 2009;11(11): 1826−31.

[56] Bumann AA, Papadokonstantakis S, Sugiyama H, Fischer U, Hungerbühler K. Evaluation and analysis of a proxy indicator for the estimation of gate-to-gate energy consumption in the early process design phases: the case of organic solvent production. Energy 2010;35(6):2407−18.

[57] Albrecht T, Papadokonstantakis S, Sugiyama H, Hungerbühler K. Demonstrating multi-objective screening of chemical batch process alternatives during early design phases. Chem Eng Res Des 2010;88(5−6A):529−50.

[58] Köhler A, Hellweg S, Recan E, Hungerbühler K. Input-dependent life-cycle inventory model of industrial wastewater-treatment processes in the chemical sector. Environ Sci Technol 2007;41(15):5515−22.

[59] Seyler C, Hofstetter TB, Hungerbühler K. Life cycle inventory for thermal treatment of waste solvent from chemical industry: a multi-input allocation model. J Clean Prod 2005;13(13−14):1211−24.

[60] Jaksland CA, Gani R, Lien KM. Separation process design and synthesis based on thermodynamic insights. Chem Eng Sci 1995;50(3):511−30.

[61] Hackl R, Andersson E, Harvey S. Targeting for energy efficiency and improved energy collaboration between different companies using total site analysis (TSA). Energy 2011;36(8):4609−15.

[62] Hackl R, Harvey S. Framework methodology for increased energy efficiency and renewable feedstock integration in industrial clusters. Appl Energy 2013;112:1500−9.

[63] Wernet G, Conradt S, Isenring HP, Jiménez-González C, Hungerbühler K. Life cycle assessment of fine chemical production: a case study of pharmaceutical synthesis. Int J Life Cycle Assess 2010;15(3):294−303.

[64] Rockström J, Steffen W, Noone K. A safe operating space for humanity. Nature 2009;461:472−5.

[65] Gall SC, Thompson RC. The impact of debris on marine life. Mar Pollut Bull 2015;92(1−2):170−9.

[66] Sandin G, Royne F, Berlin J, Peters GM, Svanström M. Allocation in LCAs of biorefinery products: implications for results and decision-making. J Clean Prod 2015;93.

[67] Aaron D, Tsouris C. Separation of CO_2 from flue gas: a review. Sep Sci Technol 2005;40(1−3): 321−48.

[68] Papadopoulos AI, Badr S, Chremos A, Forte E, Zarogiannis T, Seferlis P, et al. Efficient screening and selection of post-combustion CO_2 capture solvents. In: Varbanov PS, Klemes JJ, Liew PY, Yong JY, Stehlik P, editors. Pres 2014, 17th conference on process integration, modelling and optimisation for energy saving and Pollution reduction. Chemical engineering transactions, vol. 39 (Pts 1−3); 2014. p. 211−6.

[69] Lepaumier H, Picq D, Carrette P-L. New amines for CO_2 capture. I. Mechanisms of amine degradation in the presence of CO_2. Ind Eng Chem Res 2009;48(20):9061−7.

[70] Lepaumier H, Picq D, Carrette P-L. New amines for CO_2 capture. II. Oxidative degradation mechanisms. Ind Eng Chem Res 2009;48(20):9068–75.

[71] De Benedetto L, Klemes J. The Environmental Performance Strategy Map: an integrated LCA approach, to support the strategic decision-making process. J Clean Prod 2009;17(10):900–6.

[72] Weidema B. Avoiding co-product allocation in life-cycle assessment. J Ind Ecol 2001;4(3):11–33.

[73] UOP light olefin solutions for propylene and ethylene production (28.05.2015). Available from: http://www.uop.com/?document=uop-olefin-production-solutions-brochure&download=1.

[74] Simapro LCA software. Available from: http://www.pre.nl/simapro/default.htm.

[75] Ecoinvent V.2. Available from: http://www.ecoinvent.org/database/ecoinvent-version-2/ecoinvent-version-2.html.

[76] Grillo Reno ML, Silva Lora EE, Escobar Palacio JC, Venturini OJ, Buchgeister J, Almazan O. A LCA (life cycle assessment) of the methanol production from sugarcane bagasse. Energy 2011;36(6): 3716–26.

[77] Munoz I, Flury K, Jungbluth N, Rigarlsford G, Canals LMI, King H. Life cycle assessment of bio-based ethanol produced from different agricultural feedstocks. Int J Life Cycle Assess 2014;19(1):109–19.

[78] Panichelli L, Dauriat A, Gnansounou E. Life cycle assessment of soybean-based biodiesel in Argentina for export. Int J Life Cycle Assess 2009;14(2):144–59.

[79] Uihlein A, Schebek L. Environmental impacts of a lignocellulose feedstock biorefinery system: an assessment. Biomass Bioenergy 2009;33(5):793–802.

[80] Guinee J. Handbook on life cycle assessment - operational guide to the ISO standards. Int J Life Cycle Assess 2001;6(5):255.

[81] Japan Petrochemical Industry Association (JPIA) (28.05.2015). Available from: http://www.jpca.or.jp/english/index.htm.

[82] Plastics – the facts 2014/2015 (28.05.2015). Available from: http://issuu.com/plasticseuropebook/docs/final_plastics_the_facts_2014_19122.

[83] Kikuchi Y, Hirao M, Ookubo T, Sasaki A. Design of recycling system for poly(methyl methacrylate) (PMMA). Part 1: recycling scenario analysis. Int J Life Cycle Assess 2014;19(1):120–9.

[84] Kikuchi Y, Hirao M, Sugiyama H, Papadokonstantakis S, Hungerbuehler K, Ookubo T, et al. Design of recycling system for poly(methyl methacrylate) (PMMA). Part 2: process hazards and material flow analysis. Int J Life Cycle Assess 2014;19(2):307–19.

[85] Sugiyama H, Fischer U, Antonijuan E, Hoffmann VH, Hirao M, Hungerbuehler K. How do different process options and evaluation settings affect economic and environmental assessments? A case study on methyl methacrylate (MMA) production processes. Process Saf Environ 2009;87(6):361–70.

[86] Displaysearch (28.05.2015). Available from: http://www.displaysearch.com/cps/rde/xchg/displaysearch/hs.xsl/130516_flat_panel_public_display_market_expected_to_grow_annually_through_2017.asp.

[87] Kaminsky W, Franck J. Monomer recovery by pyrolysis of poly(methyl methacrylate) (PMMA). J Anal Appl Pyrolysis 1991;19:311–8.

CHAPTER FOURTEEN

Life Cycle Sustainability Assessment: A Holistic Evaluation of Social, Economic, and Environmental Impacts

L.Q. Luu, A. Halog
The University of Queensland, Brisbane, Australia

INTRODUCTION

Since the first introduction of the Brundtland report in 1987, sustainable development has become a common journey that humans need to consider when decoupling the environmental, economic, and societal pillars [1,2]. In order to move successfully toward sustainable industrial development, we need to know how we can measure our progress using environmental, social, and economic indicators over time. As a result, a range of methodologies and tools has been developed for assessing environmental indicators for sustainability achievement across products and technologies. Among these methodologies and tools, life cycle sustainability assessment (LCSA) is introduced here to broaden the tool of environmental life cycle assessment (E-LCA) by accounting economic and social sustainability [3]. This chapter will start with a discussion on methodologies for assessing life cycle sustainability and the need of LCSA. It is then followed by a case study that assesses rice husk-based bioelectricity in Vietnam over its life cycle to verify the practical application of LCSA methodology.

METHODOLOGIES FOR ASSESSING LIFE CYCLE SUSTAINABILITY

Life cycle assessment (LCA) is defined in Horne [4] as the "compilation and evaluation of inputs and outputs and the potential impacts of a product system throughout its life cycle." In E-LCA, all input materials, waste, and emissions are accounted for at all stages: raw material extraction and processing; product and/or service manufacturing; use and disposal; and finally transportation. The comprehensive data requirement of LCA makes it a particularly effective mechanism for systematic assessment of environmental impacts when designing chemical engineering processes to produce chemicals, fuels, and other product systems [4].

The seminal definition of sustainable development was introduced as "development that meets the needs of the present without compromising the ability of future generations to meet their own needs" [1]. This definition requires the consideration of sustainable development under the lens of system thinking, or, in other words, the system over time and space, with an acknowledgment of the needs of human beings and the limitations of natural resources.

LCSA extends the environmental boundaries of traditional LCA in an attempt to incorporate the concept of sustainable development. It is defined as a method of addressing environmental, economic, and social sustainability of a product system over its life cycle, indicated through the measurement of either positive or negative impacts [3]. LCA has been implemented through an integration of E-LCA, life cycle costing (LCC), and social life cycle assessment (S-LCA) [3]. Brief definitions of E-LCA, LCC, and S-LCA are described in Table 14.1.

E-LCA is a well-developed methodology, with internationally standardized detailed guidelines for implementation in practice (i.e., ISO 14040:2006, ISO 14044:2006, and other standards in the ISO 14040 range). LCC has not been widely applied in practice as a component of LCSA, despite being considered as the oldest of three life cycle approaches, with origins reaching back as far as the 1930s [5]. In most of the cases, the traditional LCC applications in engineering systems are not that comprehensive, as they do not cover the whole life cycle (as understood in E-LCA) and the use of additional cost data estimated by expert opinions [6]. Further reasons for limited application of traditional LCC in an LCSA context are that it needs first to have an equivalent system boundary as in E-LCA. However, this traditional LCC has now evolved into what we call "environmental" LCC [5], which needs to be standardized too. For S-LCA, a standardized methodology (like E-LCA) is yet to be fully developed, which has been a major barrier to the implementation of S-LCA in practice [7]. The slow progress in the S-LCA methodology development may be because of its requirement of quantifying qualitative indicators such as child labor, fair salary, working hours, health and safety, social benefits, etc. to quantitative values. There has been a growing body of S-LCA studies focusing on social "hotspots," such as those surrounding the mining industry and production of electronic devices. However, until now, most available S-LCAs have been limited in their methodology regarding how to perform an S-LCA [8,9], in which most social indicators are expressed qualitatively.

Although the research community accepts the life cycle sustainability concept, there is no current consensus on how to implement an LCSA. There are a number

Table 14.1 Three Pillars of Life Cycle Sustainability

E-LCA	(Potential) environmental impacts over a product system's life cycle [4].
LCC	All costs and benefits directly related to the product system over its life cycle with some consideration on the external relevant costs and benefits [5].
S-LCA	Social and socioeconomic impacts of the product system throughout its life cycle, which causes directly/indirectly and positively/negatively affected stakeholders [3].

methodologies that support LCSA such as the United Nations Environment Program Life Cycle Sustainability Analysis (UNEP LCSA), Co-ordination Action for innovation in Life-Cycle Analysis for Sustainability (CALCAS), Advancing Integrated Systems Modeling Framework for Life Cycle Sustainability Assessment (AISMF LCSA), and the Prospective Sustainability Assessment of Technologies (PROSUITE) [3,10−12]. These methodologies share a common foundation, developed based on the three frameworks of E-LCA, LCC, and S-LCA [3,10−12].

The most common methodology to assess the life cycle sustainability of a product system (which can also be the basis for designing/improving chemical processes) is following UNEP LCSA [3]. According to this methodology, the results of E-LCA, LCC, and S-LCA are integrated with a set of weighting indicators to obtain a single common life cycle sustainability result [3]. The set of weighting indicators can be presented in the form of a Life Cycle Sustainability Dashboard, with different scores and colors, or the Life Cycle Sustainability Triangle [10].

CALCAS is a framework proposed by the EU 6th Framework Co-ordination Action. The development of LCSA methodology is based on the ISO 14040−14044 frameworks for E-LCA by enhancing E-LCA with LCC and S-LCA. It expands the concept of E-LCA to include physical, social, economic, cultural, institutional, and political aspects, and broadens the boundary of a product system to assess the sustainability of a product system in which one product system can induce impacts on other product systems over time and space, and eventually impact the whole economy [11].

Halog and Manik [12] developed AISMF LCSA through the combination of the E-LCA, LCC, and S-LCA frameworks, incorporated with multistakeholders' analysis. The authors used multicriteria decision analysis to obtain the key indicators for LCSA, which were then used as critical variables for agent-based and/or system dynamics (e.g., use of causal loop relationships) modeling to ascertain the final results of sustainability decisions.

Most recently, another methodology development with an emphasis on causal relationships, PROSUITE, was introduced under the EC 7th framework program. On the foundation of the (ISO) E-LCA framework, PROSUITE sustainability assessment is based on evaluating 16 midpoint and five endpoint impacts as shown in Fig. 14.1 [13].

Being different from other approaches to assess life cycle sustainability, PROSUITE methodology utilizes five endpoint impacts, which reduce the risk of overlapping between three pillars of sustainability. For example, the indicator of income, which can be used as a pathway to social sustainability, brings better quality of life. At the same time, income indicator can be considered under the economic sustainability point of view [14].

The five endpoint indicators are aggregated into one single score of sustainability by applying three approaches: graphical representation, weighted sum, and outranking

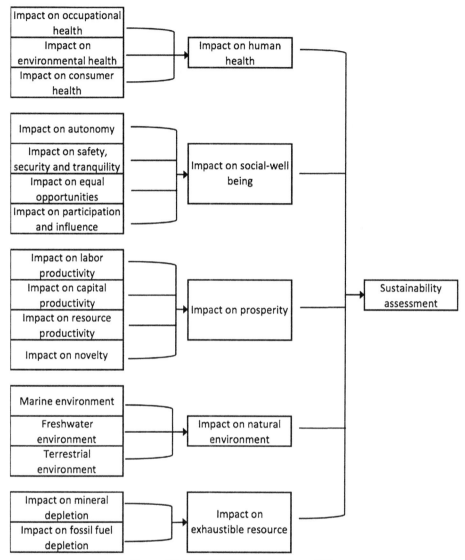

Figure 14.1 PROSUITE methodology [13].

analysis [13]. For graphical representation, five endpoint impacts will be presented in the form of a "five-armed star" chart. Each arm of the star illustrates one endpoint impact. For impacts of social well-being and prosperity, the longer the arms are, the better the product system is. However, for the impacts on human health, natural environment, and exhaustible resource, the shorter the arms are, the less negative the impacts are, and the product system causes and uses up less resource consumption, which means the product system is relatively more sustainable than the others [14].

The weighted sum results are calculated on the principle that the positive impacts can compensate for the negative impacts. In other words, impact on social well-being and impact on prosperity can be subtracted from the impacts on human health, ecosystem quality, and natural resource. As a result, the higher the scores are, the worse the product systems are [14].

The outranking analysis result is obtained by comparing the difference in each pair of categories of two product systems. It first starts with five endpoint impacts, which are followed by different midpoint impacts. The product system, which outranks the other, will receive a "win." The more "wins" a product system receives, the more sustainable it is [14].

The UNEP LCSA methodology is transparent and easily replicable in terms of scoring; however, it ignores the interconnections among E-LCA, LCC, and S-LCA and the cause—effect relationships between midpoint and endpoint indicators. The two methodologies of CALCAS and AISMF LCSA are more suitable to a complex system (e.g., global supply chains in producing and consuming chemicals); however, we did not apply these more advanced methods because of our temporal and budgetary constraints. The last methodology—PROSUITE methodology—is the most updated one (which is supported by the well-developed OpenLCA software, available at http://www.openlca.org/), which suggests the consideration of human health impacts as one of the factors for social sustainability. However, there are not many documents and studies on the validity of this LCSA methodology yet. This chapter, through a case study on assessing the life cycle sustainability of rice husk-based bioelectricity and coal-fired electricity in Vietnam, verified the application of the PROSUITE method.

As this methodology is developed under the E-LCA framework, it generally follows the ISO 14040 standards for implementing an LCA. The framework starts with defining the goal and scope, which requires setting up the goal, the process technology, the product, their functional units, system boundary of the study, and other pertinent background and procedural information. Being different from conventional LCA, PROSUITE assesses the sustainability of two product systems: a reference system and a prospective system. The reference system is normally the product system that uses existing technology and plays the role of comparable product [14]. Meanwhile, the prospective system is produced as an alternative or substitute for the reference system. The prospective system is frequently produced by the new and emerging technologies [14].

The goal and scope definition stage is then followed by a life cycle inventory analysis. In the life cycle inventory analysis stage, data on raw materials and energy consumption, emissions, waste, monetary flows, and social issues are collected. While data on raw material inputs, environmental outputs, and monetary flows are quantitative, data on social issues are in forms of both quantitative (e.g., number of working hours, amount of child labor) and qualitative (e.g., risk of public resistance to the prospective product, risk of misusing the prospective product, etc.). Some of the data of general processes are integrated in the model [14]. For example, data on occupational health issues are collected from the World Health Organization and provided in the model [14]. Other data of specific processes must be collected by the authors [14].

These data are then entered into the PROSUITE Decision Support System (DSS) for calculating the impacts and assessing the sustainability of the process technology and (or) product system [14], which are implemented by combining with normalization and weighting sets. Normalization shows the contribution of an endpoint impact to the overall sustainability, while weighting is used to aggregate normalized results into a single sustainability score [14]. The normalization set is developed by calculating the impact of the product's life cycle and comparing it to the European sustainability value; therefore the normalization set is fixed and built in the PROSUITE model. The weighting set is more flexible than the normalization set, as it can be altered according to the researcher's subjective opinion on the relative contribution among endpoint impacts [14]. Table 14.2 represents the normalization and weighting sets of PROSUITE.

The impact on the natural environment is measured in number of species loss over a year/person/year, which is rather common for LCA users. Different pathways contributing to "natural environment" impact (e.g., acidification, climate change, etc.) are finally converted into this unit. However, for the impact on exhaustible resources, instead of using "kg" of consumed metal/mineral and "MJ" of consumed fossil fuels, the unit of "US $2010/person/year" is utilized to express both consumed resources. This unit is based on the quantity of resources used in combination with its market value in 2010 [16]. For the economic pillar of sustainability, the methodology used "€/person/year" to measure the prosperity [16], which is similar to the unit used in LCC.

The social pillar of sustainability, nevertheless, requires creating a unit system to convert qualitative data into quantitative data. The human health impact is measured on the number of "healthy life" years lost because of premature mortality and disability/person/year, or so-called disability-adjusted life year (DALY)/person/year [15]. For social well-being, the result is obtained based on 11 midpoint impacts, which are divided

Table 14.2 Normalization and Weighting Sets for Five Impacts [14]

Impact	Normalization Set	Unit	Weighting Set
Natural environment	7.30E − 05	Species loss year/person/year	0.25
Human health	3.31E − 02	Disability-adjusted life year (DALY)/person/year	0.3
Exhaustible resources	1.04E + 02	US $2010/person/year	0.1
Prosperity global GDP—economy wide	9.24E + 03	€/person/year	0.1
Social well-being	N/A	N/A	0.25

Table 14.3 Normalization Reference for Social Well-Being Midpoint Indicators [16]

Sustainability Indicators	Normalization References	Unit
Total employment	8.98E + 02	h/person/year
Knowledge-intensive jobs	1.95E + 02	h/person/year
Child labor hazardous activities	1.96E + 01	h/person/year
Forced labor total	2.22E + 00	h/person/year
Regional income inequalities	−2.19E + 03	€/person/year
Global inequalities	N/A	N/A

into four endpoint impacts. Among the 11 midpoint impacts, six are quantitative indicators, including knowledge-intensive jobs, total employment, child labor, forced labor, regional income inequities, and global inequities [14]. These quantitative data are calculated with different normalization references for each impact [16] (Table 14.3). Therefore the obtained result of social well-being is already normalized before being aggregated and there is no need for a common normalization set for the social well-being impact. Five remaining qualitative indicators, including possibility of misuse, risk perception, stakeholder involvement, trust in risk information, and long-term control functions, are collected by consultation with an expert board, and will provide extra information for the social well-being impact and are not aggregated in the final result.

THREE PILLARS OF SUSTAINABILITY AND THE NEED FOR LIFE CYCLE SUSTAINABILITY ASSESSMENT

For a sustainable manufacturing process, it is suggested that the three pillars of environment, economy, and society should be given equal attention. In practice the environmental and economic aspects are normally given more emphasis than the social aspect of sustainability, which can be proved through the out-number of E-LCA compared to S-LCA studies conducted in general chemical processing and manufacturing industries, and the energy industry for a particular instance.

Studies on environmental sustainability of energy systems showed that the renewable energies, including hydropower, solar power, and wind power, have less negative environmental impacts than conventional power, which is generated from fossil fuels [17]. Coal-fired electricity has the highest greenhouse gas (GHG) emissions while it requires much less direct land use (land for power plant and mining activities) than renewable energy; and biomass has the highest direct land requirement (mostly for biomass plantation) per energy production units [17].

In addition, biomass energy has been gaining more attention, with a range of LCAs comparing the environmental impacts of bioelectricity and bioethanol [18–23]. Prasara and Grant [20] and Delivand et al. [21] compared the environmental impacts of bioelectricity and bioethanol from rice residue and conventional electricity, which identified that rice residue-based energy contributes smaller impacts of fossil fuel depletion and GHG emission than its conventional counterpart. Unfortunately, using rice straw leads to a slightly higher amount of particulate matters than conventional electricity and fuels [20].

For rice husk-based electricity, Pham et al. [24] conducted a life cycle inventory analysis study to evaluate the GHG emissions of different methods of rice husk treatment. In the baseline scenario, rice husk is mainly used for cooking and brick making, and the excess rice husk is openly burned [24]. This baseline scenario is compared with seven main scenarios in which the excess (all) rice husks are used for cooking and brick making; generating power through combustion or gasification; and producing bio-oil by pyrolysis [24]. A research conducted in Vietnam identified that open burning of rice husks is the main contributor of GHG emission, while using rice husk for generating electricity either by direct combustion or by gasification can reduce GHG emission [24]. However, the research is limited in the GHG emission impacts of rice husk-based electricity, which requires further investigation for other environmental impacts of rice residue-based electricity, either positive or negative.

Most of the current studies on economic sustainability of energy systems are in fact analyzing its economic feasibility studies. Several studies have been conducted on the foundation of cost and benefit analysis to compare different types of biomass feedstock as well as energy conversion technologies. As early as the 1990s, Mitchell et al. [25] compared different technologies (including pyrolysis, gasification, and combustion), and concluded that the main costs and benefits include cost for feedstock, cost for conversion technology, and sale of energy. In addition, the International Renewable Energy Agency (IRENA)'s study on bioenergy, with a concentration on agricultural residues, showed that cost for feedstock such as rice straw, rice husk, wheat straw, and sugar cane bagasse is lower than other types of biomass, at 40–50% of total electricity generation cost [26]. Cost for either procurement or transportation of feedstock is identified to be the main barrier for promoting biomass-based energy [27]. This leads to the weak economic sustainability of biomass-based energy compared with that of fossil fuels [27].

In contrast with the available studies on environmental and economic sustainability, very few studies were conducted on social sustainability of biomass-based electricity systems [28]. Most available studies evaluate bioenergy on its socioeconomic impacts such as land use, labor use, and health issues separately rather than integratively, covering all impacts to show its social sustainability.

As biomass-based electricity systems utilize the same feedstock as second-generation biofuel, its social impacts may be similar to that of biofuel. Studies on social impacts of biofuels indicate that the largest social concern might be the tradeoff of socioeconomic benefits of bioenergy and its impacts on food security. On the one hand, bioenergy leads to competition on land use, natural resources, and other assets for biomass growing and food growing [28–30]. This competition increases food prices [29], which is a big threat for the livelihood and food security of the poor. On the other hand, bioenergy is believed to bring additional income for farmers, which can then contribute to the national economy and living standards [31].

In addition, some impacts such as employment, working conditions, and health and gender issues are estimated. For example, direct labor for woody biomass-based energy is identified to be two or three times higher than that of conventional energy [28]. As a result the potential of labor injury is higher in biobased energy than in coal, oil, or gas [32]. This is even more serious in developing countries where more women, with limited social security and medical support, are employed in the feedstock growing industry [33].

Several health impacts including physical, mental, and social health problems were identified in biobased power plants [34]. In the study of Juntarawijit et al. [34], half of the surveyed residents, who lived near the power plants, agreed that power plants caused health problems. The problems related to pollutant particles and noise seem to be the most serious, for example, air pollution from power plants, noise from power plants, respiratory problems caused by dust, and frustrations caused by cleaning the house. However, the methodology they used in their research is subjective and site specific, which requires caution when applying their results.

Because of the imbalance of available studies on social, economic, and environmental aspects of sustainability, it is essential to create a common framework to include all three pillars with equal importance. The contribution of the common framework does not only lie in providing a comprehensive point of view of a product sustainability over its life cycle, but also assists the designers, policymakers, and planners in promoting a socially inclusive product/service.

CASE STUDY OF RICE HUSK-BASED ELECTRICITY AND COAL-FIRED ELECTRICITY

In Vietnam, renewable energy is mostly deployed in the form of hydropower with 37.6% of national electricity generation in 2011. Other types of renewable energy contribute a very small share, at 3.5% of the national electricity system [35]. However, it is planned that the shares of other types of renewable energy, including wind energy and biomass-based energy, will gradually increase in the near future to 4.5% by 2020 and 6% by 2030 [35]. For biobased electricity, it is expected that the installed capacity will increase to 500 MW by 2020 and 2000 MW by 2030, being equal to 1.1% of total electricity generation [35].

Vietnam is the second largest rice exporting country in the world where the potential of biomass residues from rice is therefore large. The national rice production yield has increased over years at 39.99 million tonnes in 2010 and 49.27 million tonnes in 2013 [36,37]. With the residue to product ratio of 0.2 for rice husk [38], the amount of rice husk, which can be used for electricity generation, is about 0.8 million tonnes. The amount of rice husk is therefore technically feasible for small-scale electricity generation.

In addition, the on-site study showed that about 20% of rice residue is kept for fertilizing the next crop, and the remaining is burned on the field [38]. The open burning of rice residue causes incomplete combustion, which emits CO, N_2O, CH_4, and polycyclic aromatic hydrocarbons [39]. As a result of this, it is not only harmful for human health but also negatively impacts the environment. If a large amount of rice residue, which is openly burnt on the field, is used to generate electricity, this would partly sustain the national energy sector. Moreover, it may help to reduce the negative social and environmental impacts of open burning of rice residue and burning fossil fuels to generate electricity.

In order to compare the relative sustainability of different energy product systems, it is necessary to construct a baseline product system, which can be used as reference for other products. This reference product system should be the common basis for the whole energy system. In this case study, electricity from coal is selected as the reference system because of its large contribution to the Vietnamese national energy grid and its developing trend over the next 20 years. The remaining product systems whose sustainability is required to be assessed are called prospective systems. These can be any product system that may or may not contribute to the long-term sustainability of the country. In this case study, rice husk-based electricity is selected as the prospective system to verify its sustainability, which will support the decision maker in planning the national energy system development.

Reference System: Electricity From Coal

The coal-fired electricity life cycle starts with extracting and processing the raw materials including coal, limestone, and crude oil. These raw materials are then transported to the power plant to prepare for the electricity generation process. In the power plant, coal is crushed into pulverized coal and fed into the combustion boiler [40]. At the same time, limestone is injected into the boiler to decrease the amount of SO_2 [41]. At the bottom of the boiler, there are a number of air nozzles to inject a mixture of hot air and gas so as to increase the efficiency of the combustion process [41]. The studied power plant has a circulating fluidized bed (CFB) combustion technology [42] with the advantage of two cyclones, which collect the dust particles from flue gas and return them to the boiler bed to mix with the input material mixture of coal and limestone [41]. The combustion process heats the feed water to generate high pressure steam. The steam vapor is then transported to the steam turbine to generate electricity. The steam, after being used to drive the turbine, is collected, condensed, and fed back to the boiler for reuse [41].

The generated electricity is transformed into high voltage, which is then transmitted via the national grid network. This high-voltage electricity is transformed back into low voltage when it is near the consumption locations [40]. Although the transformer process helps to increase the efficiency of the transmission, the loss on the transmission line is unavoidable. Because of the age of materials and technology used for transmission lines, the transmission loss may account for up to 10% of the total output [43]. As a result the input of this process (generated electricity in a power plant) is not equal to the output (consumed electricity to consumers).

In our research the Thai Binh thermal power plant is selected as it uses CFB technology, which is the same as that of the theoretical rice husk-based power plant and the most updated technology in Vietnam [42]. Technical characteristics of the Thai Binh thermal power plant can be found in Table 14.3.

Prospective System: Bioelectricity From Rice Husk

The Mekong river delta is selected as the study site for the prospective system as it is the largest area for the cultivation of rice in Vietnam, with more than 50% of rice yield, and a rice production yield of 24.29 million tonnes in 2012 [37]. Because of high production yield, the amount of rice husk is therefore large, which makes this area technically suitable for constructing a bioelectricity plant with rice husk as raw material.

The rice husk-based bioelectricity life cycle starts with the production of agricultural inputs such as rice seed, fertilizer, and pesticide, which then are used for rice farming. After the rough rice grains are harvested, rice husk is separated from brown rice [44]. While the brown rice is normally processed to obtain white rice for commercial purpose, the rice husk is directly transported to the power plant for generating electricity. The operation of rice husk-based power plants is similar to that of coal-based power plants. Nevertheless, as rice husk tends to contain less sulfur than coal, limestone consumption is omitted when considering the rice residue-based power plant.

As this type of bioelectricity is not available in Vietnam, the studied system used the technical characteristics of a theoretical bioelectricity power plant, based on the feasibility study of Nguyen and Nguyen [45]. Details on the technical characteristics of the rice husk-based power plant are summarized in Table 14.4.

GOAL AND SCOPE DEFINITION

- Goal

  This study aims at assessing the sustainability of rice husk-based bioelectricity from raw material extraction to its consumption; and compares it with that of coal-fired

Table 14.4 Technical Characteristics of Coal-Fired Power Plant and Rice Husk-Based Power Plant [42,45]

Technical Characteristics	Coal-Fired Power Plant (Reference System)	Rice Husk-Based Power Plant (Prospective System)
Operation lifetime	30 years	20 years
Operating hours	6000 h/year	4800–7968 h/year
Installed capacity	2 units * 600 MW = 1200 MW	11 MW
Electricity own use	6.2%	10%
Efficiency	38.5%	21%
Technology	Circulating fluidized boiler	Circulating fluidized boiler
Input material consumption	4,151,160 tonne/year	109,500 tonne/year
Heavy fuel oil for start-up	10,800 tonne/year	2400 tonne/year
Limestone for neutralizing SO_2	216,240 tonne/year	N/A

electricity. The implementation and results of the study are the basis to review the applied methodology in assessing the life cycle sustainability of an energy system. Moreover, this study can be used as a basis for sustainably developing the Vietnamese energy system with a higher renewable energy content.

- Product systems

  There are two product systems compared here: the reference system (coal-fired electricity) and the prospective system (rice husk-based bioelectricity).

- Function of the product

  The two product systems have the same function of providing electricity, but slightly different functional units, which are 1 MWh of coal-fired electricity and 1 MWh of rice husk-based bioelectricity.

- System boundaries

  Both systems are assessed from cradle to gate on three aspects of sustainability (Fig. 14.2) and for specific boundaries for reference and prospective systems (Figs. 14.3 and 14.4). The system boundaries are slightly different for three pillars of sustainability. For the environmental impacts, the study excludes impacts from the production of capital goods, which are typical to any previous LCA studies conducted because of its added complexity when collecting data. Nevertheless, the economic and social impacts are assessed on an economy-wide scale, which includes all sectors directly or indirectly related to the electricity generation process.

  The stages of the bioelectricity life cycle include: (1) raw material process and production; (2) raw material transportation; (3) R&D; (4) plant construction; (5) bioelectricity generation; and (6) bioelectricity transmission. These stages are similar for conventional electricity. Fig. 14.2 illustrates six stages of bioelectricity's life cycle and its system boundaries for the three pillars of sustainability, which is similar for conventional electricity (reference system).

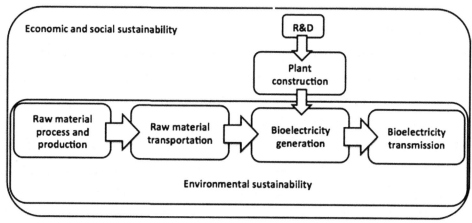

Figure 14.2 Life cycle system boundaries of bioelectricity.

- Allocation

  The environmental, economic, and social impacts are allocated for main products and coproducts (e.g., rice and rice husk in the rice husk preparation process, electricity and ash in the electricity generation process) according to their economic values. It should be noted that in this case study the main product (electricity) is quantified in MWh while its coproducts (bottom ash, fly ash, etc.) are quantified in tonne. The differences in quantification units make it impossible to allocate environmental, social, and economic impacts by "weight" basis. Among the available allocation methods, though the economic-based allocation is limited because of changing market values, it is, in this case study, the most suitable choice for assessing the sustainability of the product system.

- Assumptions

  This study assumed that both the coal-fired power plant and the bioelectricity power plant started to operate in 2013, although none of these power plants is currently operating. In addition, during the whole lifetime of these power plants, it is assumed that there is no change in their operating efficiency. In other words, the power plant investors spend an increasing amount of money on maintenance of boilers, turbines, generators, and transmission lines so that there is no need to increase input materials over the years.

- Impact assessment method

  This study used the PROSUITE endpoint impact assessment method in which all impacts are categorized into five main impacts, including impact on human health, impact on social being, impact on prosperity, impact on natural environment, and impact on exhaustible resources. These impacts are then combined with normalization and weighting to obtain a single sustainability score (see the section Methodologies for Assessing Life Cycle Sustainability).

Table 14.5 Data Quality and Data Sources for Each Unit Process

Unit Process	Impact on Natural Environment	Impact on Exhaustible Resources	Impact on Human Health	Impact on Prosperity	Impact on Social Well-Being
Raw material process and production	Data in the last 8 years from the Ecoinvent database and other literature	Data in the last 8 years from the Ecoinvent database and other literature	Calculated by impact on natural environment, combined with data from World Health Organization, integrated in PROSUITE Decision Support System (DSS)	Calculated by THEMIS economic model, integrated in PROSUITE DSS	Calculated by THEMIS economic model, combined with data from International Labor Organization, integrated in PROSUITE DSS
Raw material transportation	Data from feasibility report	Data from feasibility report			
Bioelectricity generation and transmission	Data from feasibility report	Data from feasibility report		Data in 2013 from literature	

- Data quality

  Data for three aspects of sustainability are gathered for a period of 8 years, from 2006 to 2014, from the Ecoinvent database, social hotspot database, feasibility reports of the power plant, and other types of literature. Because both of these power plants are not currently operating, their emissions are approximately calculated on the basis of the National Pollutant Inventory Emission estimation technique, developed in Australia. Relevant details of data quality and data sources can be found in Table 14.5 and explained in the section Inventory Analysis.

INVENTORY ANALYSIS

Data for the reference system—coal-fired electricity—are adapted from Loi et al.'s study [46], the Ecoinvent database for rest of world (RoW) [47], technical reports of PECCI [42], and GIZ-GDE/MOIT's report [48], in combination with calculation results following the guidelines of the Australian National Greenhouse Accounts and the National Pollutant Inventory Emission estimation technique manual for Fossil Fuel Electric Power Generation [49,50]. Fig. 14.3 illustrates different processes of generating coal-fired electricity.

Similarly, data for the prospective system—rice husk-based bioelectricity—are taken from the Ecoinvent database and literature [45,47,51—57], in combination with calculations based on emission factors of rice husk burning for energy purposes in Vietnam [39], the IPCC Guidelines for National Greenhouse Gas Inventories,

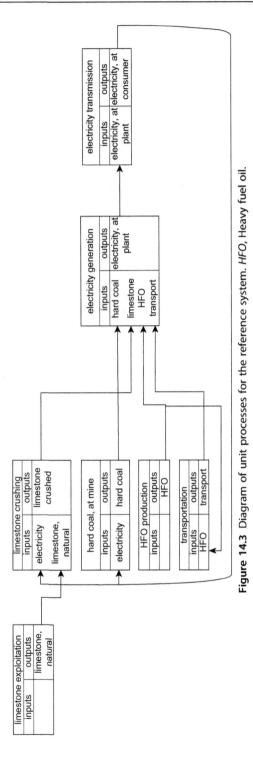

Figure 14.3 Diagram of unit processes for the reference system. *HFO*, Heavy fuel oil.

the Australian National Greenhouse Accounts, and the National Pollutant Inventory Emission. Fig. 14.4 illustrates different processes of generating rice husk-based bioelectricity.

RESULTS

Life Cycle Inventory Results

For the environmental inventory analysis, most inputs are the consumptions of fossil fuel and water (including natural gas, crude oil, hard coal, water salt sole, water for cooling, and water from the river). Consumptions of metals and minerals are insignificant compared to those of fossil fuels and water. In general, the total energy consumption for generating 1 MWh of coal-fired electricity is higher than that of rice husk-based electricity, at 5.69E + 04 MJ and 8.33E + 03 MJ, respectively. For environmental output, while most of the emissions from the bioelectricity product system are higher than those of coal-fired electricity, sulfur dioxide is the only exception as it is much lower in bioelectricity.

In terms of economic sustainability, the total costs for producing two types of electricity are not that different; however, the components that make up the costs are much different. While the cost for coal-fired electricity mostly comes from the high price of input material, the majority of bioelectricity's costs originates from high capital investment and labor cost. Considering how the PROSUITE model is designed, the software did not show our results for direct capital investment. Also the PROSUITE applied methodology did not present the result for raw material cost.

The results of social sustainability are not much different between the two product systems, ranging from 1.23 to 1.28 labor hours, 5.30E − 06 to 4.86E − 06 DALYs for 1 MWh. The number of DALYs, hours of child labor, and forced labor will be reduced if bioelectricity replaces coal-based electricity. Meanwhile, the hours of total labor and high-skilled labor needed for bioelectricity are higher than those for coal-fired electricity (Table 14.6). This is a good result for bioelectricity as we would like to increase employment and, at the same time, would not want to exploit child labor for the energy industry in particular or for any industry in general.

Impact Assessment Results

As the human health impacts are calculated based on environmental human health (number of DALY caused by the environmental condition), occupational human health (number of DALY caused by working condition), and consumer health (number of DALY caused to consumers), the results of human health impact in impact assessment should be different from the number of DALY in inventory analysis. Details of human health impact can be found in Table 14.7.

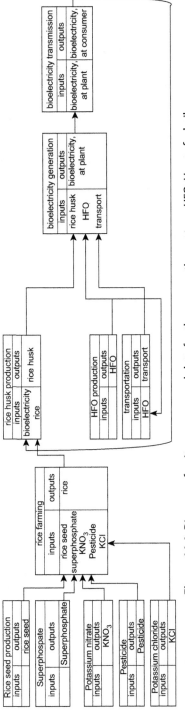

Figure 14.4 Diagram of unit processes and data for the prospective system. *HFO*, Heavy fuel oil.

Table 14.6 Main Environmental, Economic, and Social Inputs and Outputs for Two Product Systems per Functional Unit (FU)

	Electricity From Coal	Bioelectricity From Rice Husk
Inputs		
Natural gas, 44.1 MJ/kg	6.9886E + 02	3.95E + 03
Crude oil, 42.3 MJ/kg	6.69E + 03	2.72E + 03
Hard coal, 26.3 MJ/kg	4.96E + 04	1.66E + 03
Water, cooling, unspecified natural origin, m^3	1.4154E + 02	1.23E+ 03
Water, river, m^3	1.3044E + 02	0
Total cost (€ per FU)	58.86	57.91
Direct compensation of employees (€ per FU)	0.21	0.74
Labor (hours per FU)	1.23	1.28
Labor high skilled (hours per FU)	0.2	0.21
DALY per FU	5.3E − 06	4.86E − 06
Child labor total (hours per FU)	3.28E − 02	3.21E − 02
Forced labor total (hours per FU)	2.25E − 03	2.15E − 03
Outputs		
Waste water/m^3	1.65	6.16
Methane, kg	1.07	2.95
Nitrogen oxides, kg	1.2	1.64
Sulfur dioxide, kg	2.03	1.45E − 02

Table 14.7 Five Impacts of Two Product Systems

Impacts		Electricity From Coal	Bioelectricity From Rice Husk	Unit
Human health, area of protection per 1 FU		0.27	0.98	DALY
Impact on social well-being per FUs	Total employment	7.53E + 12	7.53E + 12	h
	Knowledge-intensive jobs	1.63E + 12	1.63E + 12	h
	Child labor	2.64E + 11	2.64E + 11	h
	Forced labor	2.00E + 10	2.00E + 10	h
	Regional income inequities	0	0	€
	Global inequities	0	0	€
Impact on prosperity per FUs	Capital productivity	6.74	6.74	€/€
	Labor productivity	2.1	2.1	€/€
	Resource productivity	7.97E + 05	7.97E + 05	
Ecosystem quality, area of protection per 1 FU		9.22E − 06	5.26E − 06	Species loss year
Natural resource, area of protection per 1 FU		2.54E04	1.86E03	US $2000

DALY, Disability-adjusted life year; *FU*, functional unit.

Although the occupational human health impact of bioelectricity is smaller than that of coal-fired electricity, the overall human health impact of bioelectricity is higher than that of electricity because of the large contribution of environmental human health impact (Table 14.7). In both systems, the environmental human health impacts come from the emission of carbon monoxide, methane, and sulfur oxides. The hotspots are also similar for the two product systems, because most of these emissions come from the electricity and bioelectricity generation stages. However, the contributions of these emissions are a little bit different, as sulfur dioxide from coal contributes mostly to human health impacts of electricity, while in the bioelectricity product system there is a small difference in contribution of carbon monoxide and sulfur oxides to human health impact. Fig. 14.5 represents the contributions of main emissions to environmental human health impact for the two product systems.

Impacts on social well-being are aggregated based on 11 pathways, six of which are quantitative indicators. The five remaining indicators, which are qualitative information collected by consultation with an expert board, are skipped in this study because of limited time. The quantitative indicators are analyzed at an economy-wide scale for the purpose of sustainability assessment. The economy-wide scale of $9.46E + 07$ MWh of electricity is selected in this study, as it is equal to about 500 MW of installed capacity of bioelectricity, which is planned to be developed by the Vietnamese government by 2020 [35].

The result of PROSUITE calculation shows that there is no difference between impacts of coal-fired electricity and rice husk-based electricity on social well-being at an economy-wide scale, while there is a small difference at a functional unit scale. This can be explained by the small contribution of the product system on the whole economy and society. Moreover, it is surprising that both product systems do not contribute to the social inequity at regional as well as global levels. Details of social well-being impacts can be found in Table 14.7.

Impact on prosperity is analyzed on the basis of impacts on labor productivity, impacts on capital productivity, and impacts on resource productivity. These impacts are assessed at an economic-wide scale, with a functional unit of $9.46E + 07$ MWh.

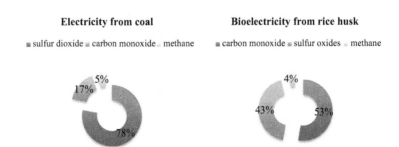

Figure 14.5 Contribution of emissions to environmental human health impact.

Being similar to the impact on social well-being, the result of PROSUITE shows that there is no difference between coal-fired electricity and rice husk-based electricity because of the small share of the product's system to the economy. In other words, replacing coal-based electricity with rice husk-based electricity does not bring any positive change (in terms of prosperity) to the economy. Details of prosperity impacts are presented in Table 14.7.

For the environmental pillar, bioelectricity is better than the conventional electricity in both natural resource and ecosystem quality. The impacts of bioelectricity on natural resource are smaller than those of conventional electricity, at 1.86E03 and 2.54E04, respectively (Table 14.7). Similarly, the results from PROSUITE show that bioelectricity impacts on ecosystem quality are smaller than those of conventional electricity (Table 14.7). Most of the ecosystem quality impacts of electricity come from carbon dioxide, sulfur dioxide, and methane, accounting for 90.96%, 5.5%, and 2.31%, respectively. While the majority of carbon dioxide and methane originates from electricity generation, sulfur dioxide mainly comes from transportation. For bioelectricity, carbon dioxide and methane are the main causes of negative impacts on the natural ecosystem, with shares of 68.37% and 11.10%, respectively. Being different from the coal-fired electricity product system, carbon dioxide from bioelectricity is mostly emitted from the pesticide production stage. Therefore the hotspot of coal-fired electricity on ecosystem quality may lie in the electricity generation process, while that of rice husk-based bioelectricity lies the pesticide production process.

INTEGRATED RESULTS OF SUSTAINABILITY ASSESSMENT

In general, except for human health impact in which bioelectricity shows its weakness over conventional electricity, impacts of bioelectricity are either more positive (for ecosystem quality and natural resource) or no big difference (for social well-being and prosperity) compared to that of electricity at full scale of development. At a functional unit scale, bioelectricity is better than conventional electricity in terms of ecosystem quality, natural resource, social well-being, and prosperity. Therefore when a single score of sustainability is aggregated, it is expected that bioelectricity is more sustainable than conventional electricity. Figs. 14.6—14.8 present sustainability assessments of coal-fired electricity and rice husk-based bioelectricity in the form of a graphical chart, weighted sum, and outranking, respectively.

The graphical chart shows the difference between the two product systems in five main impacts. However, because the impact on exhaustible resources by electricity is much larger than other impacts in both product systems, other impacts including impact on human health, impact on social well-being, impact on prosperity, and impact on ecosystem quality do not seem to have been considered and are invisible on the graphical chart.

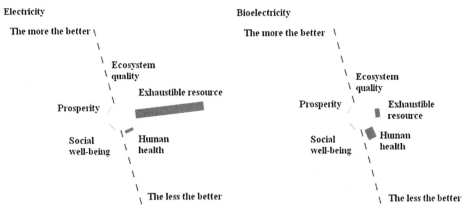

Figure 14.6 Graphical chart: sustainability assessment of electricity from coal and bioelectricity from rice husk over their life cycle.

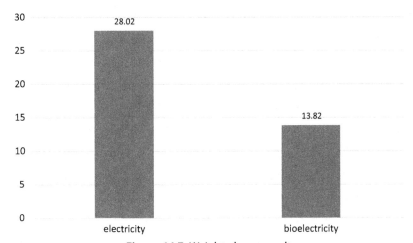

Figure 14.7 Weighted sum result.

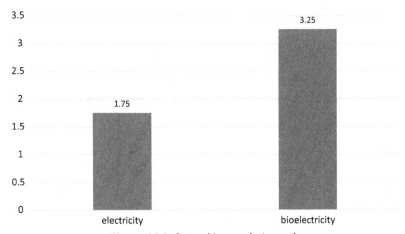

Figure 14.8 Outranking analysis result.

The weighted sum results are calculated on the principle that the positive impacts can compensate for the negative impacts. In other words, impact on social well-being and impact on prosperity can be subtracted from the impacts on human health, ecosystem quality, and natural resource. As a result, the higher the scores are, the worse the product systems are. The weighted sum result of rice husk-based bioelectricity impacts on sustainable development is half as much negative as that of coal-fired electricity, at 13.82 and 28.02, respectively.

An outranking analysis result is obtained by comparing the difference in each pair of categories of the two product systems considered. It first started with five endpoint impacts, which were followed by different midpoint impacts. The product system, which outranks the other, will receive a "win." The more "wins" a product system receives, the more sustainable it is. Moreover, as it is combined with the weighting factor for different endpoint impacts, the "win" of the impact with high weighting factor (e.g., human health, social well-being, and natural environment) will be more influential than the "win" of the impact with low weighting factor (prosperity and exhaustible resources). With outranking analysis, rice husk-based bioelectricity appears to get more "wins" than coal-fired electricity, with 3.25 and 1.75, respectively.

SOME REMARKS ON THE METHODOLOGY

PROSUITE DSS is a streamlined methodology to assess the sustainability of a technology or a product. With this methodology the decision maker only needs to import environment and cost inflow/outflow data. PROSUITE DSS calculates the five endpoint impacts and sustainability score on the basis of its built-in economic and social databases. This saves a lot of time and labor, which is a common hindrance when performing an LCA.

Conversely, this is also one of the weaknesses of the software. As PROSUITE has its own economic and social databases, it only differentiates between European countries and other countries in the world. Therefore the users should expect that results are purely indicative and it could be site specific. Also it is designed to be used in European countries; therefore the sustainability is assessed in creating a welfare, equity, and healthy environment for European communities (domestic) rather than for the place where the technology/product is applied. Meanwhile, pathways to sustainability vary among regions and countries, and are subjectively dependent on the definition of *win* and the measure of *success*. It is expected that further research on multicriteria decision analysis/multiobjective decision analysis would be conducted to improve the relevance and utility of the results.

Another limitation is the lack of comprehensiveness in the PROSUITE impact methodology as well as on its calculated results. With regard to the impact methodology, not all inventory flows of the foreground data are included. For instance, the flow of the transformation of land use is not included in impact methodology, while it is available in the Ecoinvent database with reference to foreground processes. It is highly possible that social well-being is influenced by transformation of land use. For example, transformation of agriculture land to urban (for power plant construction) and transformation of forest to agriculture land (for biomass plantation) can cause varying impacts on society. The former may negatively impact the incomes of farmers, who are normally poor in developing countries like Vietnam and hinder the potential of other socially unsustainable impacts. The latter may bring some short-term incomes for farmers but is a big threat to environmental sustainability.

The calculated results are not all encompassing, particularly the results for economic and social pillars. There is no result on cost components contributing to the total cost (e.g., cost for raw material). In other cases the results are not displayed for capital cost and income equalities. Because of the role of cost components in identifying the economic efficiency of product/technology in different stages and for different types of expenses, and the importance of equity in income distribution to social well-being, it is recommended that further research should be done on the PROSUITE methodology as well as the DSS platform to further elaborate the results for these two pillars.

CONCLUSION

In summary, LCSA is a holistic approach to assess the life cycle sustainability of a product/service, which is inclusive for three pillars of sustainability. It overcomes the difficulty in quantifying social indicators to provide relatively indicative quantitative results.

The case study is an illustrative application of LCSA in the energy industry, which suggested that bioelectricity in Vietnam is more sustainable than coal-fired electricity. At a functional unit scale, bioelectricity in Vietnam yields less negative impacts on ecosystem quality, natural resource, social well-being, and prosperity, but more negative impacts on human health compared to coal-fired electricity. Meanwhile, at an economy-wide scale, bioelectricity elicits less negative impacts on ecosystem quality and natural resource, no difference to social well-being and prosperity, and more negative impacts on human health compared to electricity.

With regard to social sustainability, although bioelectricity brings the same positive impacts on employment, high-skilled employment, as well as negative impacts on child labor and forced labor, it can cause more health issues than coal-fired electricity. Moreover, the human health impact is the drawback in both product systems as it is even larger than the positive social well-being impact. It is for this reason that the electricity/bioelectricity generation processes become the social hotspots for both product systems, as these processes are the main causes of human health impacts.

It is indicated, through existing literature and the case study, that the methodology to assess sustainability of a product/technology over its life cycle is not inclusive, with a built-in concentration on and prioritization of environmental impacts and a relative lack of attention on socioeconomic dimensions. This disparity in prioritization among the three pillars of sustainability is also shared within the broader framework of sustainability and feasibility assessment of energy systems. Therefore it is expected that further research would be conducted to complete/enhance the methodology of LCSA and broader energy assessment techniques, especially for economic and social pillars, which will consequently encourage the development of new products and technologies targeted at sustainable development in its intended form.

REFERENCES

[1] UN. Report of the World Commission on Environment and Development: our common future. 1987 [cited 2014 Aug 10]. Available from: http://www.un-documents.net/wced-ocf.htm.

[2] UN. Report of the World Summit on Sustainable Development. 2002 [cited 2014 Aug 10]. Available from: http://www.un.org/jsummit/html/documents/summit_docs/131302_wssd_report_reissued.pdf.

[3] UNEP/SETAC Life Cycle Initiative. Towards a life cycle sustainability assessment making informed choices on products. 2011 [cited 2014 Aug 10]. Available from: http://www.unep.org/pdf/UNEP_LifecycleInit_Dec_FINAL.pdf.

[4] Horne RE. Life cycle assessment: origins, principles and context. In: Horne RE, Grant T, Verghese K, editors. Life cycle assessment — principles, practice and prospects. CSIRO Publishing; 2009.

[5] Lichtenvort K, Rebitzer G, Huppes G, Ciroth A, Seuring S, Schmidt WP, et al. History of life cycle costing, its categorization and its basic framework. In: Hunkeler D, Lichtenvort K, Rebitzer G, editors. Environmental life cycle costing. New York: CRC Press and SETAC Press; 2008.

[6] Korpi E, Ala-Risku T. Life cycle costing: a review of published case studies. Manag Audit J 2008;23(3):240−61.

[7] Martínez-Blanco J, Lehmann A, Muñoz P, Anton A, Traverso M, Rieradevall J, et al. Application challenges for the social life cylce assessment of fertilizers within life cycle sustainability assessment. J Clean Prod 2014;69:34−48.

[8] Tsurukawa N, Prakash S, Manhart A. Social impacts of artisanal cobalt mining in Katanga. Democratic Republic of Congo; 2011 [cited 2014 Aug 10]. Available from: http://www.oeko.de/oekodoc/1294/2011-419-en.pdf.

[9] Franze J, Ciroth A. Social and environmental LCA of an eco labeled notebook. 2011 [cited 2014 Aug 10]. Available from: http://www.greendelta.com/uploads/media/LCA_laptop_final.pdf.

[10] Finkbeiner M, Schau EM, Lehmann A, Traverso M. Towards life cycle sustainability assessment. Sustainability 2010;2(10):3309−22.

[11] Zamagni A, Buttol P, Buonamici R, Masoni P, Guinée JB, Huppes G, et al. D20 blue paper on life cycle sustainability analysis deliverable 20 of work package 7 of the CALCAS project. 2009 [cited 2014 Aug 10]. Available from: http://www.leidenuniv.nl/cml/ssp/publications/calcas_report_d20.pdf.

[12] Halog A, Manik Y. Advancing integrated systems modelling framework for life cycle sustainability assessment. Sustainability 2011;3(2):469−99.

[13] Blok K, Huijbregts M, Roes L, van Haaster B, Patel M, Hertwich E, Wood R, et al. A novel methodology for the sustainability impact assessment of new technologies. 2013 [cited 2014 Aug 10]. Available from: http://www.prosuite.org/c/document_library/get_file?uuid=bdbb04e9-1a34-434b-85a8-44bafb28155b&groupId=10136.

[14] Blok K, Huijbregts M, Roes L, van Haaster B, Patel M, Hertwich E, Wood R, et al. Handbook on a novel methodology for the sustainability impact assessment of new technologies. 2013 [cited 2014 Sep 15]. Available from: http://prosuite.org/c/document_library/get_file?uuid=31404a5b-b716-4a65-8d4e-1ac991a6dd79&groupId=12772.

[15] WHO. Metrics: disability-adjusted life year (DALY); n.d. [cited 2015 Mar 22]. Available from: http://www.who.int/healthinfo/global_burden_disease/metrics_daly/en/.

[16] Laurent A, Hauschild MZ, Golsteijn L, Simas M, Fontes J, Wood R. Normalization factors for environmental, economic and socio-economic indicators. 2013 [cited 2014 Sep 15]. Available from: http://prosuite.org/c/document_library/get_file?uuid=750ef6d0-4e9d-4a00-913c-3f4cfd632782&groupId=12772.

[17] Gagnon L, Belanger C, Uchiyama Y. Life-cycle assessment of electricity generation options: the status of research in year 2001. Energy Policy 2002;30(14):1267–78.

[18] Campbell JE, Lobell DB, Field CB. Greater transportation energy and GHG offsets from bioelectricity than ethanol. Science 2009;324(5930):1055–7.

[19] Clarens AF, Nassau H, Resurreccion EP, White MA, Colosi LM. Environmental impacts of algae-derived biodiesel and bioelectricity for transportation. Environ Sci Technol 2011;45(17):7554–60.

[20] Prasara AJ, Grant T. Comparative life cycle assessment of uses of rice husk for energy purposes. Int J Life Cycle Assess 2011;16(6):493–502.

[21] Delivand MK, Barz M, Gheewala SH, Sajjakulnukit B. Environmental and socio-economic feasibility assessment of rice straw conversion to power and ethanol in Thailand. J Clean Prod 2012;37:29–41.

[22] Farine DR, O'Connell DA, Raison JR, May BM, O'Connor MH, Crawford DF, et al. An assessment of biomass for bioelectricity and biofuel, and for greenhouse gas emission reduction in Australia. GCB Bioenergy 2012;4(2):148–75.

[23] Luk JM, Pourbafrani M, Saville BA, Maclean HL. Ethanol or bioelectricity? life cycle assessment of lignocellulosic bioenergy use in light-duty vehicles. Environ Sci Technol 2013;47(18):10676–84.

[24] Pham TMT, Kurisu KH, Hanaki K. Greenhouse gas emission mitigation potential of rice husks for an Giang province, Vietnam. Biomass Bioenergy 2011;35(8):3656–66.

[25] Mitchell CP, Bridgwater AV, Stevens DJ, Toft AJ, Watters MP. Techno-economic assessment of biomass to energy. Biomass Bioenergy 1995;9(1):205–26.

[26] IRENA. Cost analysis of biomass for power generation. Renewable energy technologies: cost and benefit analysis series. Volume 1: Power sector. Issue 1/5: Biomass Power Gener 2012. [cited 2014 Sep 12]. Available from: http://costing.irena.org/media/2793/re_technologies_cost_analysis-biomass.pdf.

[27] Ikonen T, Asikainen A. Economic sustainability of biomass feed stock supply. TR2013:01. IEA Bioenergy Task 43; 2013 [cited 2014 Sep 12]. Available from: http://ieabioenergytask43.org/wp-content/uploads/2013/09/IEA_Bioenergy_Task43_TR2013-01i.pdf.

[28] Evans A, Strezov V, Evans TJ. Sustainability considerations for electricity generation from biomass. Renew Sustain Energy Rev 2010;14(5):1419–27.

[29] FAO. Sustainable bioenergy in Asia: improving resilience to high food prices and climate change. In: Damen B, Tvinnereim S, editors. 2012.

[30] Acheampong E, Campion BB. The effects of biofuel feedstock production on farmers' livelihoods in Ghana: the case of Jatropha curcas. Sustainability 2014;6(7):4587–607.

[31] Achterbosch T, Bartelings H, van Berkum S, van Meijl H, Tabeau A, Woltjer G. The effects of bioenergy production on food security. In: Rutz D, Janssen R, editors. Socio-economic impacts of bioenergy production. Springer; 2014.

[32] Abbasi SA, Abbasi N. The likely adverse environmental impacts of renewable energy sources. Appl Energy 2000;65:121–44.

[33] Farioli F, Dafrallah T. Gender issues of biomass production and use in Africa. In: Janssen R, Rutz D, editors. Bioenergy for sustainable development in Africa. Springer Science, Business Media; 2012.

[34] Juntarawijit C, Juntarawijit Y, Boonying V. Health impact assessment of a biomass power plant using local perceptions: cases studies from Thailand. Impact Assess Proj Apprais 2014;32(2):170–4.

[35] Decision 1208/QD-TTg dated 21 July 2011 on approving the master plan of national electrical power system in the period 2011–2020, with consideration to 2030 (in Vietnamese).

[36] GSO. Socio-economic status of Vietnam from 2001 to 2010 (in Vietnamese). Statistics Publishing. Hanoi: General Statistics Office; 2011.

[37] GSO. Statistics on agriculture (in Vietnamese). General Statistics Office; 2014 [cited 2014 Aug 10. Available from: http://www.gso.gov.vn.

[38] Do DT. Assessment of potential biomass energy from rice residues in Vietnam (in Vietnamese). In: Paper presented at forum on energy and petrol: investment and sustainable development; May 09, 2013 [Hanoi, Vietnam].

[39] Bhattacharya SC, Salam PA, Sharma M. Emissions from biomass energy use in some selected Asian countries. Energy 2000;25(2):169−88.

[40] Speight JG. Electric power generation, coal-fired power generation handbook. John Wiley & Sons Inc.; 2013.

[41] Rayaprolu K. Boilers: a practical reference. CRC Press; 2012.

[42] PECCI. Thai Binh Thermal power plant project: general technical provisions. Power Engineering Consulting Joint Stock Company I; 2010.

[43] Koskinen M, Haarla L, Hirvonen R, Labeau PE. Transmission grid security: a PSA approach. London: Springer-Verlag London Ltd; 2011.

[44] Yossapol C, Nadsataporn H. Life cycle assessment of rice production in Thailand. In: Paper presented at LCA Food; November 12−14, 2008 [Zurich, Switzerland].

[45] Nguyen VH, Nguyen VS. Clean development mechanism project design document (CDM-PDD) for pilot grid connected rice husk fueled bio-power development projects in Mekong delta, Vietnam. Second technical report. Economy and Environment Program for Southeast Asia (EEPSEA); 2006.

[46] Loi DC, Gheewala SH, Bonnet S. Integrated environmental assessment and pollution prevention in Vietnam: the case of anthracite production. J Clean Prod 2007;15(18):1768−77.

[47] Ecoinvent Database, http://www.ecoinvent.org.

[48] GIZ-GDE/MOIT. Summary report research on supporting mechanism for developing on-grid bioelectricity in Vietnam. 2014.

[49] NGAR. Australian national greenhouse accounts national greenhouse accounts factors. Department of Climate Change and Energy Efficiency; 2012.

[50] NPI. National Pollutant Inventory. Emission estimation technique manual for Fossil Fuel Electric Power Generation version 3.0. Department of Sustainability, Environment, Water, Population and Communities; 2012.

[51] GIZ. Identification of biomass market opportunities in Vietnam. 2011.

[52] Huan NH, Thiet LV, Chien HV, Heong KL. Farmers' participatory evaluation of reducing pesticides, fertilizers and seed rates in rice farming in the Mekong Delta, Vietnam. Crop Prot 2005;24(5):457−64.

[53] Kasmapragruet S, Paengjuntuek W, Saikhwan P, Phungrassami H. Life cycle assessment of milled rice production: case study in Thailand. Eur J Sci Res 2009;30(2):195−203.

[54] Chapagain AK, Hoekstra AY. The green, blue and grey water footprint of rice from production and consumption perspectives. Ecol Econ 2011;70(4):749−58.

[55] Pham VT. Pesticide use and management in the MeKong delta and their Residues in surface and drinking water [Ph.D. thesis]. Institute for Environment and Human Security - United Nations University in Bonn; 2011.

[56] IEA. Key world energy statistics. 2013 [cited 2014 Oct 10]. Available from: http://www.iea.org/publications/freepublications/publication/KeyWorld2014.pdf.

[57] MOIT. Rice price in some provinces (in Vietnamese). 2014 [cited 2014 Oct 10]. Available from: http://www.vinanet.com.vn/tin-thi-truong-hang-hoa-viet-nam.gplist.289.gpopen.237210.gpside.1.gpnewtitle.gia-gao-tai-mot-so-tinh-ngay-10-10-2014.asmx.

CHAPTER FIFTEEN

Embedding Sustainability in Product and Process Development—The Role of Process Systems Engineers

C. Jiménez-González

GlaxoSmithKline, NC, United States

INTRODUCTION

Engineers and others involved in designing and implementing new processes and products need to address a range of questions. For example: Which materials would be best to use in a particular formulation? Which unit operations are needed for the separation train? What is the optimal design for the flowsheet? When should a bioprocess, rather than a chemical process, be implemented? Are there any environmental, health, or safety hazards that need to be addressed?

Process systems engineering (PSE) has been defined by Grossman, in a wider sense, as the discipline "concerned with the improvement of decision making processes for the creation and operation of the chemical supply chain. It deals with the discovery, design, manufacture and distribution of chemical products in the context of many conflicting goals" [1]. Over the last few decades, PSE has already developed many of the appropriate tools to address the product and process development questions such as those presented earlier. These tools include computer modeling, process integration, process synthesis, and process optimization, to name a few. These tools and techniques are pivotal for the design and optimization of products and processes, and also lend themselves to incorporate environmental, health, and safety (EHS) considerations as part of the design work.

In the framework of PSE, embedding sustainability concepts means that when one is developing novel chemical processes, selecting reactors or separations, building and running plants, and so on, a process systems engineer should strive to follow three principles [2]:

- Maximize resource efficiency (mass and energy);
- Eliminate and minimize EHS hazards; and
- Design systems holistically and using life cycle thinking.

Process systems engineers have at their disposal a series of tools and techniques that have evolved, or were designed, to precisely maximize the use of materials and energy, reduce hazards, and incorporate life cycle thinking and holistic design. A testament of this is

Sustainability in the Design, Synthesis and Analysis of Chemical Engineering Processes

the top key green engineering challenges identified by the American Chemical Society Green Chemistry Institute Pharmaceutical Roundtable, which include some of quintessential PSE tools, such as process intensification, mass and energy integration, and life cycle thinking [3].

This chapter intends to present the PSE tools, techniques, and areas of opportunities for embedding sustainability during material selection, process design, process and product modeling, and supply chain implications. It also provides a perspective of the challenges to this task and the skill sets that will need to be developed to overcome those challenges successfully.

MATERIAL SELECTION

As part of product and process design, some materials are predefined by the chemical reactions involved or by the type of properties of the products. However, scientists and engineers will have to make several decisions regarding materials as part of the process and product design. Some of these materials include solvents, excipients, formulation materials, catalysts, and reagents. Material selection would have impacts not only on the intrinsic EHS hazards of the process, but also on supply chain impacts and mass and energy usage. Although the preferred timing for material selection is at the design stage, materials with EHS issues can be substituted after a process or a product has been established. Substitution has been done routinely as part of normal continuous improvement processes, and oftentimes as part of customers' and other pressures.

Process systems engineers play an active role in ensuring that these decisions are technically sound and integrate environmental sustainability aspects. Some examples of PSE tools used for key materials such as solvents, reagents, and formulation materials are described next.

Solvents

Solvents are used in all types of industries for processing, manufacturing, and formulation. For instance, solvents are used as mass separation agents, acting as a reaction medium, in equipment cleaning. Solvents are also used within product formulations, such as in paints, adhesives, and cleaners (Table 15.1). At the same time, many commonly used solvents have EHS hazards, including human toxicity, process safety, and waste management issues. In addition, solvents are subject to increasing and continuous regulatory scrutiny [4,5].

Green chemistry and engineering principles [6,7] suggest that the use of ancillary materials not integrated in the product should be avoided, and when organic solvents and other ancillary materials must be employed, their use must be minimized and optimized to enhance reactions with minimal environmental and operational concerns. Process

Table 15.1 Some Selected Examples of Solvent Use

Application	Function	Industrial Examples
Process	Separation • Dissolve • Extract Synthesis • React • Dissolve	Chemical Pharmaceuticals
Formulated products	Active ingredient Excipient Encapsulation Carrier	Pharmaceuticals Food Paint Food Skin care Pesticides
Cleaning	Equipment Surfaces	Chemical Pharmaceutical Skin care

systems engineers designing a process are then faced with that challenge. The appropriate selection of solvents depends to a large extent on the application, along with other considerations such as solvent recovery and solvent release. There are several PSE tools and techniques that have been used to address the solvent selection (Table 15.2), most of them can be described following a generic process:

1. Problem identification—to identify the actual problem and the direction required to solve it. This includes asking whether the use of a solvent is even necessary, or whether the objective can be met by some other means, such as physical separation or solvent-free reaction.

Table 15.2 Illustrative Examples of Existing PSE Tools for Solvent Selection [12]

Databases	Data-Driven Benchmarking	Predictive Computer Models
• CAPEC-Database [13] • CambridgeSoft, ChemFinder [14] • DECHEMA Chemistry Data Series [15] • CRC Handbook of Chemistry and Physics • DETHERM • CHEMSAFE • The NIST Webbook	• Solvent Selection guides from GlaxoSmithKline, Pfizer, AstraZeneca, ACS GCI Pharmaceutical Roundtable, and others • SAGE (solvent alternative guide) • Pharmaceutical Roundtable Solvent Selection Guide • Britests Ltd	• ProCAMD (ProPred, PDS, SoluCalc, SMSwin) [13] • Aspen Properties, Aspen Plus: NRTL-SAC and eNRTL-SAC [16] • COSMO-RS; COSMO-SAC • Principal component analysis techniques

2. Search. In this part, several of the PSE tools can be used, including computational modeling, molecular design, principal component analysis, databases, and benchmarking, among others.

3. Verification. This could include a computational simulation, using a structured design of experiments for laboratory verification.

In general, the properties and functions needed for a specific solvent selection problem include pure—solvent properties, EHS properties, solvent—solute properties, and solvent—function properties. Part of the role of the process systems engineer is to utilize the data on solvents and make selection decisions based on the application and function. This can be done working directly with the data, using a computer algorithm to guide the decisions, or using a heuristic model for benchmarking the solvent. When using a computer algorithm, it is also advisable to incorporate in the modeling a risk-based modeling approach for a more holistic assessment. A risk-based approach would incorporate the assessment of both the hazards and probability of occurrence to get a better estimation of how severe and likely potential issues would be.

Since there are several options for solvent selection tools, there will be tradeoffs between completeness and the time and information necessary to make a decision. Table 15.2 presents the examples of tools from simple to more sophisticated approaches. One can perform a quick screen using few data points, but after the initial screening, one may need to do some advanced computer modeling before the experimental validation.

In addition to solvent selection from the known molecules, there is a series of approaches for designing solvents tailored to a set of desired characteristics [8]. One of these approaches is the computer-aided molecular design (CAMD) methodology developed by Gani and coworkers at CAPEC, Technical University of Denmark [9]. This approach is the reverse of that used by property prediction methodologies, which start with the identity of the molecule and/or the molecular structure, and then a set of target properties is calculated. CAMD may be performed at various levels of size and complexity of molecular structure representation. Some application examples include solvent design for organic reactions [10,11].

Reagents

Similar to the solvent selection problem, a process systems engineer will have the need to select specific reagents for a given transformation. Sometimes the choice of reagent is limited to only a few, but very often there are choices [17]. For instance, for O-dealkylation reactions, some of the reported reagents include thiols/methionine [18], metal/hydrogen for hydrogenolysis [19,20], acids [21], metals [22], amine/alkoxides [23], and hydrides [24], among others. Reagents will also exhibit different EHS properties, which coupled with the type of process and the potential controls, would contribute to a sound reagent selection. To facilitate the selection of reagents with fewer

EHS issues such as toxicity, decomposition, footprint, and others; some efforts have been made to summarize and rank commonly used reagents for certain industries. For instance, Pfizer [25] and GSK [26] have developed reagent selection guides for the most commonly used transformations. In the case of Pfizer the potential reagents for a given transformation are evaluated in terms of utility, scalability, and "greenness" and set into a region within a Venn diagram as shown in Fig. 15.1A. Pfizer describes utility as the ability to work in good yield in a wide variety of drug-like molecules; scalability as the ability to be used to prepare multikilogram batches; and defines greenness as an assessment based on the available data on worker safety, ecotoxicity, and atom economy. A qualitative assessment is performed by a group of chemists that, after reviewing the literature, place the reagent in a specific part of the Venn diagram. A link to the literature consulted is provided in the Venn diagram. In this approach the goal is to find a reagent in the intersection of the three circles, although very often this is not possible, and tradeoffs will happen, as the next choice are reagents in the intersection of two circles.

In the case of GSK's guide the potential reagents for a given transformation are evaluated with an overall EHS score and a clean chemistry score, which are combined into an overall greenness score through a geometric mean. Individual EHS scores are estimated based on EU risk phrases [27] and known issues in each area, as shown in Table 15.3. The overall EHS score combines those three using a weighted average. The reagent guide also includes a life cycle score that provides an evaluation of the environmental life cycle impacts of the reagent using a streamlined life cycle assessment tool [28].

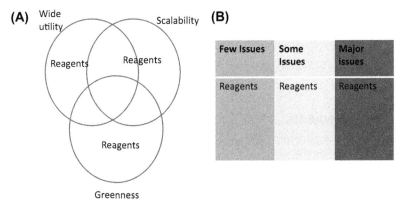

Figure 15.1 Examples of the framework to help select reagents for commonly used transformations in the reagent selection guides within (A) Pfizer and (B) GlaxoSmithKline (GSK). In the guide from Pfizer the goal is to select the reagent closer to the center of the Venn diagram, as placed by a group of chemists based on literature data. The GSK guide ranks and color codes the reagents based on the environmental, health, and safety (EHS) issues (based on EU risk phrases) and chemistry issues (based on stoichiometric calculations and assessment by a group of chemists).

Table 15.3 Calculation of the Environmental, Health, and Safety Score Components of the GSK Reagent Guide

Score	Environment	Health	Safety
1	Risk phrases R50 and R53, or one of R40, R45, R49, R60, R61, R64, or R68	Risk phrases R26, R27, R28, R45, R49, R60, R61, R64, or R68 or OEL < 0.5 ppm or < 0.01 mg/m^3	Risk phrases R1, R2, R3, R4, R5, R6, R7, R9, R16, R17, or R44
4	Risk phrases R51 and R53, or any of R50, R62 or R63,	Risk phrases R23, R24, R25, R34, R35, R39, R40, R41, R42, R43, R48, R62, or R63, or OEL ≤5 ppm or ≤ 0.1 mg/m^3	Risk phrases R8, R11, R12, R14, R15, R18, R19, R29, R30, R31, or R32
7	Risk phrases R52 and R53	Risk phrases R20, R21, R22, or R33, or OEL ≤ 50 ppm or ≤1 mg/m^3	Risk phrase R10
10	All others	All others	All others

The chemistry score evaluates atom efficiency, stoichiometry, work-up, coreagents, and other issues. Atom efficiency and stoichiometry were evaluated using direct algorithms, whereas the other components were given a score by a group of chemists. A full description of the methodology to generate the score is provided in the literature [27]. An overall summary provides an "at-a-glance" classification that provides a quick view of which reagents will have more issues to be managed (Fig. 15.1B). This type of tool of course comes with the caveat that the initial assessment that identifies and ranks issues does not substitute a risk-based approach. Ideally the issues identified would include life cycle or supply chain issues, or at the very least some life cycle thinking, as in the GSK tool presented earlier. The approach is to first eliminate hazards, and when that is not possible, to minimize the risks. Thus, after the issues are identified, it is necessary to couple the solvent selected (or the top choices) with an evaluation of the risks and identify any potential mitigation measures for those risks.

Formulation Ingredients

When talking about consumer products the selection of ingredients to use in formulations is a particular safety interest given the potential direct contact with consumers, for instance, in skin care or personal hygiene products. In this case we are not reacting chemicals, but we are forming emulsions or other types of mixtures that would become a formulated product such as a cream or an ointment, for example, diaper rash cream, moisturizing lotions, antibacterial ointments, sunscreens, and others.

The selection of the appropriate ingredients is also of commercial significance, given the increased pressure of consumer groups and retailers. Some of the big box retailers have designed frameworks to rank certain sustainability aspects of the products sold in their stores (e.g., Walmart's Sustainability Index) [29]. The same type of EHS information of reagents can be used to determine the suitability of formulation ingredients. In addition to the typical databases, there are some specialized systems that would provide information on the ingredients and at the same time provide a ranking on the EHS aspects of a given ingredient, such as the safer chemical ingredients for use in design-for-environment (DfE) labeled products (Fig. 15.2) [30]. This is built by evaluating each ingredient in a formulation against master and functional-class criteria that define the characteristics and toxicity thresholds for ingredients that are acceptable in safer choice products. The criteria are based on US Environmental Protection Agency expertise in evaluating the physical and toxicological properties of chemicals.

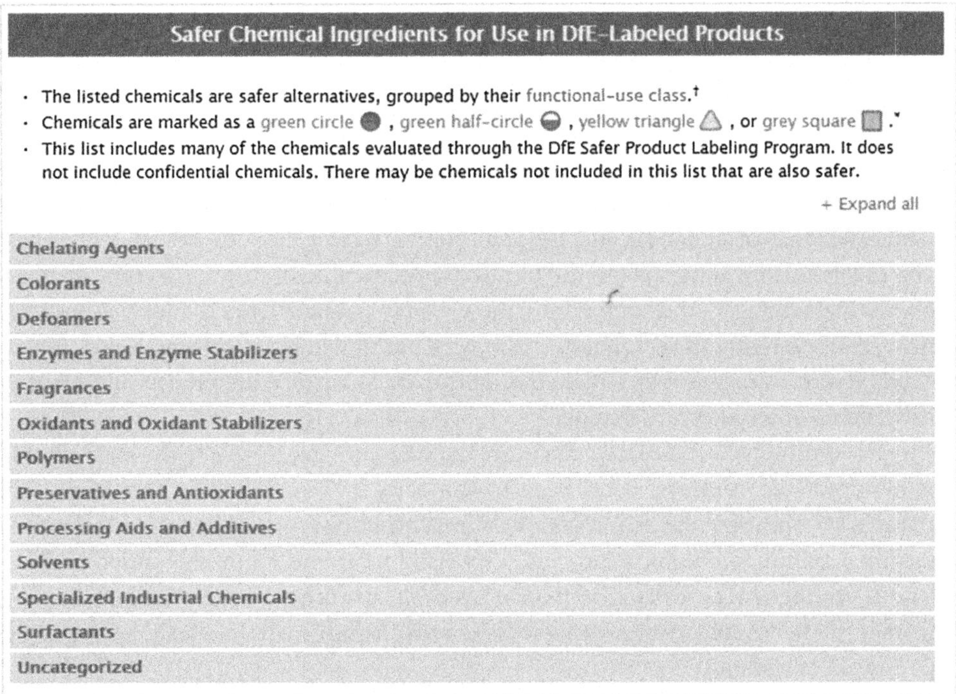

Figure 15.2 Screenshot of the Safer Chemical Ingredients for DfE-Labeled Products. The chemicals are marked with a *green circle* (dark gray in print versions) or a *green half circle* (dark gray in print versions). *Green circle* (dark gray in print versions): the chemical has been verified to be of low concern based on experimental and modeled data. *Green half circle* (dark gray in print versions): the chemical is expected to be of low concern based on experimental and modeled data and additional data would strengthen our confidence in the chemical's safer status. *Yellow triangle* (light gray in print versions): the chemical has met safer choice criteria for its functional ingredient class, but has some hazard profile issues. Specifically, a chemical with this code is not associated with a low level of hazard concern for all human health and environmental endpoints.

Some of the challenges faced when selecting formulation ingredients is that new ingredients are being generated sometimes faster than the rate at which the necessary information to make a sound assessment is generated. The other challenge is that the initial assessment on the properties is not a risk-based approach. Further assessment would be needed to estimate the extent of the possible risks related to the properties of a particular ingredient.

PROCESS DESIGN, SYNTHESIS, AND INTEGRATION

The dream situation for a process systems engineer would be to have a chemical process that is one single unit operation that receives reactants and outputs the desired product at the desired purity with no waste, but this does not happen in reality. Most chemical processes are complex systems in which a series of reactors and separators are intimately related and interrelated to each other through pumps, transport pipes, valves, utilities, heat exchangers, controls, recirculation loops, waste streams, and pollution control devices, among others. The reactor can be seen as the center of a chemical process design, but without separations the end product would not be possible. Separation processes generally account for between 40% and 70% of the chemical plant cost [31–33] and are therefore identified as part of the areas of focus for energy optimization [34,35]. The goal of chemical and process systems engineers is to design a chemical process that is as close as possible to the idealized process where the product self-separates with no waste produced, thus optimizing the use of mass and energy as much as possible. Utilizing process synthesis techniques is also very useful when performing some process redesign or retrofitting, as it provides an opportunity to analyze the process and identify improvements in a systematic manner.

Process synthesis, a systematic decision-making process to select the unit operations required for a process, dates back to the early 1970s [36]. The aim of process synthesis is to optimize the sequence of processing steps (e.g., reactions, distillations, extractions) within a chemical process, especially as it relates to the order and interactions among the processing steps (e.g., recycle streams). Process synthesis can be performed to revise an existing process (i.e., retrofitting) or to design a new process (e.g., flowsheet design).

Process synthesis is closely aligned with the aims of sustainable design, as a more efficient process will utilize energy and mass resources more effectively and minimize waste. This involves many decisions that need to be taken at multiple levels. Douglas [37] estimated that for a typical design the number of alternatives that might accomplish the same goal can be over 1 billion. Designers need to select the one alternative closest to the objectives.

A series of different approaches to process synthesis has been developed to facilitate and optimize the design of a process. These techniques and tools have been utilized and

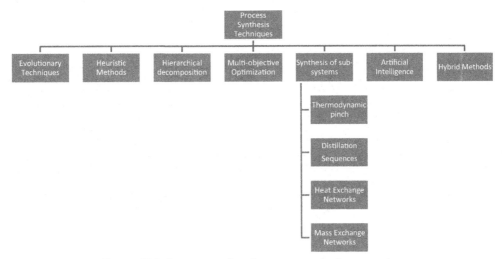

Figure 15.3 Some examples of process synthesis approaches.

tailored directly to produce process designs that will optimize material and energy requirements and minimize environmental impacts. Fig. 15.3 shows some examples of process synthesis approaches [38]. These approaches are primarily tools such as software and mathematical algorithms, to simplify and automate the calculations. Some brief description is given in the following paragraphs.

- Evolutionary techniques. In this approach a first flowsheet is produced and then refined incrementally to the desired endpoint that is set based on the performance metrics. This can be done using computer systems such as Aspen [16] or ICAS [13] and evolutionary rules. At each step, performance metrics for the flowsheet are calculated and the incremental improvements are repeated until no further significant improvements can be measured. One very good example of integrating sustainability and green engineering into evolutionary process synthesis is the methodology that Carvalho et al. have developed for retrofitting or designing new processes once the first flowsheet is available [39] (In addition, see chapter 11).
- Heuristic methods. This approach uses rules of thumb generated by observing the behavior of systems when solving many problems of that type. Many rule-of-thumb books have been published to condense the empirical knowledge for chemical processes, such as Woods's *Rules of Thumb in Engineering Practice* [40], Walas's *Chemical Process Equipment: Selection and Design* [41], or Branan's *Rules of Thumb for Chemical Engineers* [42].
- Hierarchical decomposition. This method breaks the design down in such a way that "high-level" decisions are taken first in the design phase (e.g., identifying reaction—separation train). After the high-level decisions are made, decisions on each unit operation are refined. Linnhoff et al. [43] represented this approach in an onion

diagram—akin to peeling the layers of an onion. Douglas developed a hierarchical decomposition approach and modified his hierarchy to account for waste minimization in the process synthesis, as shown in Table 15.4 [44].

- Multiobjective optimization/superstructure optimization. In this approach all the possible flowsheets are combined into one, normally called a superstructure. The superstructure contains more streams and unit operations than would be needed in reality, but it can be reduced to a practical flowsheet. The objective of the optimization is to remove those aspects of the flowsheet that are not going to render the best options. To solve this optimization, mixed–integer nonlinear problems are commonly used, as proposed by Grossmann [45]. Some examples of application include supply chain modeling, optimization of retrofit of existing systems, and optimization of process synthesis, among others. Many examples of current applications of multiobjective optimization can be found in the PSE literature [46–48].
- Synthesis of subsystems. The main objective of this approach so far has been the reduction and optimization of energy, including thermal pinch, distillation sequences, and mass and heat exchange networks, which will be described in a little more detail later in the chapter (In addition, see chapter 4).
- Artificial intelligence. There are numerous software packages for computer-aided chemical process design. The design of chemical plants is arguably a complex problem, as there are several competing objectives with their associated tradeoffs. Artificial intelligence algorithms have been used for modeling and simulation of chemical systems, discrete event simulation to analyze malfunctions, process control, and process design [49].

Although process synthesis techniques have been used as tools to design more sustainable processes, this was not always the case, and there has been an evolutionary approach to

Table 15.4 Douglas' Hierarchical Decomposition Approach [33]

Level	Decision
1	Input information—type of problem
2	Input—output structure of the flowsheet
3	Recycle—structure of the flowsheet
4	Specification of the separation system
	a. General structure—phase splits
	b. Vapor recovery system
	c. Liquid recovery system
	d. Solid recovery system
5	Energy integration
6	Evaluation of alternatives
7	Flexibility and control
8	Safety

process synthesis and design that has increasingly incorporated EHS aspects into process synthesis. For example, initially, chemical process design was limited to reaction and separation steps. However, with the energy crisis of the 1970s, energy integration was included. As environmental costs rose, waste treatment was added to the scope for process synthesis, and more recently the trend is to include materials integration, risk analysis, and life cycle assessment. Cano-Ruiz and McRae [50] and Stephanopoulos et al. [51] have presented an evolutionary framework for chemical process synthesis in the last 40 years or so, from the viewpoint of integrating environmental aspects into process design, as presented in Fig. 15.4.

Process Integration

The synthesis of process subsystems was mentioned briefly in the previous section. This practice is most commonly known as process integration, and has evolved to be a full PSE area in itself. Process integration was initiated primarily as a way of optimizing energy and then expanded to optimization of mass flows, and it can be applied both for new process design and during the implementation of changes and modifications in an existing process. Process systems engineers need to spend considerable time optimizing mass and energy—and more application of process integration is needed to design more efficient, sustainable processes. Some of the application examples include energy optimization through pinch analysis [52], energy optimization through heat exchange networks (HENs) [53–55], pollution prevention through combined process integration [56], distillation system design [57], mass exchange networks (MENs) for waste

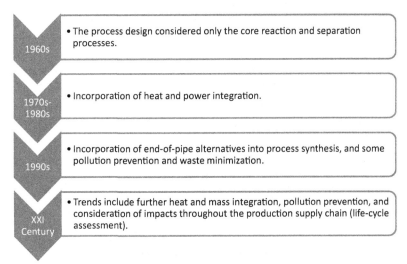

Figure 15.4 Evolution of chemical process synthesis from the perspective of integrating environmental considerations into chemical process design.

reduction [58,59], mass integration on batch processes [60], reactive MENs [61,62], synthesis of heat-induced separation networks [63], synthesis of crystallization-based separations [64], wastewater minimization [65], water minimization [66], and wastewater minimization [67], among others.

In general, process integration involves the following activities:

- Task identification. In other words, why the process intensification work is being conducted (goal setting) and then define a specific task to achieve the goal. One example is to have the goal of improving overall energy efficiency, and the specific task is to reduce net energy requirements, which might involve energy integration.
- Target setting. In this step the desired performance levels are identified ahead of the detailed design: for example, a 25% reduction of net energy requirements.
- Alternatives generation. This is step 1 of process synthesis (see earlier). In PSE there are often multiple solutions to achieve a certain target; therefore it is necessary to follow a methodology that would allow identifying the potential alternatives that would achieve the desired target. There are many approaches that have been used for alternative generation, such as physical rules, matrix methods, neuron networks, and heuristics, to mention a few easily found in the literature [68,69].
- Alternative analysis and selection. This involves evaluating the alternatives based on a set of desired performance metrics (e.g., efficiency, economic, environmental, safety) and determining the best alternative for the given target. This is performed using mathematical programming techniques in conjunction with optimization algorithms [70,71].

Heat Exchange Networks and Mass Exchange Networks

Mass and energy are of course intimately related in any production process, since, for instance, a chemical process can be described as the way by which we convert mass and energy into a more valuable product. In a chemical process, we start with raw materials (mass) and then require steam (energy) to carry out chemical and physical transformations. As a result we typically will have several hot streams that need to be cooled and several cold streams that need to be heated. To achieve this, one could cool all the hot streams with, say, cold water or refrigerants and then heat all the cold streams with steam; but by doing this the total energy requirements would be maximized. To minimize the amount of utilities that need to be used, and therefore minimize costs and environmental impacts, we use process integration—specifically energy integration.

Energy integration can be simply described as using hot streams to heat cold streams, and vice versa, before any additional utilities are used, with the result of reducing the overall use of utilities. The simpler example is a heat exchanger—but this could be very complex, depending on the system, and one would need an entire network of those heat exchangers in the plant for effective energy integration. These networks are known as HENs (Fig. 15.5). There are many examples of successful heat integration in a wide variety of industries, such as refineries, petrochemical, chemical, food and drink, pulp and paper, and metallurgical. The net results of heat integration applied successfully are cost savings, increased throughput, and reductions in emissions and environmental impacts.

MENs are the parallel concept for using process streams to integrate mass. MENs are very similar to HEN, but instead of exchanging energy using process streams, we exchange mass. For instance, very often we need to separate unreacted species from our desired product [72]. We could use an additional material (say, a solvent) to remove the unreacted material and send it back to the reactor. However, that comes with additional cost and environmental impacts. By using MEN, we take advantage of lean streams (streams with low concentrations of a given material) to separate and recover mass from rich streams. The idea is to use as much of the lean streams to recover and potentially recycle the materials before we have to use an external agent, thus reducing environmental and life cycle impacts in the process. For instance, a reported simultaneous HEN and MEN integration of a three-component system reported profits of $19 million/year with close to 32,000 kW of heat exchanged and increased conversion [73].

Of course, designing HEN and MEN is easier said than done—the mass and energy transfer is governed by thermodynamic and equilibrium laws, and thus will require process systems engineers to perform some calculations and graphs to produce the design, but that is where the fun begins.

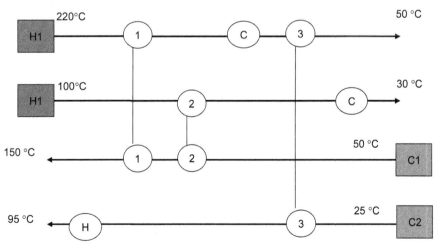

Figure 15.5 Simplified representation of a heat exchange network, showing hot streams, cold streams, heat exchanges (numbered pairs), and external utilities such as heating (H) or cooling (C).

PROCESS INTENSIFICATION

Process intensification was first defined by Ramshaw as a strategy to make dramatic reductions in the size of a chemical plant (100- to 1000-fold), either by reducing the size of the equipment or by combining multiple operations into fewer pieces of equipment [74]. The concept of process intensification has since been expanded to include increases in production capacity within a given piece of equipment, multifunctional systems, significant decreases in energy consumption, a marked cut in waste or by-product formation, and other advantageous outcomes [75]. Process intensification can be achieved through the use of large forces, such as increased pressure, smaller geometry, microfluidic interactions, high pressures, and different types of energy (e.g., magnetic fields, ultrasound, oscillatory forces).

Table 15.5 shows some of the potential advantages of process intensification from a sustainability perspective.

Although there are many advantages of applying process intensification, there are some aspects that will invariably need to be managed. Some of these issues to be managed are a potential for decreased inherent safety as there will be less incentive to eliminate hazardous materials, given that the inventories are minimized; a potential for reaction runaway if heat transfer is not sufficient, given a reduction of solvent used; a potential for

Table 15.5 Potential Sustainability Advantages and Disadvantages of Process Intensification

Area	Potential Sustainability Advantages	Potential Sustainability Disadvantages
Improved mass transfer	• Reduction of by-product formation and thus a reduction in waste • Increased throughput • Faster reactions: equipment tailored to intrinsic kinetic rate • Possibility of higher selectivity, mass efficiency, yield, and quality	• Potential for reaction runaway if heat transfer is not sufficient
Improved energy transfer	• Reduction of hot spots, improved control of temperature distribution • Reduction of energy requirements and related life cycle impacts • Potential reduction in secondary reactions	• Potential clogging issues if precipitates form because of rapid cooling
Reduced size	• Reduced inventories of potentially hazardous materials • Reduction or elimination of scale-up issues • Might not need frequent cleaning if equipment is dedicated	• Potential for blockages during operation • Less incentive to eliminate hazardous material • Potentially more difficult to clean and maintain

Table 15.6 Examples of Process Intensification Areas. Both Techniques and Technologies Are Intimately Interrelated, but Presented Separately for Easy of Discussion

Area	Examples of Process Intensification
Techniques	Hybrid systems • Hybrid reactors: reverse flow, reactive distillation, reactive extraction, reactive crystallization, chromatographic reactions, membrane reactions, fuel cells • Hybrid separators: membrane absorbers, adsorptive distillation, membrane distillation Other: supercritical fluids, dynamic reactors, alternative energy (e.g., solar, plasma)
Technologies	• Reactors: spinning disk reactor, static mixer reactor, microreactors, heat exchange reactors, monolithic reactors, oscillatory flow reactors, trickle bed reactors • Separators: centrifugal absorbers, rotating packing beds • Mixing and processing: intensive granulation, static mixers, vortex mixers, rotor mixers

increased blockages during the operation, given reduced size; and having equipment that may be more difficult to clean and maintain given the smaller size and increased specialization.

Process intensification uses only engineering principles to achieve the desired results, which means that the changes needed to intensify a process will be in "how" the chemistry performs, not in "what" of the chemistry itself. Process intensification techniques use different ways of processing to intensify production, such as the use of multifunctional reactors (e.g., reaction plus separation in one reactor), hybrid separations (e.g., membrane absorption and stripping), and use of alternative forms of energy (e.g., centrifugal force instead of gravity for phase separations). Process intensification technology includes microreactors, static mixers, and compact heat exchanger reactors, among others. Table 15.6 shows some examples of process intensification areas.

Although it is true that process intensification may lead to smaller, cleaner, safer, more energy-efficient chemical plants, and therefore has significant sustainability advantages, the main drive for process intensification has been the need for innovation and the economic and competitive advantages that innovation brings to a company.

HAZARD ASSESSMENTS AND INHERENT SAFETY

Although the previous section highlighted significantly the optimization of resources, the importance of minimizing health and safety risks has been largely integrated into the work of developing chemical processes. The importance of preventing accidents and other health and safety hazards has brought attention to the development of methodologies, processes, and regulations to eliminate or minimize the potential for such events. To identify hazards related to a chemical process, a process hazard analysis can be conducted [76]. This is a thorough, orderly, systematic approach to identifying, evaluating, and controlling the hazards of processes. In other words it is a careful review of what could go wrong in a process and what types of controls must be implemented to prevent accidents. Hazard assessments and inherent safety reviews are imperative when designing a new process, but are also part of standard reviews of existing processes, particularly as equipment ages and when changes or retrofitting are needed. There are several methodologies for hazard identification and analysis [77,78], some examples are:

- Checklists. These have a varied degree of detail and are often used to verify compliance with standards. The output of this analysis is mainly only as good as the checklist it is based on, and the use of checklists is frequently combined with other hazard analysis methods.
- What–if. This is an approach whereby a group of experienced people who are very familiar with a system brainstorm about its potential undesired effects. This technique is not structured by nature, but it is used widely to identify hazards and undesired effects.
- Hazard and operability study. This is a structured methodology to investigate each element of a system systematically for all the potential deviations that can cause hazards and operability issues. The effects caused by deviations from design conditions cases are examined for each key process parameter (e.g., flow, volume, pressure, temperature) using phrases such as "more of," "less of," "none of," to describe potential deviations and causes of failure. Finally, an assessment is made weighing the consequences, causes, and protection requirements.
- Failure mode and effects analysis. This is a methodical study of component failures. It normally starts with a process diagram showing all the components that could fail. The components are listed and analyzed individually to document potential causes of failure (e.g., leak), the consequences of the failure (e.g., fire, explosion), the severity (high, moderate, and low), the probability of the failure, the detection modes, and the mitigation measures.
- Fault tree analysis. A fault tree analysis is a quantitative assessment of all the undesirable outcomes, such as a toxic gas release or explosion, that could result from a specific initiating event. It begins with a graphic representation (using logic symbols) of all possible sequences of events that could result in an incident. Probabilities are assigned to each event using failure rate data and then used to calculate the probability of occurrence of the undesired event.

However, many of these considerations have been made by the use of controls after a process is designed (e.g., globe boxes for dispensing highly toxic materials, explosion suppression systems). The sustainability challenge is to ensure that the health and safety considerations are, as much as possible, inherent to the process design as a built-in feature, not as a bolt-on control. Chemical engineers and process systems engineers have the mission to eliminate the need for materials of concern, designing systems where the remaining materials are contained by design, and designing inherently safer processes through improved heat and mass transfer.

The safety of a process can be achieved by inherent (internal) and external means. Inherent safety focuses on the intrinsic properties of a process and attempts to "design out" hazards rather than trying to control hazards through the application of external protective systems. Inherently safer processes rely on chemistry and physics (properties of materials, quantity of hazardous materials) instead of control systems (interlocks, alarms, procedures) to protect workers, property, and the environment. It would be inappropriate to talk about an inherently safe process, as an absolute definition of safe is difficult to achieve in this context since risk cannot be reduced to zero. However, one can talk about a process or chemical being inherently safer than other(s). For instance, water can be an extremely hazardous chemical under certain conditions (e.g., floods), but in the context of a chemical process, water is an inherently safer solvent than other chemicals. Trevor Kletz has postulated some basic principles of inherent safety [79,80] that process systems engineers can follow when designing or retrofitting chemical processes. Kletz's inherent safety principles can be summarized as follows:

- Intensification. Reduce or eliminate hazardous materials or hazardous conditions.
- Substitution. Replace hazardous materials or conditions with safer ones.
- Attenuation or moderation. Use a hazardous material under less hazardous conditions.
- Simplification. Design plants as simply as possible so they provide fewer opportunities for error and less equipment that can fail.
- Limitation of effects. Change the design or the reaction conditions to reduce the severity of adverse effects.
- Avoiding knock-off effects. Design to avoid domino effects if adverse events do occur.
- Preventing incorrect assembly. Design equipment so that improper construction or reassembly is impossible or very difficult.
- Making status clear. Design and use equipment that indicates clearly if it is on, off, open, shut, or assembled correctly or incorrectly.
- Error tolerance. Design and use more robust equipment that endures poor operation or poor conditions without failing.
- Ease of control. When possible, control by physical principles instead of add-on systems.
- Computer control. Design simple, user-friendly, thoroughly tested software by people who understand the process.
- Instructions. Write clear and short instructions.

- Life cycle thinking. Consider construction and demolition phases as well as the operation phase.
- Passive safety. If hazards cannot be avoided and there is a need to add safety measures, use passive protective measures instead of active or procedural measures.

Inherent safety principles applied during the design stage of a process provide a larger return on investment than when the process has been operating for some time, as it is more difficult and costlier to retrofit inherent safety concepts (i.e., elimination of a hazardous material). For instance, the process used by the former Union Carbide in Bhopal, India, to make the pesticide carbaryl had a contamination of the methyl iso-cyanate with water that cost the lives of 2000 people, caused many injuries, and contributed heavily to the company going out of business. An inherently safer route has been developed and is now available, which would have avoided all these costs if it had been implemented at the design stage.

In the earliest phases of design, process systems engineers have greater freedom to consider changes in basic process chemistry or technology that can result in an inherently safer plant. Although these principles seem to be primarily common sense, they have not been put into practice in the industrial world as widely as have engineering or procedural measures [81]. This is probably related to the fact that designing processes that are inherently safer require both a fundamental change in the way chemists and engineers approach process development and very close collaboration among disciplines.

IMPACTS THROUGHOUT THE SUPPLY CHAIN—LIFE CYCLE THINKING

The true measure of whether a new or retrofitted process is more sustainable needs to encompass the assessment of the impacts of the process throughout its entire supply chain and life cycle. There are many examples of changes in technology or chemistry that on the surface appear to be an improvement of the sustainability performance when looking at a single stage of a process. However, when expanding the boundaries beyond what is done in the plant, one might discover that they are merely transferring impacts from one place to another instead of solving the problem. One of the typical examples of this is removing a contaminant from a waste stream using a solvent. On the surface we have prevented pollution, but when we account for the impacts related to generating the solvent and disposing of the pollutant-rich stream, we have to ask if that is really an overall improvement.

Life cycle inventory and assessment is a methodology that allows one to more pre-cisely estimate the cumulative environmental impacts associated with manufacturing all the chemicals, materials, and equipment used to make a product or deliver a service across the supply chain; thereby providing a comprehensive view of the potential tradeoffs in environmental impacts (For more details, see chapters 13 and 14). Life cycle assessment is the framework that would allow us to address the wider aspects of sustainability, since the design of new processes and products needs to account for the impacts that extend beyond the factory boundaries.

The minimization of life cycle impacts has become one of the areas of work for process systems engineers. Ten years ago the terms "carbon footprinting" or "eco-footprinting" were relegated to academic settings and now they are mostly mainstream. Minimizing life cycle impacts is closely related to resource efficiency and health and safety aspects. For instance, a chemical process with enhanced mass and energy efficiency will have a reduced environmental footprint given the reduction of resource consumption (i.e., mass and energy) and thus the reduction of waste. Using life cycle assessment (or at the very least life cycle thinking) in the design stage allows process systems engineers to take into consideration the entire supply chain and life cycle of a product or process. For instance, the following questions can be addressed through this:

- Are the raw materials produced from renewable sources? Do they need to be?
- How would the product or waste be disposed of?
- Can packaging be minimized?
- Do we use any materials in the supply chain that can cause major impacts/accidents?
- Can we design the product in a way that it would degrade after its useful life?
- What are the procurement impacts on this material decision?

There have been some efforts aimed at integrating the life cycle considerations within the design of process and products. Examples of this include GlaxoSmithKline's technology selection frameworks [82], incorporating life cycle assessment metrics for process retrofitting [83], incorporating streamlined life cycle assessment in tools and guides used in material selection and process development [29], the use of life cycle thinking in Johnson & Johnson Earthwards [84], and incorporating life cycle assessment in supply chain optimization [85], among others. For instance, Fig. 15.6 shows the output of the streamlined life cycle assessment tool developed by the American Chemical Society Green Chemistry Institute Pharmaceutical Roundtable to incorporate life cycle considerations into process design [86]. However, there is much work that needs to be done in this regard, particularly as related to the integration of life cycle thinking as a standard component of process and product design. Process systems engineers have the necessary skills to integrate this type of assessment seamlessly into the typical tools of process and product design.

MODELING AND COMPUTER-AIDED TOOLS

Many of the tools and techniques described so far are underpinned by the ability to perform fast computational simulations and modeling. This includes both process simulations and modeling, product design, and enterprise-wide modeling and simulations.

PMI/LCA Summary Table				
Step Name/Number	1	2	3	TOTALS
Mass Substrate (kg)	0.00	1.00	1.50	2.50
Mass Reagents (kg)	2.00	1.53	0.50	4.03
Mass Solvents (kg)	109.30	27.47	30.00	166.77
Mass Aqueous (kg)	0.00	10.00	0.00	10.00
Step PMI	111.30	44.44	35.56	
Step PMI Substrate, Reagents, Solvents	111.30	33.33	35.56	
Step PMI Substrates and Reagents	2.00	2.81	2.22	
Step PMI Solvents	109.30	30.52	33.33	
Step PMI Water	0.00	11.11	0.00	N/A
Cumulative PMI	111.30	167.00	312.22	
Cumulative PMI Substrate, Reagents, Solvents	111.30	155.89	293.70	
Cumulative PMI Substrates and Reagents	2.00	3.92	7.09	
Cumulative PMI Solvents	109.30	151.97	286.61	
Cumulative PMI Water	0.00	11.11	18.52	
Mass Net [kg]	63.46	55.39	62.28	181.14
Energy [MJ]	66.93	156.90	38.59	262.41
POCP [kg of ethylene equivalents]	0.04	0.06	0.24	0.34
GWP [kg CO2 equivalents]	86.59	76.91	91.04	254.55
Acidification Potential [kg SO2 equivalents]	0.19	0.41	0.28	0.87
Eutrophication Potential [kg phosphate equivalents]	0.05	0.09	0.10	0.24
TOC [kg]	0.07	0.39	0.24	0.70
Oil/Gas Depletion [kg]	98.32	46.05	59.21	203.57
Water [kg]	251.14	304.43	316.75	872.32

NOTE: DO NOT MODIFY THIS TABLE. All cells have been set to summarize the information in the individual templates. As each template is completed, the appropriate information will be displayed. If major modifications are made to a given template below, it is recommended that the embedded cell references are still valid. THIS TABLE IS ONLY VALID FOR THIS STRING OF 11 TEMPLATES. Additional steps or convergent synthetic pathways will not be accurately depicted in this table.

Figure 15.6 Example output of the process mass intensity and life cycle assessment tool developed by the American Chemical Society Green Chemistry Institute Pharmaceutical Roundtable. The data presented in the columns provide the metrics for steps 1, 2, 3, and the total of the synthesis. Some of the instructions in the tool are included as an illustration.

Here, process systems engineers work in the interfaces of first principle sciences, information technology, statistics, biology, and operation research.

The simulations and models can be as simple as management of databases performing some statistical analysis, or as complicated as a full supply chain simulation.

With the increasing development of computing environments and speed, it will be possible to accommodate more flexible modeling environments that would run the gambit from simple algebraic and statistical systems (e.g., a principal component analysis for material selection), to the more complex world of partial differential equations for modeling dynamic systems (e.g., supply chain modeling with dynamic demand across 20 markets). Process systems engineers need to be able to manage this range, using pure mathematical equations, numerical analysis, and mixed algorithms. CAMD for property prediction [87], integrated process design and control [88], uncertainty impacts in process design [89], material selection in formulations [90], computer-aided solvent screening for biocatalysis [91], computer-aided product design [92], integration of process synthesis and intensification [93], enterprise-wide modeling [94] and enterprise-wide optimization [95], and life cycle assessment of products [96] are a few of the examples in which process systems engineers are making inroads in the areas of modeling and computer-aided design.

However, there are many additional challenges and gaps that process systems engineers will need to develop and fill, working closely with information technology, mathematics, biology, and operation research experts. These challenges include large-scale differential methods for multiple-scale simulations, uncertainty and sensitivity modeling, mixed integer and logic modeling that may mix quantitative and qualitative variables, modeling of heuristics systems, improved optimization methods for nonlinear variables, and dynamic systems for planning and schedule modeling, to mention a few.

The capability of process systems engineers to leverage computer-aided techniques and modeling will be intimately linked to the extent the previous tools and fields can be leveraged across the enterprise.

FUTURE OUTLOOK

Process systems engineers have a pivotal role in embedding sustainability into new and existing processes and products. To this extent it is necessary to leverage the existing skill set of chemical and process systems engineers with an integrated sustainability mindset.

It all starts with innovation.

Arguably, engineers and scientists belong to the most creative disciplines, and the design of cleaner, safer processes and products can be seen as another very intriguing creative challenge. Thus there is a series of key areas that process systems engineers need to integrate actively to design more sustainable processes. These areas have different levels of development and thus different levels of innovation would be needed. Some of these areas are by no means new, such as process intensification, energy integration, and hazard assessment; but these good old friends may need to be adapted and adopted to become an integral part of process design. Some of these areas have not been explored in much detail by process systems engineers, such as life cycle assessment, process intensification, and material selection, so they will require implementation of different approaches on design, scale-up, and operation of manufacturing processes. Some areas are well developed in the chemical space, but less so in other emerging areas, such as bioprocessing, electronics, and nanotechnology. In addition, there are some areas that need to be addressed in an interrelated or combined fashion, such as considering an intensified process that uses hybrid reaction/separation unit operations to enhance mass and energy transfer.

There are already many examples of advances of PSE embedding sustainability aspects as part of the process development cycle. Continuous microreactors from Corning utilized for fine chemicals and pharmaceuticals [97], continuous formulation processes, process intensification demonstrated in several processes [98], and life cycle assessment evaluations integrated into process development [99], to mention a few. However, a more widespread uptake is needed, so they are used routinely.

The challenge for process systems engineers will be to continue to advance the integration of PSE tools and techniques with the objectives of optimizing resources, minimizing hazards, and designing systems using life cycle thinking. This will need process systems engineers to continue fostering their engineering skills and develop other skills that may not be identified yet. Some of the skills and practices that will need development or enhanced adaptation are:

- Mastery of the scientific process understanding to leverage opportunities for intensification and optimization.
- Better understanding of enzymatic, fermentation, and other bioprocesses. This would need to include fundamental design parameters to design more resource efficient bioprocesses.
- More extensive use of process integration techniques, especially at the development phase and with bioprocesses, where this approach has not been widely used.
- More routine application of multiobjective optimization techniques for process design, product design, sustainability assessments, and supply chain modeling
- Development of quantitative decision-making tools and rapid simulations that would integrate both process design and sustainability principles.
- Development of better and more sophisticated computer-aided tools, including property prediction packages for chemical and biochemical processes.
- Increased understanding of the uncertainties in modeling processes, from the viewpoint of both process design and sustainability assessment.
- Better techniques for enterprise-wide modeling to optimize the supply chain through planning, scheduling, real-time optimization, and inventory control.

- Understanding of life cycle assessment and the development of improved consistency and transparency of streamlined and easy-to-use life cycle assessment methodologies.
- Ability to collaborate closely with other disciplines such as information technology, biology, operations research, mathematicians, and statisticians, to mention a few.

Addressing these skills and process understanding needs will make it possible to integrate sustainability principles into process design and development in a far more rigorous manner. In pursuing this approach to process development and manufacturing, process systems engineers may be in a better position to deliver on the promise of cleaner, safer, more efficient manufacturing processes.

ACKNOWLEDGMENTS

Much appreciation goes to the many process systems engineers who are currently designing processes and products that are more sustainable. Many thanks particularly go to Angus Freeman, Steve Makin, and other leaders in GlaxoSmithKline who continue to support the development of strong technical leaders in the pharmaceutical and consumer health care industries.

REFERENCES

[1] Grossman I. Research challenges in process systems engineering. AIChE J 2000;46(9):1700−3.
[2] Jimenez-Gonzalez C, Ponder CS, Hannah RE, Hagan JH. Green engineering in the pharmaceutical industry. In: Zhang W, Cue BW, editors. Green techniques for organic synthesis and medicinal chemistry. UK: John Wiley & Sons; 2012. Chichester.
[3] Jiménez-González C, Poechlauer P, Broxterman QB, Yang BS, am Ende D, Baird J, et al. Key green engineering research areas for sustainable manufacturing —a perspective from pharmaceutical and fine chemicals manufacturers. OPRD 2011;15(4):900−11.
[4] European Commission. Council Directive 1999/13/EC. March 11, 1999.
[5] Toxic Chemical Release Reporting: Community Right-to-Know. 40 CFR part 372.
[6] Anastas P, Warner J. Green chemistry: theory and practice. New York: OUP USA; December 1998.
[7] Abraham MA, Nguyen N. Green engineering: designing the principles — results from the SanDestin conference. Environ Prog 2003;22(4):233−6.
[8] Venkatasubramanian V, Chan K, Caruthers JM. Computer-aided molecular design using genetic algorithms. Comput Chem Eng 1994;18(9):833−44.
[9] Harper PM, Gani R, Kolar P, Ishikawa T. Computer aided molecular design with combined molecular modelling and group contribution. Fluid Phase Equilibria 1999;158(160):337−47.
[10] Gani R, Jiménez-González C, Ten Kate A, Crafts PA, Jones M, Powell L, et al. A modern approach to solvent selection. Chem Eng 2006;113(3):30−43.
[11] Samudra AP, Sahinidis NV. Optimization-based framework for computer-aided molecular design. AIChE J 2013;59(10):3686−701.
[12] Jimenez-Gonzalez C, Constable DJC. Green chemistry and engineering, a practical design approach. 1st ed. Wiley and Sons; 2011.
[13] CAPEC database, http://www.capec.kt.dtu.dk/Software/ICAS-and-its-Tools/ [last accessed 16.11.14].
[14] ChemFinder, http://www.cambridgesoft.com/databases/login/?serviceid=128 [last accessed 16.11.14].
[15] DECHEMA data series, http://www.dechema.de/en/CDS.html [last accessed 16.11.14].
[16] Aspen Plus, http://www.aspentech.com/products/aspen-plus.aspx [last accessed 16.11.14].
[17] ACS GCI Pharmaceutical roundtable reagent guides — version for members, http://www.acs.org/content/acs/en/greenchemistry/industry-business/pharmaceutical.html [last accessed 16.05.15].
[18] Chakraborti AK, Sharma L, Nayak MK. Demand-based thiolate anion generation under virtually neutral conditions: influence of steric and electronic factors on chemo- and regioselective cleavage of aryl alkyl ethers. J Org Chem 2002;67:6406−14.

[19] Perosa A, Tundo P, Zinovyevb S. Mild catalytic multiphase hydrogenolysis of benzyl ethers. Green Chem 2002;4:492–4.

[20] Sergeev AG, Hartwig JF, Selective. Nickel-catalyzed hydrogenolysis of aryl ethers. Science 2011;332(6028):439–43.

[21] Dunn PJ, Hughes ML, Searle PM, Wood AS. The chemical development and scale-up of sampatrilat. Org Proc Res Dev 2003;7(3):244–53.

[22] Yang H, White PS, Brookhart M. Scope and mechanism of the iridium-catalyzed cleavage of alkyl ethers with triethylsilane. J Am Chem Soc 2008;130:17509–18.

[23] Hwu JR, Wong FF, Huang JJ, Tsay SC. Sodium bis(trimethylsilyl)amide and lithium diisopropylamide in deprotection of alkyl aryl ethers: α-effect of silicon. J Org Chem 1997;62(12):4097–104.

[24] Majetich G, Zhang Y, Wheless K. Hydride-promoted demethylation of methyl phenyl ethers. Tetrahedron Lett 1994;35(47):8727–30.

[25] Alfonsi K, Colberg J, Dunn PJ, Fevig T, Jennings S, Johnson TA, et al. Green Chem 2008;10:31–6.

[26] Adams JP, Alder CM, Andrews I, Bullion AM, Campbell-Crawford M, Darcy MG, et al. Green Chem 2013;15:1542–9.

[27] Council directive 67/548/EEC of 27 June 1967 on the approximation of laws, regulations and administrative provisions relating to the classification, packaging and labeling of dangerous substances. Official J 16/08/1967;196:1–98.

[28] Curzons A, Jiménez-González C, Duncan A, Constable D, Cunningham V. Fast life-cycle assessment of synthetic chemistry tool, FLASC™ tool. Int J LCA 2007;12(4):272–80.

[29] Walmart Sustainability Index, http://corporate.walmart.com/global-responsibility/environment-sustainability/sustainability-index [last accessed 16.11.14].

[30] US EPA, http://www.epa.gov/dfe/saferingredients.htm [last accessed 16.11.14].

[31] Humphrey J, Keller G. Separation process technology. New York: McGraw-Hill; 1997.

[32] Purdue School of Chemical Engineering. Separations Research group website, https://engineering.purdue.edu/Separations/[last accessed 17.05.15].

[33] Humphrey JL, Fair JR. In: Proceedings from the fifth industrial energy technology conference volume II, Houston, TX; April 17–20, 1983.

[34] Neelis M, Worrell E, Masanet E. Energy efficiency improvement and cost saving opportunities for the petrochemical industry. LBNL-964E. Berkeley, CA: Berkelye National Laboratory; June 2008.

[35] US Department of Energy. Energy intensive processes portfolio: addressing the key energy challenges across the US industry. DOE/EE-0389. Industrial Technology Program; March 2011.

[36] Rudd DF, Powers GJ, Siirola JJ. Process synthesis. Englewood Cliffs, NJ: Prentice Hall; 1973.

[37] Douglas JM. Conceptual design of chemical processes. New York: McGraw-Hill; 1988.

[38] Johns WR. Process synthesis: poised for a wider role. Chem Eng Prog 2001;97(4):59–65.

[39] Carvalho A, Gani R, Matos H. Systematic methodology for process analysis and generation of sustainable alternatives. In: Braunschweig B, Joulia X, editors. 18th European Symposium on Computer Aided Process Engineering (ESCAPE 18). Amsterdam: Elsevier; 2008.

[40] Woods DR. Rules of thumb in engineering practice. Weinheim, Germany: Wiley-VCH; 2007.

[41] Walas S. Chemical process equipment: selection and design. Oxford, UK: Butterworth-Heinemann; 2002.

[42] Branan C. Rules of thumb for chemical engineers. Houston, TX: Gulf Professional Publishing; 2005.

[43] Linnhoff B, Townsend DW, Boland D, Hewitt GF, Thomas BEA, Guy AR, et al. A user guide on process integration for the efficient use of energy. Rugby, UK: Institute of Chemical Engineers; 1982.

[44] Douglas JM. Process synthesis for waste minimization. Ind Eng Chem Res 1992;31:238–43.

[45] Grossmann IE. A modelling decomposition strategy for MINLP optimisation of process flowsheets. Comput Chem Eng 1989;13:797.

[46] Severson K, Martín M, Grossmann IE. Optimal integration for biodiesel production using bioethanol. AIChE J 2013;59(3):834–44.

[47] Baliban RC, Elia JA, Misener R, Floudas CA. Global optimization of a MINLP process synthesis model for thermochemical based conversion of hybrid coal, biomass, and natural gas to liquid fuels. Comput Chem Eng 2012;42:64–86.

[48] Mallapragada DS, Tawarmalani M, Agrawal R. Synthesis of augmented biofuel processes using solar energy. AIChE J 2014;60(7):2533–45.

[49] Mavrovouniotis ML, editor. Artificial intelligence in process engineering. San Diego, CA: Academic Press Inc.; 1990. 367 pp.

[50] Cano-Ruiz JA, McRae GJ. Environmentally conscious chemical process design. Ann Rev Energy Environ 1998;23:499—536.

[51] Stephanopoulos G, Reklaitis GV. Process systems engineering: from Solvay to modern bio- and nanotechnology: a history of development, successes and prospects for the future. Chem Eng Sci 2011;66(19):4272—306.

[52] Linnoff B. Pinch analysis: a state of the art overview. Chem Eng Res Des, Trans IChemE, A 1993;71:503—22.

[53] Papoulias S, Grossmann IE. A structural optimization approach in process synthesis—II: heat recovery networks. Comput Chem Eng 1983;7(6):707—21.

[54] Zhu XX, O'Neill BK, Roach JRC, Wood RM. A method for automated heat exchanger network synthesis using block decomposition and non-linear optimization. Trans IChemE 1995;73(8):919—30.

[55] Linnoff B, Hindmarsh E. The pinch design method for heat exchange networks. Chem Eng Sci 1983;38(5):745—63.

[56] El-Halwagi MM. Pollution prevention trough process integration: systematic design tools. San Diego, CA: Academic Press; 1997.

[57] Shah P, Kokossis A. Systematic optimization technology for high-level screening and scoping of complex distillation systems. Comput Chem Eng 1999;23(Suppl.):S137—42.

[58] El-Halwagi MM, Hamad AA, Garrison GW. Synthesis of waste interception and allocation networks. AIChE J 1996;42(11):3087—101.

[59] Srinivas BK, El-Halwagi MM. Optimal design of pervaporation systems for waste reduction. Comput Chem Eng 1993;17(10):957—70.

[60] Foo CY, Manan ZA, Yunus RM, Aziz RA. Synthesis of mass exchange network for batch processes: I. Utility targeting. Chem Eng Sci 2004;59:1009—26.

[61] El-Halwagi MM, Srinivas BK. Synthesis of reactive mass exchange networks. Chem Eng Sci 1992;47(8):2113—9.

[62] Srinivas BK, El-Halwagi MM. Synthesis of reactive mass exchange networks with general non-linear equilibrium functions. AIChE J 1994;40(3):463—72.

[63] El-Halwagi MM, Srinivas BK, Dunn RF. Synthesis of optimal heat-induced separation networks. Chem Eng Sci 1995;50(1):81—97.

[64] Dye SR, Berry DA, Ng KM. Synthesis of crystallisation-based separation scheme. AIChE Symp Ser 1995;91(304):238—41.

[65] Wang YP, Smith R. Wastewater minimisation. Chem Eng Sci 1994;49:981—1006.

[66] Hallale N. A new graphical targeting method for water minimisation. Adv Environ Res 2002;6(3): 377—90.

[67] Wang YP, Smith R. Wastewater minimisation. Chem Eng Sci 1994;49:981—1006.

[68] Shah VH, Agrawal R. A matrix method for multicomponent distillation sequences. AIChE J 2010;56(7):1759—75.

[69] Henao CA, Maravelias CT. Surrogate-based superstructure optimization framework. AIChE J 2011;57(5):1216—32.

[70] Tawarmalani M, Sahinidis NV. Convexification and global optimization in continuous and mixed-integer nonlinear programming: theory, algorithms, software, and applications, vol. 65. Springer Science & Business Media; 2002.

[71] Biegler LT. Nonlinear programming: concepts, algorithms, and applications to chemical processes. (MPS-SIAM Series on Optimization, book 10). SIAM-Society for Industrial and Applied Mathematics; September 2010.

[72] El-Halwagi MM, Manousiouthakis V. Synthesis of mass exchange networks. AIChE J 1989;35(8):1233—44.

[73] Duran MA, Grossmann EI. Simultaneous optimization and heat integration of chemical processes. AIChE J 1986;32(1):123—38.

[74] Ramshaw C. The incentive of process intensification. In: Proceedings of the first international conference on process, intensification for the chemical industry. London: BHR Group; 1995.

[75] Stankeiwicz AI, Moulijn AJ. Process intensification: transforming chemical engineering. Chem Eng Prog 2000;96(1):22—34.

[76] US Occupational Safety and Health Administration. Process safety management. OSHA 3121. Washington, DC: OSHA; 2000 [reprint].

[77] American Institute of Chemical Engineers. Center for chemical process safety. Guidelines for hazard evaluation procedures. 2nd ed. New York: AIChE; 1992.

[78] Venkatasubramanian V, Vaidhyanathan R. A knowledge-based framework for automating HAZOP analysis. AIChE J 1994;40(3):496—505.

[79] Kletz TA. Inherently safer plants. Plant Oper Prog 1985;4(3):164—7.

[80] Kletz TA. Process plants: a handbook of inherently safer design. Boca Raton, FL: CRC Press; 1998.

[81] Amyotte PR, Khan FI, Dastidar AG. Solids handling: reduce dust explosions the inherently safer way. Chem Eng Prog Oct. 2003:36—43.

[82] Jiménez-González C, Constable DJC, Curzons AD, Cunningham VL. Developing GSK's green technology guidance: methodology for case-scenario comparison of technologies. Clean Technol Environ Policy 2002;4:44—53.

[83] Carvalho A, Gani R, Matos H. Design of sustainable chemical processes: systematic retrofit analysis generation and evaluation of alternatives. Process Saf Environ Prot 2008;33(12):2075—90.

[84] Earthwards®life cycle impact areas. Johnson & Johnson website, http://www.jnj.com/caring/citizenship-sustainability/strategic-framework/lifecycle-impact-areas [last accessed 21.11.14].

[85] Fengqi Y, Tao L, Graziano DJ, Snyder SW. Optimal design of sustainable cellulosic biofuel supply chains: multiobjective optimization coupled with life cycle assessment and input—output analysis. AIChE J 2012;58(4):1157—80.

[86] Jiménez-González C, Ollech C, Pyrz W, Hughes D, Broxterman QB, Bhathela N. Expanding the boundaries: developing a streamlined tool for eco-footprinting of pharmaceuticals. OPRD 2013;17:239—46.

[87] CAPEC, http://www.capec.kt.dtu.dk/[last accessed 25.11.14].

[88] Mansouri SS, Huusom JK, Woodley J, Gani R. Systematic process design and operation of intensified processes. In: Presentation in the AIChE annual meeting, Atlanta Georgia; 2014.

[89] Cheali PA, Quaglia KV, Gernaey GS. Effect of market price uncertainties on the design of optimal biorefinery systems—a systematic approach. Ind Eng Chem Res 2014;53:6021—32.

[90] Mattei M, Kontogeorgis GM, Gani R. A comprehensive framework for surfactant selection and design for emulsion based chemical product design. Fluid Phase Equilibria 2014;362:288—99.

[91] Abildskov J, van Leeuwen MB, Boeriu CG, van den Broek LAM. Computer-aided solvent screening for biocatalysis. J Mol Cat B Enzym 2014;85(86):200—13.

[92] Yunus NAB, Gernaey KV, Woodley JM, Gani R. A systematic methodology for design of tailor-made blended products [ESCAPE-23 special issue] Comput Chem Eng 2014;66(4):201—13.

[93] Lutze P, Babi DK, Woodley JM, Gani R. Phenomena based methodology for process synthesis incorporating process intensification. Ind Eng Chem Res 2013;52:7127—44.

[94] Grossmann I. Enterprise-wide optimization: a new frontier in process systems engineering. AIChE J 2005;51(7):1846—57.

[95] Quaglia A, Sarup B, Sin G, Gani R. A systematic framework for enterprise-wide optimization: synthesis and design of processing network under uncertainty. Comput Chem Eng 2013;59:47—62.

[96] P&G Environmental Brochure, 2008, http://www.pg.com/en_US/downloads/sustainability/pov/EnvBrochure2008.pdf [last accessed 17.05.15].

[97] Sutherland J. Efficient processing with corning® advance flow™ glass technology. In: Presentation during the 13th annual green chemistry and engineering conference. MD: College Park; June 23—25, 2009.

[98] Poechlauer P, Braune S, Reintjens R. Continuous processes in small-scaled reactors under cGMP conditions: towards efficient pharmaceutical synthesis. In: Presentation during the 13th annual green chemistry and engineering conference. MD: College Park; June 23—25, 2009.

[99] Gebhard R. Sustainable production of pharmaceutical intermediates and APIs—the challenge for the next decade. DSM Webminar; 2009.

INDEX

'Note: Page numbers followed by "f" indicate figures and "t" indicate tables.'

A

ABB algorithm. *See* Accelerated branch and bound algorithm (ABB algorithm)
Absorption, 44—47
Accelerated branch and bound algorithm (ABB algorithm), 205—206, 210—211
  software implementations, 211
  in software PNS Studio, 218—219
Adsorption, 44—47
Advanced control approach, 120—122, 121f
Advancing Integrated Systems Modeling Framework for Life Cycle Sustainability Assessment (AISMF LCSA), 328—329
AIChE sustainability metrics, 228
Air pollution control district (APCD), 79—80
Ambiguity, 170
2-Aminochromes, 28
Ammonia (NH_3), 277—278
  chemical compounds producing from, 278t
  downstream products derived from, 279f
Ammonification, 277
Ants' rule of pursuit, 120—121, 120f
APCD. *See* Air pollution control district (APCD)
APEA. *See* Aspen process economic analyzer (APEA)
Arbitrary weights, 151
Artificial intelligence, 362
Aspen process economic analyzer (APEA), 156—158
"At-a-glance" classification, 358
Augmented property (AUP), 94
Average metrics, 170, 179

B

Balsas watershed system, 106—107, 107f
Base-case designs, 68, 70
BASF, 228
Batch chemical processing, 15
Batch MENs, 91
bcf. *See* Billion cubic feet (bcf)

Benchmarking
  cogeneration targets, 97, 98f
  tools, 88—89
Billion cubic feet (bcf), 276—277
Biocatalysis implications, 26—27
Bioethanol process
  corn dry grind *vs.* corn wet milling, 151—154
  economic performance, 158t
  economic sustainability evaluation, 157f
  parameters for economic evaluations and prices, 158t
  process modeling, 154—156
  sustainability indicators, 156—163
  values of social indicator, 161t
Bioethanol production system, fermentation for, 127—134
  closed-loop simulation, 129f, 131f
  open-loop simulations, 128f—129f, 131f, 133f
  Radar plot with GREENSCOPE indicators, 130f, 132f—133f, 135f
Biogeochemical cycles, 275—276
Biomass, 309—310
  biomass-based electricity systems, 334—335
  biomass-derived compounds, 291t—292t
Brundtland Commission, 67—68
Bulk stream, 45

C

CALCAS. *See* Co-ordination Action for innovation in Life-Cycle Analysis for Sustainability (CALCAS)
Calculus of variation, 196
CAMD. *See* Computer-aided molecular design (CAMD)
Carbon
  chemical industry profile for
    for carbon emissions, 286—287
    for carbon sequestration, 283—285, 284f
    chemical sectors with NAICS code, 285t
  cycle, 276—277
    flows in Eco-LCA inventory, 281t—282t

Carbon (*Continued*)
  emissions, 173, 286—287
  footprinting, 371
  LCA for chemical industry interaction with
      carbon and nitrogen cycles
    carbon cycle flows in Eco-LCA inventory,
        281t—282t
    direct and indirect impact/dependence,
        280
    Eco-LCA inventory, 280
    nitrogen cycle flows in Eco-LCA inventory,
        283t
  sequestration, 283—285, 284f
Carbon, hydrogen, oxygen symbiosis network
    (CHOSYN), 108, 109f—110f
Carbon dioxide (CO_2)
  capture, 54—56
  utilization, 303—304
Cash criterion, 143
Catalysis, 25—26
CEI calculation. *See* Chemical exposure index
    calculation (CEI calculation)
12-Cell ecological model, 182—183
CFB combustion technology. *See* Circulating
    fluidized bed combustion technology
    (CFB combustion technology)
Characterization factors, 71—72
Checklists, 368
Chemical engineers, 368—369
Chemical exposure index calculation (CEI
    calculation), 148—149
Chemical feedstock implications, 18—19
Chemical industry
  and biogeochemical cycles, 275—276
  biomass-derived compounds, 291t—292t
  and carbon cycle, 276—277
  LCA for interaction with carbon and nitrogen
      cycles, 279—280
    carbon cycle flows in Eco-LCA inventory,
        281t—282t
    direct and indirect impact/dependence, 280
    Eco-LCA inventory, 280
    nitrogen cycle flows in Eco-LCA inventory,
        283t
  and nitrogen cycle, 277—279
  profile for carbon
    for carbon emissions, 286—287
    for carbon sequestration, 283—285, 284f
    chemical sectors with NAICS code, 285t

  profile for nitrogen
    nitrogen emissions profile, 289—290
    nitrogen mobilization profile, 287—288
    nitrogen product profile, 288—289
  techno-ecological approach and sustainability,
      290—293
Chemical processes, sustainability approach for,
    70—72
Chemical reactivity, 12—13
Chemical—biochemical processes, 252, 258
Chemically mediated membrane processes,
    48—49, 49f
Chitin, 20, 24
Chlor-alkali production with human toxicity
    potential analysis, 72—81
  human toxicity potential analysis,
      75—81
    PEI profiles, 78f
    process emission and toxicity characterization
        factors, 76t
    process release profiles, 77f
  process economics analysis, 72—75
    input—output flowsheet structure, 73f
    recycle—reactor system structure flowsheet
        for, 74f
    separation system structure flowsheet, 75f
Chlorine gas (Cl_2 gas), 73
CHOSYN. *See* Carbon, hydrogen, oxygen
    symbiosis network (CHOSYN)
CHP. *See* Combined heat and power (CHP)
Circulating fluidized bed combustion technology
    (CFB combustion technology), 336
Cl_2 gas. *See* Chlorine gas (Cl_2 gas)
Cleaning and rinse operation optimization
    technology, 239—241
Co-ordination Action for innovation in
    Life-Cycle Analysis for Sustainability
    (CALCAS), 328—329
Coal-fired electricity, 333, 335
  prospective systems, 336—337
  reference system, 336—337
  technical characteristics, 338t
Coefficient of performance (COP), 42
Cogeneration, 97—98, 102f
Collaborative profitable pollution prevention
    (CP3), 228
Combined heat and power (CHP), 97
Combustion process, 336
Compensating surplus (CS), 144—145

Competitiveness evaluation of emerging
    technologies, 221–222
Composite sustainability indices, 233–234
Computer-aided molecular design (CAMD),
    306–307, 356
Conceptual chemical process design,
    67–70
  chlor-alkali production with human toxicity
      potential analysis, 72–81
  sustainability approach for chemical processes,
      70–72
Conditional-value-at-risk metrics (CVaR metrics),
    170, 178–179
Constraint vector, 171
Continuous extraction process, 44f
Control theory, 189–191
Controllability analysis, 189–195. See also
    Sustainable system dynamic models;
    Techno-socio-economic policies, optimal
    control for
  results of network and controllability analysis,
      195t
COP. See Coefficient of performance (COP)
Corn dry grind ethanol production process,
    151–154
  stages, 156t
Corn wet milling ethanol production process,
    151–154, 153f
  stages, 157t
Corn-to-ethanol processes
  component balance, 155t
  material balance, 155t
  reactions in, 156t
  utility consumption, 155t
Cost, 46
  function, 203–205
Coupling of process integration with MFA,
    105, 106f
CP3. See Collaborative profitable pollution
    prevention (CP3)
"Cradle-to-gate" approach, 298
"Cradle-to-grave" approach, 298,
    300–301
Crude oil, 276–277
CS. See Compensating surplus (CS)
CSRM. See Specific raw material cost (CSRM)
CVaR metrics. See Conditional-value-at-risk
    metrics (CVaR metrics)

D
DALY. See Disability-adjusted life year (DALY)
Danger characteristics. See Hazard class
DCFROR. See Discounted cash flow rate of
    return (DCFROR)
DDGS. See Distillers dried grains with solubles
    (DDGS)
Decision making, 150–151, 169, 179
  procedure, 295
  setting, 173
Decision Support System (DSS), 332
Definition of solid waste (DSW), 14
1-D metrics, 151
2-D metrics, 151
Denitrification, 278
Density operator, 93
Desalination, 53–54
Design-for-environment (DfE), 358–359
Destroyed exergy, 40–41
DfE. See Design-for-environment (DfE)
Diaphragm process, 72–73, 76–77, 79
  cell cost, 74
Direct impact/dependence, 280
Direct recycle, 88, 90
Disability-adjusted life year (DALY),
    332–333
Disagreement, 169, 171–172
  measured stakeholder, 170
Discounted cash flow rate of return (DCFROR),
    144, 156–158
Discounted payback period (DPBP), 143
Distillation, 41–44, 41f
Distillers dried grains with solubles (DDGS),
    151–154
DPBP. See Discounted payback period (DPBP)
DSS. See Decision support system (DSS)
DSW. See Definition of solid waste (DSW)

E
E-factor, 7
E-LCA. See Environmental life cycle assessment
    (E-LCA)
Eco-industrial park (EIP), 107–108, 108f
Eco-LCA. See Ecologically based life cycle
    analysis (Eco-LCA)
Eco-LCA inventory, 280
  carbon cycle flows in, 281t–282t
  nitrogen cycle flows in, 283t

"Ecofootprinting", 371

Ecologically based life cycle analysis (Eco-LCA), 278–279

ECON, 268

   economic analysis—ECON interface, 268f

Economic indicators, 126–127

Economic input—output LCA (EIO-LCA), 279–280

Economic pillar, 252–255

Economic potential (EP), 126

Economic sustainability, 71, 230–231. *See also* Environmental sustainability; Social sustainability

   analysis, 142–145

   indicators, 231t

Economic value of environmental quality, 144–145, 145t

Economical and ecological analysis, 217–219

   default measurement units in, 219f

   single optimal solution of the mathematical programming model, 221f

EF. *See* Emission Factor (EF)

Efficiency indicators, 123–125

EFRAT method. *See* Environmental fate and risk assessment tool method (EFRAT method)

EHS. *See* Environmental, health, and safety (EHS)

EIO-LCA. *See* Economic input—output LCA (EIO-LCA)

EIP. *See* Eco-industrial park (EIP)

Electricity cost, 173

Electroplating industry, 227

Electroplating process, 227, 238f

   case study, 237–246

      result of system sustainability assessment, 239t

      results of sustainability improvement, 246t

      selected sustainability metrics, 238t

      sustainability assessment of technologies, 243–244, 244t–245t

      technology candidate selection, 239–243

      technology recommendation, 244–246

   fundamentals for process sustainability, 229–230

   industry-centered supply chain, 230f

   P3 and CP3, 228

   sustainability assessment framework, 233–237

   sustainability metrics system, 230–232

Electroplating system sustainability (ESS), 229

ELR. *See* Environmental loading ratio (ELR)

Emergy, 147

Emergy sustainability index (ESI), 147

Emergy yield ratio (EYR), 147

Emission Factor (EF), 159t

End-of-pipe technologies, 297

Energy integration, 96–98, 364–365. *See also* Mass integration; Property integration

Energy producer (EP), 186–187

Energy source (ES), 186–187

Energy—comfort management in buildings, 175–179

Environmental, health, and safety (EHS), 14, 353

Environmental fate and risk assessment tool method (EFRAT method), 256

Environmental hazard, water hazard (EH$_{water}$), 125

Environmental indicators, 125–126

Environmental life cycle assessment (E-LCA), 327–328

Environmental loading ratio (ELR), 147

Environmental pillar, 255–258

Environmental quality level, 144–145

Environmental quotient (EQ), 125

Environmental sustainability, 28–29, 231. *See also* Economic sustainability; Social sustainability

   analysis, 145–148

   indicators, 232t

Environmentally conscious dynamic hoist scheduling technology, 243

EP. *See* Economic potential (EP); Energy producer (EP)

EPA. *See* US Environmental Protection Agency (US EPA)

EQ. *See* Environmental quotient (EQ)

Equilibrium distribution coefficient, 45

Equivalent surplus (ES), 144–145

ES. *See* Energy source (ES); Equivalent surplus (ES)

Escherichia coli (*E. coli*), 265

ESI. *See* Emergy sustainability index (ESI)

ESS. *See* Electroplating system sustainability (ESS)

Ethanol, 118

Ethanol/water separation, 58, 60f

Evaluation of design alternatives, 151

Evolutionary methods, 205

Evolutionary techniques, 361
Exergy, 39—40
    analysis, 146—147
    destruction, 39—40
    of system, 146
Expenditure function, 144—145
Extraction, 44—47
EYR. *See* Emergy yield ratio (EYR)

F

Failure mode and effects analysis, 368
Fault tree analysis, 368
Fermentation
    for bioethanol production system, 127—134
        closed-loop simulation, 129f, 131f
        open-loop simulations, 128f—129f, 131f, 133f
        Radar plot with GREENSCOPE indicators,
            130f, 132f—133f
    process model, 118—119, 120t
Fermentor unit, 118f
FI. *See* Fisher information (FI)
FineChem tool, 307—308
Fisher information (FI), 181—182, 187—189, 195
"Five-armed star" chart, 329—330
Flammability Subindex, 162t
Flowsheets for processes, 70
Flux, 48—49
Food web model equation, 183
Formulation ingredients, 358—360
Framework, 170
    molecules, 19—25
Frequency class, 148—149
Frequency of use. *See* Frequency class
FU. *See* Functional unit (FU)
Fuel specifications, 57
Functional unit (FU), 208, 344t

G

GA. *See* Genetic algorithms (GA)
β-Galactosidase (β-gal), 265
    bottleneck identification, 266
    defining level of analysis, 267—268
    economic, environmental, and social indicators
        selection, 268—269
    reporting assessment, 270
    retrofit action classification, 266—267
Gas stripping, 44—47
Gaseous waste, 94—96

Gasification, 309—310
Gate-to-gate processes, 71
Gauging reaction effectiveness for the
        environmental sustainability of
        chemistries with a multi-objective
        process evaluator (GREENSCOPE tool),
        116—117
    assessment tool, 122—123
    indicators, 124t
    methodology, 295—296
    sustainability assessment tool, 127
    taxonomy, 300
    tool, 71
GCC. *See* Grand composite curve (GCC)
Generation expansion, 173—175
Generic separation process, 40f
Genetic algorithms (GA), 97—98, 100f
GHG emission. *See* Greenhouse gas emission
        (GHG emission)
Ghosh inverse, 280
Gibb's free energy of mixing, 37—38
GlaxoSmithKline's technology, 371
Global warming potential (GWP), 125—126
Goal setting, 235—236
Grand composite curve (GCC), 96—97, 96f,
        101f
Graph theoretic method, 205—206
Green chemistry, 296
    underpinnings, 1—5
        hazard and risk, 3—4
        waste and hazard, 4—5
Green engineering, 5, 12
    principles, 296
Greenhouse gas emission (GHG emission), 311,
        333
GREENSCOPE tool. *See* Gauging reaction
        effectiveness for the environmental
        sustainability of chemistries with a
        multi-objective process evaluator
        (GREENSCOPE tool)
GWP. *See* Global warming potential (GWP)

H

Haber—Bosch process, 277, 288—289
Hazard, 3—5
    assessments, 367—370
    class, 148—149
    and operability study, 368

Heat exchange networks (HENs), 96—97,
    364—365, 365f
Heat induced MENs, 91
Heat integration, 69—70
    distillation column, 43—44
HENs. *See* Heat exchange networks (HENs)
Heuristics, 205, 361
HH. *See* Human compartment (HH); Human
    household (HH)
Hicksian consumer's surplus, 144—145
Hicksian measures, 144—145
Hierarchical decomposition, 361—362
Hierarchical design approach, 98, 102f
"HiGee" technology, 43—44
Human compartment (HH), 185—186
Human household (HH), 183—187
Human toxicity potential analysis, 75—81
Hybrid vapor stripping-vapor permeation process,
    57, 59f
Hybrids, 52—53
Hydrogen gas (H_2), 277

I

IA. *See* Impact assessment (IA)
IChemE metrics, 269
IChemE sustainability metrics, 228
Impact assessment (IA), 11—12
    method, 339
    model-based calculations for, 310—317
    results, 342—346
In-degree distribution, 193—194
Inaccessible resource pool (IRP), 183—187
Indirect impact/dependence, 280
Industrial sector (IS), 185—187
Inherent safety, 367—370
    index method, 160—161, 162t—163t
Integrated ecological and economic model,
    186—189, 187f, 197—198
    degree distributions, 192f
    network representation, 191f
    scenario analysis for various compartments,
        188f
    state variables, 188t
    transient Fisher information, 187—189, 189f
Integrated model, 182—183
Interception network, 88
Interest rate criterion, 143—144
Interfacial tension, 45

Intergovernmental Panel on Climate Change
    (IPCC), 256
Intermediate integrated model, 185—186, 185f
    degree distributions, 192f
    network representation, 190f
    scenario analysis for various compartments, 188f
    state variables, 188t
    transient Fisher information, 187—189, 189f
Intermediate model, 182—183, 196—197
    consumption increase scenario, 200f
    multivariable control profiles for sustainability, 198f
    multivariable optimal control profiles, 200f
    single-and multivariable control for sustainability,
        197f
Internal rate of return (IRR). *See* Discounted cash
    flow rate of return (DCFROR)
International Renewable Energy Agency
    (IRENA), 334
Investment analysis. *See* Investment assessment
Investment assessment, 235, 255, 267
IPCC. *See* Intergovernmental Panel on Climate
    Change (IPCC)
IRENA. *See* International Renewable Energy
    Agency (IRENA)
IRP. *See* Inaccessible resource pool (IRP)
IS. *See* Industrial sector (IS)

K

Kletz's inherent safety principles, 369—370

L

Lactic acid, 23
LC_{50}, 148—149
LCA. *See* Life cycle assessment (LCA)
LCC. *See* Life cycle costing (LCC)
LCD. *See* Liquid crystal displays (LCD)
LCI/A. *See* Life cycle inventory and assessment
    (LCI/A)
LCIA. *See* Life cycle impact assessment (LCIA)
LCIs. *See* Life cycle inventories (LCIs)
LCSA. *See* Life cycle sustainability assessment
    (LCSA)
LCSoft, 268
    carbon footprint—LCSoft results, 269f
LD_{50}, 71, 77—78, 159
Level-arm rule, 94—96
LGP. *See* Light guide panels (LGP)
LHV. *See* Lower heating value (LHV)

Life cycle analysis (LCA). *See* Life cycle assessment (LCA)

Life cycle assessment (LCA), 116–117, 145–146, 255, 278–279, 295–296, 327
  aspects of solvent selection postcombustion CO_2 capture
    generic flowsheet for, 306f
    life cycle inventories, 307–308
    motivation, 305
    process system and scope of LCA, 305–307
  for chemical industry interaction with carbon and nitrogen cycles
    carbon cycle flows in Eco-LCA inventory, 281t–282t
    direct and indirect impact/dependence, 280
    Eco-LCA inventory, 280
    nitrogen cycle flows in Eco-LCA inventory, 283t
  framework, 297–298
  in lignocellulosic biorefineries design
    key parameters and selected data for, 312t
    LCA metrics, 315f–316f
    life cycle inventories, 310–317
    midpoint impact categories of RECIPE Method, 316t
    motivation, 308–309
    multifunctional biorefinery system, 309f
    process system and scope of LCA, 309–310
    properties for coproducts, 313t
    values for partitioning coefficients, 314t

Life cycle costing (LCC), 328

Life cycle health and safety, 46

Life cycle impact assessment (LCIA), 299

Life cycle inventories (LCIs), 11, 298–300
  in early to basic process design stages, 300–301
    LCI data gaps and process design decisions, 301–303
    process scale, 303–305
  lignocellulosic biorefineries design, 310–317
  PMMA recycling process, 319–321
  process decomposition for extracting cradle-to-gate LCIs, 301f
  results, 342
  solvent selection postcombustion CO_2 capture, 307–308

Life cycle inventory and assessment (LCI/A), 11, 370–371

Life cycle sustainability assessment (LCSA), 327
  coal-fired electricity, 335–337
  data quality and data sources, 340t
  environmental human health impact, 345f
  goal and scope, 337–340
  integrated results, 346–348
    graphical chart, 347f
    outranking analysis result, 347f
  inventory analysis, 340–342
  life cycle system boundaries of bioelectricity, 339f
  methodologies for, 327–333
  normalization
    reference for social well-being midpoint indicators, 333t
    and weighting sets, 332t
  pillars, 328t, 333–335
  PROSUITE methodology, 330f
  remarks on methodology, 348–349
  results
    impact assessment results, 342–346
    life cycle inventory results, 342
  rice husk-based electricity, 335–337
  unit processes
    and data for prospective system, 343f
    for reference system, 341f

Light guide panels (LGP), 317

Lignocellulosic biorefineries design, LCA in
  key parameters and selected data for, 312t
  LCA metrics, 315f–316f
  life cycle inventories, 310–317
  midpoint impact categories of RECIPE Method, 316t
  motivation, 308–309
  multifunctional biorefinery system, 309f
  process system and scope of LCA, 309–310
  properties for coproducts, 313t
  values for partitioning coefficients, 314t

Limonene, 25

Liquid crystal displays (LCD), 317

Liquid–liquid extraction, 44–47

Lotka–Volterra models, 182–183

Lower heating value (LHV), 60–61

M

Macroconomic model equation, 183

Mass integration, 88–91, 89f. *See also* Energy integration; Property integration

Mass intensity (MI), 7, 7t

Mass separating agent (MSA), 44, 91
    separations, 45
    viscosity/flowability, 45
Mass-exchange networks (MENs), 91,
        364–365
Mass-exchange pinch diagram, 91, 92f
Mass-integration approach, 107–108, 109f
Material factor (MF), 149–150
Material flow analysis (MFA), 105, 105f
Material recycle pinch point, 89–90, 90f
Material selection, 71, 354
    formulation ingredients, 358–360
    reagents, 356–358
    safer chemical ingredients for DfE-labeled
        products, 359f
    solvents, 354–356, 355t
Mathematical models, 182–183
Mathematical programming methods, 205
MATLAB functions, 122
MAVS. See Membrane-assisted vapor stripping
        (MAVS)
Maximal structure generation algorithm
        (MSG algorithm), 210–211, 215f
    eliminating operating units, 214f
    group of operating units, 215f
    software implementations, 211
Mechanical vapor recompression (MVR), 42
Mekong river delta, 337
Membrane permeability, 50
Membrane separation quality, 49
Membrane-assisted vapor stripping (MAVS),
        58–60
Membrane-based separation processes,
        47–52
MENs. See Mass-exchange networks (MENs)
Mercury process, 72–73
Metal degreasing plant, 94–96
Metal discharge technology, 242–243
"Methanol to olefins" conversion, 310
Methyl methacrylate (MMA), 319–320. See also
        Poly(methyl methacrylate) (PMMA)
MF. See Material factor (MF)
MFA. See Material flow analysis (MFA)
MI. See Mass intensity (MI)
Micelles, 17
Microreactors, 27–28
Minireactors, 27–28
Mixing rule, 93
MMA. See Methyl methacrylate (MMA)

Model-based calculations for impact assessment,
        310–317
Molecular design, 100–103
Monod-type equation, 118–119
MSA. See Mass separating agent (MSA)
MSG algorithm. See Maximal structure generation
        algorithm (MSG algorithm)
Multicomponent reaction, 28
Multieffect evaporator, 43–44
Multifunctional processes, 299–300, 304–305
Multiobjective decision making
    approach, 170–173
    energy–comfort management in buildings,
        175–179
    generation expansion, 173–175
    objective prioritization, 169
    Pareto front and compromise decisions,
        177f
    Pareto front computation, 169
Multiobjective optimization/superstructure
        optimization, 362
Multiple decision makers. See Stakeholders
Multiscale approaches, 99–108
    EIP, 107–108
    process integration and molecular design,
        100–103, 103f–104f
    process integration with surrounding
        environment, 105–107
    property-based material recycle pinch diagram,
        104f
Mutual solubility, 45
MVR. See Mechanical vapor recompression
        (MVR)

N

National emissions inventory process data
        (NEI process data), 76
National Research Council (NRC), 35–36,
        67–68
NatureWorks, 22–23
Near-zero chemical technology, 242–243
NEI process data. See National emissions
        inventory process data (NEI process data)
Net present value (NPV), 143, 156–158, 255
Network Workbench (NWB), 191
Nitrification, 277
Nitrogen gas (N_2), 277
    chemical industry profile
        nitrogen emissions profile, 289–290

nitrogen mobilization profile, 287–288
nitrogen product profile, 288–289
cycle, 277–279
    flows in Eco-LCA inventory, 283t
emissions profile, 289–290
fertilizer manufacturing (325310) sector,
    287–288
fixation, 277–278
LCA for chemical industry interaction with
    carbon and nitrogen cycles
    carbon cycle flows in Eco-LCA inventory,
        281t–282t
    direct and indirect impact/dependence, 280
    Eco-LCA inventory, 280
    nitrogen cycle flows in Eco-LCA inventory,
        283t
mobilization profile, 287–288
NLP. See Nonlinear programming (NLP)
No-/low-cost strategies, 88–89
Non-random two liquid thermodynamic
    model, 154
Nonlinear programming (NLP), 194–195
Nonporous dense membranes, 47–48
NPV. See Net present value (NPV)
NRC. See National Research Council (NRC)
NWB. See Network Workbench (NWB)

O

O-dealkylation reactions, 356–357
O-type node, 213
O-type representing operating units, 208
Oak Ridge National Laboratory (ORNL), 35
OFT. See Operational time factor (OFT)
Operational time factor (OFT), 266
Optimal control
    problem, 196
    theory, 181–182
Optimal water use and reuse network design
    technology, 241, 241f
Organic Rankine cycles (ORC), 97–98, 99f,
    101f
ORNL. See Oak Ridge National Laboratory
    (ORNL)
OSBL. See Outside the battery limits (OSBL)
Out-degree distribution, 193–194
Outside the battery limits (OSBL),
    156–158
Oxidation-reduction process, 276–277

P

P-graph framework, 208–211
    MSG, SSG, and ABB algorithms, 210–211
    PNS Studio model development by P-graphs,
        213
    representing environment as resource,
        223f
    structural representation, 208, 209f
    structurally feasible process, 209–210
P3. See Profitable pollution prevention (P3)
Pareto analysis, 251–252
Pareto curve, 106–107, 108f
Pareto front computation, 169
PBT Profiler tool, 76–77
PEI. See Potential environmental impact (PEI)
Permeance, 48–49
Permselectivity, 49
Petrochemical manufacturing processes, 15
Photosynthesis, 276–277
PI. See Process intensification (PI)
PLA. See Polylactic acid (PLA)
PMMA. See Poly(methyl methacrylate) (PMMA)
PNS. See Process network synthesis (PNS)
PNS Draw software, 211–224, 212f
PNS Studio software, 211–224
    economical and ecological analysis,
        217–219
        default measurement units in, 219f
        single optimal solution of the mathematical
            programming model, 221f
    evaluation of competitiveness of emerging
        technologies, 221–222
    generate structurally feasible flowsheets by
        algorithm SSG, 216–217, 218f
        optimal and alternative suboptimal process
            networks, 220f
    model development by P-graphs, 213
    sensitivity analysis of best flowsheets, 221
    set default values, 216f
    structural analysis, 214–216
Polluted water, 183–185
Poly(methyl methacrylate) (PMMA), 317
    life cycle, 318f
    monomer recycling process, 319f
    recycling process
        life cycle inventories, 319–321
        motivation, 317
        process system and scope of LCA, 317–319

Polyester, 290–293
Polylactic acid (PLA), 22–23
Potential environmental impact (PEI), 71–72, 78f, 158–160
  component-specific, 159t
  EF and, 159t
Potential exposure. *See* Potential exposure class
Potential exposure class, 148–149
Potential impact balance, 71–72
Predator–prey interactions, 182–183
Process control, 135–136
Process design, 360. *See also* Life cycle assessment (LCA)
  decision-making procedure, 295
  Douglas' hierarchical decomposition approach, 362t
  HEN, 364–365, 365f
  LCA framework, 297–298
  LCIs, 298–300
    in early to basic process design stages, 300–305
  MEN, 364–365
  primarily tools, 360–362
  sustainability frameworks for, 296–297
  sustainability frameworks for process design, 296–297
Process economics analysis, 72–75
Process flowsheet, 88, 88f
  allocation through, 310
Process integration, 87, 100–103, 360, 363–364
  Douglas' hierarchical decomposition approach, 362t
  energy integration, 96–98
  HEN, 364–365, 365f
  mass integration, 88–91
  MEN, 364–365
  multiscale approaches, 99–108
  primarily tools, 360–362
  primary categories, 87
  property integration, 92–96
Process intensification (PI), 27, 366–367
  advantages and disadvantages, 366t
  examples of areas, 367t
Process modeling, 154–156
Process network synthesis (PNS), 206, 208
Process scale, 303–305
Process sustainability, fundamentals for, 229–230
  electroplating plant with sustainability concerns, 229f

Process synthesis, 203, 360
  activities in chemical process design, 203–205, 204f
  approaches, 361f
  classes of systematic methods, 205
  Douglas' hierarchical decomposition approach, 362t
  evolution of chemical, 363f
  HEN, 364–365, 365f
  illustrative example, 206–208
    alternative technologies and design configurations, 207f
  MEN, 364–365
  P-graph framework, 208–211
    representing environment as resource, 223f
  PNS Draw software, 211–224
  PNS Studio software, 211–224
  primarily tools, 360–362
  sustainability as alternative objective for, 222–224
Process systems engineering (PSE), 116–117, 353
Process systems engineers
  hazard assessments and inherent safety, 367–370
  impacts throughout supply chain–life cycle thinking, 370–371
  material selection, 354
    formulation ingredients, 358–360
    reagents, 356–358
    safer chemical ingredients for DfE-labeled products, 359f
    solvents, 354–356, 355t
  modeling and computer-aided tools, 371–373
  process design, 360–365
  process integration, 360–365
  process intensification, 366–367
  Process mass intensity and life cycle assessment tool, 372f
  process synthesis, 360–365
  tools for solvent selection, 355t
"Process windows", 16–17
Product and process development, 353
Product concentration (C_p), 118
Product systems, 338
  impacts, 344t
Profitable pollution prevention (P3), 228
Property
  clusters, 94
  load, 93
  operators, 94

property-based material recycle pinch diagram, 92–93, 93f

property-mixing rules, 93

Property integration, 92–96. *See also* Energy integration; Mass integration

Prospective Sustainability Assessment of Technologies (PROSUITE), 328–329, 330f, 331, 342

Prospective system, 337

    DSS, 332, 348

    unit processes and data for, 343f

Prospective systems, 336

PROSUITE. *See* Prospective Sustainability Assessment of Technologies (PROSUITE)

PSE. *See* Process systems engineering (PSE)

"Pulp mills (322110)" sector, 283–285

Q

Quantity class, 148–149

R

RadFrac model, 154–156

Reaction spaces, 15–17

Reaction yield (RY), 124

Reactive MENs, 91

Reactive-distillation solution, 70

Reactor cell, 73

Reagents, 356–358

RECIPE method, 311–313

    midpoint impact categories, 316t

Recycle—reactor system structure flowsheet, 74, 74f

Reference doses (RfD), 71–72, 78–79

Reference system, 336–337

    unit processes for, 341f

Reid vapor pressure (RVP), 94–96

Relative quantity. *See* Quantity class

Renewability-material index (RI_M), 124

Resource pool (RP), 183–187

Rest of world (RoW), 340

Retrofit design, 249

    economic assessment areas to evaluating, 253f

    environmental assessment areas, 257t

    framework for assessment of alternatives, 259

        bottleneck identification, 259–260

        defining level of analysis, 262–264

        economic, environmental, and social indicators selection, 264

        reporting assessment, 264–265

        retrofit action classification, 260–261, 261t

    β-gal production, 265–270

    SARD framework, 260f

    social assessment areas, 260f

    state of art, 250

        classification of environmental impact assessment methods, 256t

        economic pillar, 252–255

        environmental pillar, 255–258

        social pillar, 258–259

        sustainability in, 250–252

        steps, 251f

Retrofit project, 69

Return on investment (ROI), 255

RfD. *See* Reference doses (RfD)

Rice husk-based electricity, 334–335

    prospective system, 337

    prospective systems, 336

    reference system, 336–337

    technical characteristics, 338t

RI_M. *See* Renewability-material index (RI_M)

Risk, 3–4

    analysis methodologies, 296–297

    risk-based approach, 356

Robeson plots, 50, 51f

Robeson upper bound, 50

ROI. *See* Return on investment (ROI)

RoW. *See* Rest of world (RoW)

RP. *See* Resource pool (RP)

RVP. *See* Reid vapor pressure (RVP)

RY. *See* Reaction yield (RY)

S

S-LCA. *See* Social life cycle assessment (S-LCA)

Safety during operation, 160

SARD. *See* Sustainable assessment of retrofit design (SARD)

Seawater reverse osmosis (SWRO), 53–54

Second Law efficiency, 40–41

Selectivity, 44–45

Sensitivity analysis

    of best flowsheets, 221

    candidate policy options for, 199t

Separations

    alternatives, 41–53

    effect of concentration and fraction, 39f

    dilemma and imperative, 35–38

Separations (*Continued*)
   examples, 53–61
   generic separation process, 40f
   methods of analysis, 38–41
"Sherwood plot", 36–37, 37f
Short-cut models for filling in data gaps, 307–308
Short-lived climate pollutants (SLCPs), 15
Simple model, 183–185, 184f
   degree distributions, 192f
   network representation, 190f
   scenario analysis for various compartments, 188f
   state variables, 188t
   transient Fisher information, 187–189, 189f
Single objective function, 181–182
Sinks, 88–89
SLCPs. *See* Short-lived climate pollutants (SLCPs)
Social life cycle assessment (S-LCA), 259, 328
Social pillar, 258–259
Social sustainability, 229–232, 258–259. *See also*
      Economic sustainability; Environmental
      sustainability
   analysis, 148–151
   indicators, 233t
Social well-being, 345
Software COIN-OR's CBC, 221f
Solution structure generation algorithm
      (SSG algorithm), 205–206, 210–211
   generate structurally feasible flowsheets by,
      216–217, 218f
      optimal and alternative suboptimal process
         networks, 220f
   software implementations, 211
"Solution–diffusion", 47–48
Solvent selection postcombustion CO_2 capture
   generic flowsheet for, 306f
   life cycle inventories, 307–308
   motivation, 305
   process system and scope of LCA, 305–307
Solvents, 13–15, 354–356, 355t
   solvent-based CO_2 capture, 305–306
   solvent/water separation, 56–61
Solver, 211
"Sorption–diffusion", 47–48
"Sources", 88–89
Specialty chemicals industry, 21
Specific raw material cost (CSRM), 126
SPI. *See* Sustainable process index (SPI)
SSG algorithm. *See* Solution structure generation
      algorithm (SSG algorithm)

SSplit model, 154–156
Stability/reactivity, 46
Stakeholders, 169
   dissatisfaction under different compromise
      decisions, 178f
   polls, 172
Steam vapor, 336
Structural Analysis, 214–216
Structural representation of P-graph, 208, 209f
Structurally feasible process, 209–210
Substrate concentration (C_S), 118
Succinic acid, 22
Super pollutants, 15
Superposition, 89–91, 97
Surface-finished parts, 227
Surfactants, 17
Sustainability, 39–40, 47, 181–182
   as alternative objective for process synthesis,
      222–224
   analysis pillars, 141
   decision-making analysis results, 246t
   evaluator, 252, 258–259
   frameworks for process design, 296–297
   indicators, 156–163
   measures, 141–142
   metrics system, 230–232
      economic sustainability, 230–231
      environmental sustainability, 231
      social sustainability, 231–232
   multivariable control profiles for, 198f
   as path, 182f
   performance improvement, need for,
      235–236
   in retrofit design, 250–252
   single-and multivariable control for, 197f
Sustainability assessment, 122–123,
      135–136
   economic indicators, 126–127
   efficiency indicators, 123–125
   environmental indicators, 125–126
   framework, 233–237
      goal setting, 235–236
      investment assessment, 235
      need for sustainability performance
         improvement, 235–236
      technology evaluation, 234–235
      technology selection, 236–237
   GREENSCOPE indicators, 124t
   of technologies, 243–244, 244t–245t

Sustainable assessment of retrofit design (SARD), 250, 259. *See also* Retrofit design
Sustainable chemical engineering processes, 28–31
   principles and implications, 5–12
     design systems holistically and using life cycle thinking, 10–12
     eliminating and minimizing hazards and pollution, 8–10
     maximizing resource efficiency, 6–8
   problems with chemicals and reaction spaces, 12
     chemical reactivity, 12–13
     reaction spaces, 15–17
     solvents, 13–15
   sustainable chemistry and chemical manufacturing, 17, 30t
     biocatalysis, implications of, 26–27
     catalysis, 25–26
     chemical feedstock implications, 18–19
     framework molecules, 19–25
     high production volume chemicals, 18f
     reducing number of steps, 27–28
   underpinnings of green chemistry, 1–5
Sustainable engineering economic and profitability analysis
   bioethanol process, 151–163
   economic sustainability analysis, 142–145
   environmental sustainability analysis, 145–148
   social sustainability analysis, 148–151
   sustainability measures, 141–142
     evaluation of design alternatives by, 151
Sustainable process index (SPI), 145–146
Sustainable process systems, 115–116. *See also* Life cycle sustainability assessment (LCSA)
   chemical processes, 115
   fermentation for bioethanol production system, 127–134
   process systems engineering, 116–117
   proposed approach
     advanced control approach, 120–122, 121f
     fermentation process model, 118–119, 120t
     fermentor unit, 118f
     sustainability assessment, 122–127
   sustainability assessment and process control, 135–136
Sustainable system dynamic models, 182–189. *See also* Controllability analysis; Techno-socio-economic policies, optimal control for

integrated ecological and economic model, 186–189, 187f
intermediate integrated model, 185–186, 185f
simple model, 183–185, 184f
SustainPro, 251–252, 266–267
SWRO. *See* Seawater reverse osmosis (SWRO)
Syngas, 309–310
Synthesis of subsystems, 362
System boundaries, 338

T

Techno-ecological approach and chemical industry sustainability, 290–293
Techno-socio-economic policies, optimal control for, 195–198. *See also* Controllability analysis; Sustainable system dynamic models
   candidate policy options for sensitivity analysis, 199t
   integrated ecological and economic model, 197–198
   intermediate model, 196–197
Technoeconomic analysis, 141
Technology
   candidate selection, 239–243
     cleaning and rinse operation optimization technology, 239–241
     design schemes for electroplating and rinsing, 242f
     dynamics of dirt residue, 240f
     environmentally conscious dynamic hoist scheduling technology, 243
     near-zero chemical and metal discharge technology, 242–243
     optimal water use and reuse network design technology, 241, 241f
   evaluation, 234–235
   recommendation, 244–246
   selection, 236–237
Ternary cluster diagram, 94–96, 95f
Thermal storage, 98
Time criterion, 143
Total water cost ($C_{water\ tot}$), 126–127
Toxic chemicals, 227
Transient Fisher information, 187–189, 189f
"Tree-view", 213
"Truncation" error, 279–280

U

UNEP LCSA. *See* United Nations Environment
Program Life Cycle Sustainability Analysis
(UNEP LCSA)
Unit operations, 70
Unit processes
and data for prospective system, 343f
for reference system, 341f
United Nations Environment Program Life Cycle
Sustainability Analysis (UNEP LCSA),
328–329, 331
Upper bound, 50
US Department of Energy National Renewable
Laboratory top value-added chemicals,
19, 20f
US Environmental Protection Agency (US EPA),
5, 67, 122–123, 146, 252

V

Value brought by process to society, 150–151
Vapor–liquid equilibrium (VLE), 50
Volatility, 45

W

Walmart's Sustainability Index, 358–359
WAR algorithm. *See* Waste reduction algorithm
(WAR algorithm)

WAR graphical user interface (WAR GUI),
258
Waste, 4–5
Waste electrical and electronic equipment
(WEEE), 318–319
Waste reduction algorithm (WAR algorithm),
71–72, 78–79, 146, 256, 297
Water intensity (WI), 125
Watershed, 105, 106f
WC. *See* Working capital (WC)
WEEE. *See* Waste electrical and electronic
equipment (WEEE)
Weighted multiobjective optimization problem,
173–174
What-if technique, 368
WI. *See* Water intensity (WI)
Working capital (WC), 142–143
World Commission on Environment and
Development, 249
Worst-case metrics, 179

X

p-Xylene, 290–293

Printed in the United States
By Bookmasters